The Rise of the Kwantung Army

Japan's Empire in Manchuria to 1932

Pacific Atrocities Education

The Rise of the Kwantung Army

Japan's Empire in Manchuria to 1932

Quin Cho

The Rise of the Kwantung Army

Japan's Empire in Manchuria to 1932

Quin Cho

Published by Pacific Atrocities Education

Printed in the United States of America.

Paperback ISBN: 978-1-947766-71-6
E-book ISBN: 978-1-947766-70-9

Table of Contents

Acknowledgments

I am immensely grateful for the various forms of support I have received over the course of this book project. Nancy and Peter Lee both board members at Pacific Atrocities Education (PAE), provided vision and material support for the project through a generous post-baccalaureate fellowship. Jenny Chan, the director of PAE has been essential to the project, from helping to conceptualize the outline of the book to availing the project of PAE's ongoing support and assistance. Jenny's steady feedback and indispensable support for labor-intensive archival research strengthened every part of the book. Two of my former professors at University of Southern California, Drs. Jacques Hymans and Robert English, helped me turn this history into a book project, having served as advisors for my undergraduate honors thesis. Other USC faculty who helped me to develop as a historian and writer include Alice Baumgartner, Jay Rubenstein, Deborah Harkness, Benjamin Uchiyama, and Dave Kang. I'm also indebted to Kaila Philo, whose dedication and expertise as an editor vastly improved the book. Fatima Janaghir translated many critical documents from Classical Japanese to English, a skill that is exceedingly rare. Eric Van Slander and other outstanding members of the National Archives staff at College Park enthusiastically helped us locate and scan primary documents. Finally, my family members, including my parents, sister, and grandmother have provided an immense amount of love, advice, and support throughout the book-writing process. I deeply appreciate all these contributions and owe a great debt to everyone who has helped me in the process of writing this book.

Author's Note

This book primarily uses Hanyu Pinyin to Romanize Chinese names, with some exceptions. These exceptions include well-known names and places (e.g. Chiang Kai-shek) and obscure places that are difficult to transcribe from Wade-Giles romanization. For Japanese romanization, Hepburn is the primary system used. Both Chinese and Japanese names are presented with the family name first and the given name second (e.g. Zhang Zuolin). Throughout the book, "Manchuria" is referred to as "northeastern China," or the "Three Eastern Provinces" (Fengtian, Zhili, and Heilongjiang) due to the negative connotations affiliated with Manchukuo and Manchu separatism.

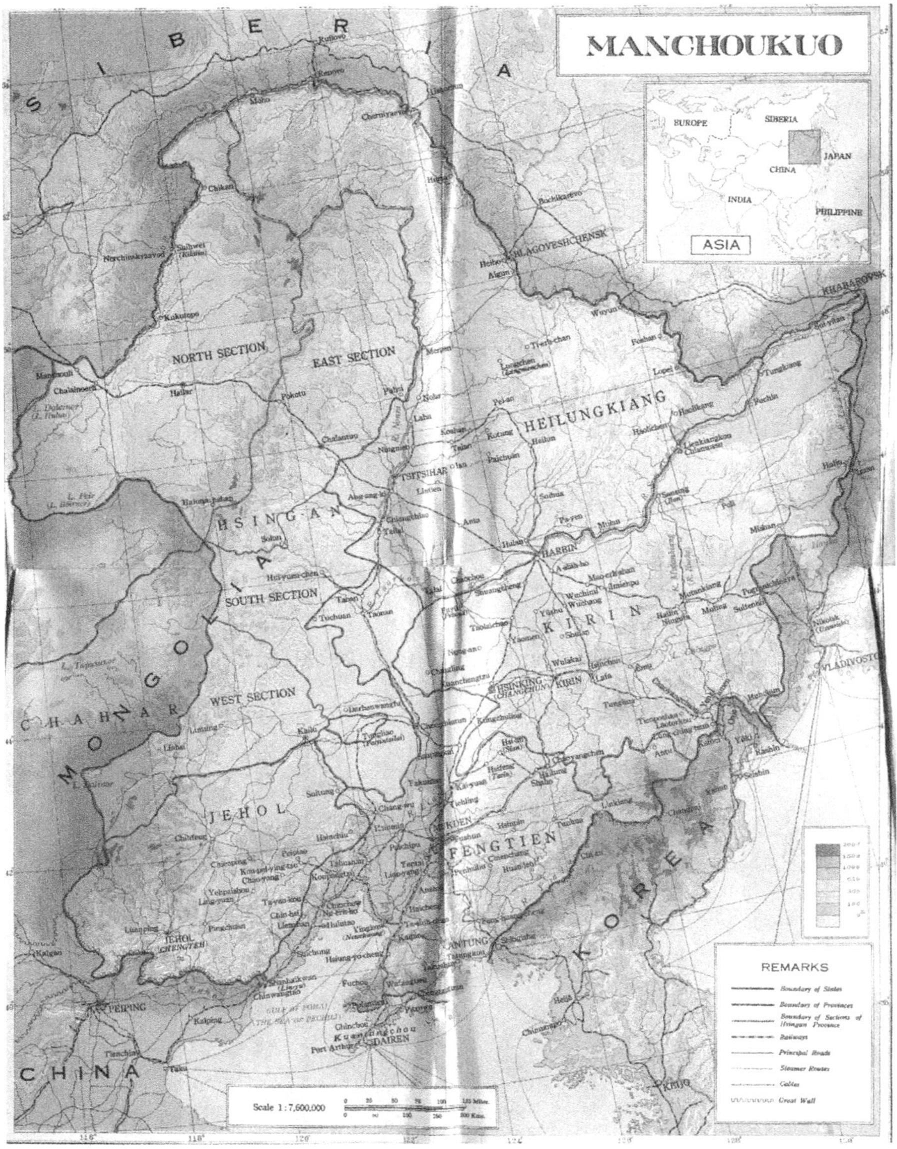

Map of Northeastern China. Note the locations of the "Three Eastern Provinces," Fengtian (Fengtien), Jilin (Kirin) and Heilongjiang (Heilungkiang) (Manchukuo Yearbook, 1934)

Introduction

On the morning of September 18th, 1931, Captain Imada Shintaro of Japan's Kwantung Army briefed his subordinate officers on an unexpected mission in northeastern China. Tensions had been high in the region, but Imada's orders surprised even the most battle ready of officers: blow up Japan's own railroad, the South Manchuria Railway (SMR) near the Beidaying Barracks and then attack the Chinese inside.[1] Simultaneously, other forces in northeastern China were to attack and seize key cities such as Mukden, Changchun, and Gongchangling. After seizing these areas, the Kwantung Army could then move on the rest of northeastern China and absorb it into the Japanese Empire. Imada's false flag attack on the SMR would serve as the justification for Japan's invasion, a necessary reprisal against insolent Chinese in what has become known in history as the Mukden Incident.

Imada and the forces under his command were well prepared to oversee his assigned task. The previous night they had prepared 42 packages of explosives.[2] The 2nd Independent Railway Guards Battalion (the force slated to assault the Beidaying Barracks) had participated in exercises and rehearsals throughout September along with the rest of the Kwantung Army.[3] Between 9 and 10 p.m. on the 18th, unscheduled railway cars delivered a "considerable number" of Japanese troops to the area near Beidaying.[4] These troops, part of the 3rd Infantry Company of the 2nd Independent Railway Guards Battalion, then moved to their assault positions.

At 10:30 p.m. on September 18th, the troops heard their signal: a large explosion occurred at the South Manchuria Railway near the Beidaying Barracks. Although the explosion did not seriously damage the railway tracks the 3rd Company, 2nd Independent Railway Guards Battalion went into action as planned at 10:50 p.m. The Japanese assault went according to plan, with the Chinese offering minimal resistance. By morning, the barracks were in Japanese hands.

A similar script played out elsewhere along the South Manchuria Railway line: The Japanese overwhelmed Chinese warlord Zhang Xueliang's armies, which were unprepared for action. The Japanese government was also unprepared for the Kwantung Army's invasion of northeastern China, as evidenced by the fact it issued no orders to invade. The Kwantung Army took full advantage of the institutional surprise that engulfed both their own government and the armies of their opponents, conquering northeastern China for the Japanese empire by the spring of 1932. This conquest culminated in the formation of the puppet state of Manchukuo, and the Kwantung Army's *de facto* administration of its puppet state.

Japan's invasion was primarily the work of two Kwantung Army staff officers, Colonel Itagaki Seishiro and Lieutenant Colonel Ishiwara Kanji. In contrast to most wars, which begin after a given country's government or head of state initiates war in pursuit of broader policy objectives, Itagaki and Ishiwara's decision to invade was taken on their own initiative. The independent decision of these officers to attack northeastern China reflected a broader Japanese phenomenon known as *gekokujo*, which literally means "domination of seniors by juniors."[5] *Gekokujo* had its roots in the civil wars of fourteenth century Japan, which were characterized by great social upheavals.[6] By the late 1800s, *gekokujo* took the form of mid-ranking Imperial Japanese Army (IJA) staff officers making unilateral decisions far above their pay-grade, in large part due to their specialized knowledge and proximity to events.[7] In general, the government and upper echelons of the IJA did not approve of these decisions, even if there were sometimes sympathizers in the Army General Staff (AGS). But Hirohito and the rest of the Japanese power structure did not discipline their unruly subordinates, despite the fact they did not approve of their extreme activities.[8] Consequently, throughout the 1930s, mid-ranking staff officers were able to hijack control over national policy and completely undermine the army's command structure, especially in northern China. In turn, these officers' actions would put Japan on the path of war with China, the United States, United Kingdom, and Soviet Union.

The Kwantung Army was a hotbed of *gekokujo* dating back to the 1910s and 1920s; its staff officers repeatedly made unilateral decisions that had great impacts on national policy. By 1931, the rise of Chinese nationalism was a great enough threat to Japan's interests in northeastern China that Ishiwara and Itagaki resolved to invade the region without permission from their superiors in Tokyo. While some members of the Army General Staff (AGS) were aware of Ishiwara and Itagaki's plans, most in Tokyo were not, and likely would not have approved if they had been aware. Indeed, on September 19th, the Japanese Cabinet issued a non-expansion order, hoping to keep the incident confined to the South Manchuria Railway (SMR) area. But Ishiwara and Itagaki managed to convince the commander of the Kwantung Army, Lieutenant General Honjo Shigeru to invade the province of Jilin, despite the government's wishes. Ishiwara also unilaterally bombed the city of Jinzhou in October and attacked the province of Heilongjiang in November. These actions directly contradicted the government's policy, which sought to minimize the scope of the incident and keep it contained to a small area. Instead, Ishiwara expanded the incident on his own initiative until it engulfed the whole of northeastern China.

Ishiwara and Itagaki's invasion of northeastern China raises a broader question: Why did they want to invade this particular region? The answer largely lies in Japan's economic troubles after the First World War, which accelerated with the onset of the Great Depression in 1929. Faced with shrinking export markets, financial failures, economic contraction, and an increasingly destitute peasantry, many Japanese believed that the only solution lay in the conquest northeastern China. Here, Japan could take the first step towards building a self-sufficient empire, as land was abundant and resources like soybeans, coal, and iron were common. Ishiwara in particular believed Japan needed to create a self-sufficient sphere in Asia because of his view that Japan and the United States were on a path of inevitable conflict. In his assessment, for Japan to emerge victorious in this "Final War," it needed to expand its existing economic base and mobilize society in its direction. He specifically advocated for the creation of a second industrial base in northeastern China so that Japanese industry would be less vulnerable to US air attacks and naval blockades. Ishiwara therefore pushed for Japan's conquest of the region as a prerequisite for the Japanese Empire's survival.

China's fragmentation and subordination to the western imperial powers were what made it so vulnerable to Japan's aggression. China's predicament was in turn the by-product of the arrival of European modernity in East Asia, which is coterminous with the "century of humiliation" which ran from 1842 to 1949. Even though the "Middle Kingdom" attempted to modernize in the last half of the nineteenth century, these efforts were dogged by a lack of centralization and corruption. By 1911, China's situation was so desperate that a revolution broke out, ending the system of dynastic rule, and nominally replacing it with a republic. But this republic was unstable and soon fractured into a series of smaller entities, as different warlords sought to unify the country. While the nationalist "Kuomintang" emerged as a centripetal force, it was unable to completely reunify China; therefore, the country remained susceptible to the machinations of other powers due to its disunity.

Unlike China, Japan was able to adapt to the West's new paradigm, incorporating western industries, technologies, and centralized modes of administration into its preexisting society. This process, which began with the so-called Meiji Restoration in 1868, achieved the "rich country and strong army" objectives of the so-called "Meiji oligarchs" by the end of the 1800s. Japan soon began flexing its muscle abroad, fighting wars with Russia and China over Korea. By 1905 not only had Japan effectively acquired dominion over Korea, but it also seized control of the Liaodong Peninsula and the South Manchuria Railway, providing a toehold in northeastern China. These territorial acquisitions were the genesis of Japan's new empire, crafted in the image of other similar European empires. The formation of the Japanese Empire therefore signaled the completion of its modernization process and ascendance as one of the world's great powers.

This book tells the story of the Mukden Incident and the rise of the Kwantung Army as rooted in the arrival of European modernity in East Asia and as fundamental to Japan's rise as an Empire. Chapters one through four discuss the arrival of European modernity in East Asia, the rise of Japan's empire, China's dissent into warlordism, and the effects of World War I. Chapters five through eight discuss the aftermath of the war, the rise of Chinese nationalism, and Japan's imperial policies in northeastern China. Chapters nine through eleven provide necessary background for the runup to the Mukden Incident, while chapters twelve through fifteen document and analyze the incident and its aftermath. The present volume is intended as the first part of a trilogy on the Kwantung Army with volumes two and three covering the years 1932-1937 and 1937-1945, respectively.

Chapter 1

European Modernity Arrives in East Asia

China's Century of Humiliation

For centuries, China was the envy of the world. It was the origin point for many of humanity's most important technologies, with a meritocratic governance system that facilitated effective administration. China's massive public infrastructure projects, perhaps best exemplified by the Grand Canal, were the fruits of this effective governance. These impressive achievements epitomized the large-scale coordination that sustained Chinese civilization. China's commodities ranging from tea and silk to porcelain and spices were desired around the world. In terms of industry, China was likely the world's most efficient iron and steel producer in the 1700s.[1] The combination of these factors, along with boasting the world's largest population, established China as a formidable leader in the global economy throughout the early 1800s.[2]

China was also a relatively peaceful civilization at the center of East Asia's Confucian international system. Two institutions defined relations between the "middle kingdom" and its smaller neighbors: investiture and tribute missions.[3] Investiture required nominally subordinate heads of state (e.g. the Korean Emperor) to prostrate themselves to the Chinese Emperor through ceremonial rites (best encapsulated through the infamous "kowtow"). In return, the Chinese Emperor—the son of Heaven in the Confucian understanding—would bestow his recognition upon these leaders, authorize trade relations, and guarantee the defense of subordinate states in the event they were invaded by a third party.[4] Symbolic gifts were exchanged between China and the lesser states as part of a broader diplomatic exchange.[5] This system of investiture and tribute established China as the legitimate Confucian hegemon in East Asia and maintained a relatively peaceful status quo in East Asia for at least a thousand years.[6] However, unlike colonies of European states, East Asian tributary states such as Korea remained autonomous in an economic, political, and diplomatic sense. In these ways, the tributary system differed greatly from the system of European and Japanese imperialism that later replaced it.

Despite the merits of its governing system, by the nineteenth century, China had fallen behind the western powers, many of whom had established globe-spanning empires (notably in the Americas, India, and Indonesia). The plundered fruits of these empires helped give rise to the Industrial Revolution beginning in Great Britain in 1750.[7] The advent of increasingly sophisticated industrial technologies afterwards (e.g. the steam engine and replaceable parts) allowed for the rapid advancement of weapons and other instruments vital to military operations. The revolution in technologies coincided with a revolution in the form of European nation-states, which consolidated, standardized, and centralized the old feudal armies and administrations of Europe into more coherent, united polities. As a result of these developments, China and Japan had to deal with increasingly bold European empires that encroached their borders and threatened the traditional governing systems in East Asia.

China first had to reckon with the West's newfound might in the First Opium War in 1839. The origins of this war date back to the 1830s, when the British East India Company began selling opium to the Chinese to reverse the flow of precious metals between China and the United Kingdom.[8] Even though the company managed to successfully reverse the outflow, its activities earned the enmity of the Qing Dynasty, which understandably did not want a foreign government poisoning its subjects. To limit the influx of opium into China, the Qing government sent commissioner Lin Zexu to Guangzhou on a mission to prevent further smuggling of the highly addictive drug.[9] While Lin was able to seize a large amount of British opium and prevent its distribution, his actions antagonized both the British Empire and the opium traders, who believed in the right of "free trade" and unfettered access to the Chinese market.[10] In the United Kingdom, these elements successfully lobbied for a war against China on the basis of their right to addict the Chinese population with the drug.[11] Pro-war forces got their wish when Britain went to war with China in 1839.

The British were able to effectively move their forces to China through one of the first fruits of industrial modernity: the steamship.[12] The Chinese, by contrast, were largely dependent on wooden ships, deploying mostly sail-dependent, flat-bottomed junks.[13] The British also had both an army and navy with a coherent, united organization and command-and-control structure. The Chinese, for their part, were saddled with a disparate array of antiquated weapons (matchlock muskets, swords, spears, crossbows, and muzzle loading cannon among them) and an even more disparate array of command structures and systems of organization.[14] These disparities allowed the British to repeatedly land along the Chinese coast and overwhelm the country's antiquated coastal defenses. The coastal landings culminated in British forces arriving at Zhenjiang and capturing the Grand Canal in July of 1842, cutting off China's means of shipping grain. Britain's capture of the Grand Canal forced China to the negotiating table.[15]

The defeated Chinese were forced to sign the Treaty of Nanjing in August 1842. In this treaty, Great Britain forced China to make a variety of concessions. These included ceding the island of Hong Kong to the British Empire, opening up a variety of ports for "free" trade with the foreigners, and allowing these foreigners to operate under their own laws and desired terms in the so-called "treaty ports."[16] China also lost control over the exercise of tariffs in these ports, where it had to confer with the British government before imposing any tariff rates.[17] These provisions enumerated in the Treaty of Nanjing marked the start of the exercise of extraterritorial rights by western powers in China, and the beginning of the end of China's previous role as the suzerain of East Asia.[18]

The First Opium War catapulted China into a prolonged period of chaos, instability, and warfare, starting with the Taiping Rebellion. The roots of the Taiping Rebellion lay with Hong Xiuquan, who claimed he was Jesus Christ's younger brother. Hong sought to build a heavenly kingdom on earth. Hong's preaching and expansive following eventually represented a great threat to the existing Qing Dynasty, which moved to suppress his movement in 1850. After twenty million deaths and fourteen years of war, the Qing were able to destroy the movement. In order to do so, they had to empower local warlords and gentry to raise their own "Yongying" militias in order to combat the Taiping kingdom. These Yongying militias numbered roughly 300,000 by the end of the 1860s and continued to grow.[19] The empowerment of these militias compromised the Qing Dynasty's monopoly on force and created parallel armed forces within the empire.

While the Taiping Rebellion raged, China became subject to yet more foreign wars and unequal treaties. The most important of these foreign wars was the Second Opium War, which took place between 1856 and 1860. In this second conflict, China faced a coalition of Britain and France. The western powers were once again able to utilize their maritime supremacy to land on the Chinese coast, seizing control of Guangzhou in southern China and the Dagu Forts further north.[20] Anglo-French forces then marched on Beijing, and seized the city on October 6th, 1860. By October 18th, the British forces had burned the Old Summer Palace to force China to acknowledge the invading foreigners' supremacy. The palace's destruction sent the Chinese a strong warning not to take up arms against Britain in the future.[21]

The seizure of Beijing forced China back to the negotiating table. At the "Peking Convention," China had to make another series of concessions which included opening seven new treaty ports, granting foreign vessels permission to navigate the Yangtze River, and ceding further control over tariffs to foreign powers.[22] The Peking Convention also affirmed the results of the 1858 Treaty of Aigun, in which China agreed to cede the land north of the Amur river and east of the Ussuri river to Russia. China's concessions to the foreign powers further cemented the system of extraterritorial rights and exposed China's internal weaknesses. To repel further foreign assaults, it was clear that China would need to modernize its military apparatus in order to build a "rich country and strong army."

China's crushing defeat in the Second Opium War was the wake-up call necessary for the Qing government to embark on a modernization program. This program, known as the "Self-Strengthening Movement," lasted for the better part of the next three decades.[23] The first wave began in 1861 under the Tongzhi Emperor, and hence is referred to as the "Tongzhi Restoration." To finance the program, the Qing government officially institutionalized the Imperial Maritime Customs Service in 1861 under the supervision of Englishman Horatio Nelson Lay. Even though Lay was a foreigner, the Maritime Customs Service was still a vital instrument for the Qing government whose efficient operation was necessary for the Qing Dynasty's program of modernization. The agency's revenues provided the capital to send Chinese students to the United States, to establish a school to teach western sciences and languages, and to build a series of key military-industrial complexes, including the Jiangnan Arsenal, the Xingu Arsenal, and the Fuzhou shipyard. These firms manufactured some of the world's most advanced weapons at that time, including rifles and rifled cannons. While inefficient and cumbersome, facilities such as these represented China's first means of producing modern weapons systems and provided the basis for the nation to build a modern army.

In the 1860s, China had two armies divided into a dual-track command system, which hindered unity of command and dissipated the country's efforts to build a large, national army. The more prominent force, known as the "Eight Banner Armies," actually consisted of twenty-four armies by the 1860s.[24] Composed of Han Chinese, Manchu, and Mongol troops, these forces were primarily tasked with defending the borders of the Qing Dynasty.[25] Those tasked with internal security were known as the "Green Standard Army," and mostly were comprised of Han Chinese soldiers. By the mid-nineteenth century, the Green Standard Army represented a greater proportion of the Chinese armed forces numbering roughly 600,000 troops.[26] However, these forces were less well-equipped and trained than their more prestigious contemporaries in the Eight Banner Armies. Both the Banner Armies and the Green

Standard Armies declined considerably in the nineteenth century, making modernization a pressing priority. It was apparent that existing forces in both armies lacked the capacity to effectively resist foreign invaders.[27]

During the period of the Tongzhi Restoration, the Qing Dynasty sought to improve the training, equipment, and command structure of both the Eight Banner and Green Standard Armies. The Qing brought in western advisors to help modernize both forces. Unfortunately, the Qing's modernization program did not go according to plan in either case. While the Qing government's "Peking Field Force,"(formed from elements of the Eight Banner Armies) was a 30,000-strong force trained and equipped to western standards, this force did not expand after the mid-1860s.[28] Similarly, even though the Qing managed to retrain a substantial number of the men in the "Green Standard Army," they maintained its decaying command structure and lacked the funds to re-train and equip 600,000 men to the standards of European armies.[29] Consequently, both the central government's armies were characterized by a divided, dual-track command system. They also remained largely dependent on old weaponry, lacked training, and were dogged by corruption and insufficient funds. These problems would only grow more pressing as the European powers made further advances and the Qing Dynasty continued to decline in the nineteenth century.[30]

The primary pockets of army modernization within China were in the Yongying militias. Less burdened by the Qing Dynasty's byzantine bureaucracy and dual track command structure, provincial leaders were able to successfully modernize their forces in some cases.[31] One of the most important of these modernized militias was the so-called "Huai Army," overseen by Li Hongzhang. Thanks to Li's patronage, the Huai Army was able to integrate cutting-edge weapons systems (such as the rifle, rifled cannon, and gatling guns) into their ranks. This militia also practiced western-style military drills, which created *esprit de corps* and inculcated prompt obedience to the command-and-control structure among the rank and file.[32] Perhaps most importantly, the Huai Army adopted a uniform organization structure from western armies, centered around 504-man battalions that were further subdivided into four companies, which in turn were subdivided into eight platoons.[33]

These reforms made the Huai Army a formidable force, with a variety of foreign observers speaking highly of its capabilities.[34]

Other modern armies also formed in China during the Self-Strengthening period. One of these modern armies, consisting of three infantry battalions, two cavalry squadrons, and artillery batteries, was constructed on the Prussian model, with a Prussian military academy in Tianjin established to supply and train its officers.[35] Another force that formed under the modernization efforts was the so-called Manchurian Army, which stood roughly 60-80,000 strong. This army had a significant minority of men who were equipped and trained to European standards.[36] By the 1890s, the "Newly Created Army" emerged, which would go on to be China's most important modernized force. While relatively small with an enlistment of 7,000 men, the New Army was under the command of Yuan Shikai, an important Qing official.[37] The New Army had a western table of organization and equipment including one infantry regiment, a cavalry squadron, three artillery battalions, and an engineer company. Its officers were trained by Prussian advisors, who instructed their Chinese understudies in the nuances of staff work, communications, and night operations.[38] Along with some updated equipment, these advantages made the New Army a cutting-edge force, and eventually a model for change within China itself.

Despite the death of the Tongzhi Emperor in 1875, China's self-strengthening program continued. Starting in the 1870s, the Chinese government invested in a variety of industries, including railroads, telegraph, textile mills, iron mines, coal mines, and steamship manufacturing.[39] Of these industries, steamships were among the most notable, as the construction of steam-powered naval vessels greatly increased China's previously lacking maritime capabilities. These ships were incorporated into a variety of different fleets, including Li Hongzhang's Beiyang Fleet, the Nanyang Fleet, the Guangzhou Fleet, and the Fujian Fleet. As in the army, divided commands hindered unity of action, which in turn limited the extent to which China could standardize its navy's organization, command structure, and logistical system. These issues would be a great hindrance to China in future conflicts, especially when its fleet was pitted against more unified, organized enemies.

China's late-nineteenth century attempts to modernize its economy, military, and industry produced mixed results. Even though the Qing Dynasty did succeed to some degree in bringing modern industry, science, technology, and infrastructure to China, large-scale changes were difficult to implement due to the fragmentation of the late Qing state and a lack of a united government-military apparatus. This fragmentation is best exemplified in the many different armies that sprouted as "modern" alternatives to the existing Eight Banner and Green Standard Armies. Whatever the merits of these forces, they did not represent a unified national army that could effectively coordinate its actions to achieve policy objectives. The disparate energies within the government and armed forces would pull the Qing government apart and make it vulnerable in the face of foreign threats. Therefore, even though China to some degree acquired greater industrial and technological capabilities, it was unable to comprehensively modernize in the nineteenth century.

The Roots of Japanese Modernity: The Boshin War and the Meiji Restoration

Japan, like China, also did not immediately have access to the technology of the Industrial Revolution. Prior to 1853, Japan was still relatively isolated from Western countries under the feudal domain of the Tokugawa Shogunate. Western modernity arrived in Japan by force when Commodore Matthew Perry's "Great Black Fleet" docked in Tokyo Bay in 1853. Commodore Perry threatened to bombard Tokyo with cannon and land troops unless the Japanese opened ports up for trade with the United States.[40] The shogun and his advisers were highly intimidated by the large steamships and the cannon onboard. They readily obliged the Commodore and subsequently signed a series of treaties with the US and European Powers.[41] These treaties, much like those China signed following the two Opium Wars, were unequal in the sense that they deprived Japan of its full sovereignty by granting Europeans and Americans special extraterritorial rights to conduct trade and commerce.[42]

Many Japanese were not happy with the direction of the country. They feared that Japan would be forced to submit to Western domination and wanted to avoid the fate of colonial or semi-colonial dependency that awaited other East Asian countries. As many of these Japanese were *daimyo* (feudal lords) from the outlying domain of Choshu, they had troops at their command and were capable of challenging the existing order. Merchants from Osaka and Kyoto supported these lords, providing them with funds.[43] Together, these *daimyo* and merchants formed a coalition that sought to transform Japan into a state that could defend its sovereign territory.

In 1864, Choshu took up arms against the Tokugawa Shogunate. Under the leadership of Takasagi Shinkasu, the domain organized units of joint samurai-commoner infantry known as the *kiheitai*, an unconventional light infantry militia, designed to strike when and where the enemy least expected attacks (often the flanks and rear).[44] In theory, they would perfectly supplement the conventional samurai, who attack enemy formations from the front. But the idea behind the *kiheitai* was not enough to compensate for poor equipment, which was a key factor in Choshu's defeat by a 150,000-man expedition from the shogunate in late 1864.[45] This expedition forced the Choshu reformers to assess what went wrong and to ascertain how to remedy the problems that arose in their first rebellion.

Omura Masujiro replaced the tubercular Takasagi as the leader of the Choshu reformers. He believed that Choshu's armies required better weapons systems to properly implement Takasagi's vision for the *kiheitai*.[46] Omura acquired his new weapons in the form of pistols, rifled muskets, cannon, and breech loading rifles from British arms traders in Shanghai. In June 1865, Choshu purchased 1,800 rifles and 2,000 pistols; that July, it purchased 4,300 pistols and 3,000 muskets. These weapons systems allowed Choshu to equip all of its frontline units with firearms.[47]

Omura also implemented new tactical reforms that increased the effectiveness of the *kiheitai*. Using Dutch military manuals, he began training the *kiheitai* in open order skirmisher tactics, which dispersed the tightly packed formations of previous eras.[48] These new skirmisher tactics also utilized the terrain, maneuver tactics and quick strikes to disrupt enemy formations and threaten their flanks and rear. The adjustments that Omura championed greatly improved the *kiheitai's* effectiveness, which by 1865 represented a force capable of taking on the shogunate's armies.

Choshu also began to adopt new systems of organization for its armed forces. In 1865, the domain required its wealthier vassals (those who had at least 1,000 koku) to send young men for service in the military.[49] Instead of these men serving under their own local leaders, they would serve under Choshu commanders in western-style platoons (*shotai*), companies (*chotai*) and battalions (*daitai*). By 1866, Choshu had assembled twenty infantry battalions in preparation for war with the shogunate.[50] The introduction of western tables of organization and equipment to Choshu's armies standardized its command-and-control structure and eliminated the clan linkages that previously defined relations in the armed forces.

In 1866, the Tokugawa Shogunate sent another expedition to suppress the Choshu reformers. The shogunate's forces sought to surround the upstarts in Choshu and annihilate them. But unfortunately for the shogunate, the neighboring domain of Satsuma (which had supported the shogunate's anti-Choshu expeditions in previous years) had signed a military alliance with Choshu.[51] Satsuma's refusal to join the shogunate's expedition bought time for Omura and the Choshu rebels, who were eventually able to employ the *kiheitai* to harass, outflank, and encircle enemy units. The *kiheitai's* attacks were in line with the skirmishing tactics Omura emphasized in training the *kiheitai* in 1864-65.[52] These skirmishing tactics left the close-order dependent shogunate troops at the mercy of the *kiheitai*. If the troops stayed in close-order formation, they were vulnerable to the *kiheitai's* firepower, but if they dispersed, they lacked the firearms, training, and doctrine for their commanders to keep control of their men across space. Under these conditions, Omura's units were able to defeat the shogunate's units in detail before going over to the offensive in October 1867.[53] The Choshu units' offensive eventually brought them across the Shimonoseki Strait to the gates of Kyoto, where the Emperor resided.

Emperor Komei's death in early 1867 led to his replacement by his 14-year-old son, Mutsuhito. Meanwhile, the Boshin War continued, as the domains of Satsuma and Choshu fought together against the shogun in the name of protecting and "restoring" the young Emperor to his rightful position as sovereign of the nation. In November 1867, the shogun resigned administrative authority over all of Japan and handed it over to the Emperor.[54] The shogun's handover of power was formalized on January 3rd, 1868, when the leaders of Satsuma and Choshu received a rescript from the emperor that stripped the shogun of all authority and restored the Emperor to his position as the rightful ruler of Japan.[55] The domains of Satsuma and Choshu then sent their armies into Kyoto to enforce the "restoration." Even though the shogun rushed troops from Osaka up the Toba Road to prevent the "restoration" from occurring, Satsuma troops blocked the road into the city and concentrated massed volleys of small arms and artillery fire on them. The shogun's men, lacking unity of command and fighting in close-order formations, were forced to retreat.[56] Subsequently, the armies of Satsuma and Choshu continued their offensive further north and unified Japan under the emperor's rule. In this manner, the "Meiji Restoration" was completed.

After the Meiji Restoration, the former Satsuma and Choshu *daimyo* attempted to modernize Japan. In 1871, they sent 108 people on the Iwakura Mission to investigate what made the West so powerful. The aims of the mission were to achieve international recognition for the new Meiji state, re-negotiate some of the unequal treaties, and study western political-economic systems.[57] The Iwakura Mission proceeded to go on a tour of the West including the United States, United Kingdom, France, Germany, Belgium, Italy, the Netherlands, Russia, and Austria.[58] Even though the mission did not succeed in re-negotiating the unequal treaties, it managed to secure recognition for the Meiji state and to study the political-economic systems that made western countries powerful. After the Mission returned to Japan, its members informed Meiji Japan's founders of the economic and political systems they found in the West, and how to best go about the modernization of the country.

The newfound Meiji government soon put the Mission's discoveries to good use, embarking on a "modernization" program to create a "rich country and strong army." The program included the creation of Japan's first industrial enterprises, the implementation of compulsory education, the utilization of foreign technical experts, and the construction of public infrastructure through government financing.[59] Crucially, Meiji oligarchs understood the importance of staying out of debt to western financial institutions in the push for Japan's industrialization.[60] They instead raised capital for industrialization by taxing the peasantry at a rate of three percent.[61] The new tax reform, alongside the import of industrial technology from overseas, allowed Japan to rapidly industrialize in the last thirty years of the nineteenth century. This industrialization was in turn the basis for the economic and military strength that the Meiji founders sought to develop in Japan.

The Formation of the Imperial Japanese Army

The first step in the process of creating a "strong army" began with the formation of the Imperial Guard. Consisting of 6,000 men from all over the nation, the Imperial Guard represented Japan's first truly national force that was not defined by provincial or clan ties.[62] Standardized ranks, pay grades, and uniforms further distinguished the Imperial Guard from the feudal and provincial armies of the past. The Imperial Guard was also the first nationwide formation to be organized on the basis of western tables of organization and equipment, with nine battalions, two artillery batteries, and two cavalry squadrons within its order of battle.[63] But perhaps most importantly, the Imperial Guard was exclusively loyal to the personage of the emperor rather than any regional domain.[64] The Imperial Guard's executive loyalty made the institution of monarchy a centripetal force for Japanese nationalism, setting the precedent for later provisions that made the armed forces exclusively responsible to the emperor. In turn, the Imperial Guard would become a template for the future Imperial Japanese Army (IJA), which would take on the national character of the Imperial Guard.

Forming a national army raised the question of conscription. Yamagata Aritomo, an officer from Choshu province, was the most prominent advocate of conscription due to his time in Prussia, which gave him an appreciation for the power and unifying potential of the European nation's conscript army.[65] In Yamagata's view, a new conscript army could do away with the peculiarities of the old order, inculcate the values of the new state, and offer a pool of ready-trained reservists in the event of foreign or domestic conflict.[66] Conscription would also disempower the samurai class, which during the time of the Tokugawa Shogunate were the only class of people permitted to bear arms. National military leaders such as Yamagata distrusted the discipline and loyalty of the samurai and wanted to loosen their monopoly on bearing arms.[67] Most importantly, conscription would act as a centripetal force for Japanese nationalism, and inculcate a sense of national belonging in Japan. Instead of seeing themselves as being from Satsuma, Choshu, Tokyo, or Kyoto, men who had been in the army would begin to see themselves as Japanese, representing the nation.[68] For these reasons, Yamagata vehemently advocated for conscription, which would accelerate the process of nation-building and make the Japanese Army a node of Japanese nationalism.

In January 1873, Yamagata finally succeeded in implementing his nationwide conscription program. The new ordinance applied to able-bodied men between the ages of 17 and 40, each with seven-year service terms. Of these seven years, three would be spent on active duty, two in the first-line reserves, and two more in the second-line reserves.[69] But despite the supposedly universal nature of Yamagata's conscription ordinance, there existed a laundry list of exemptions. These exemptions included heads of household, heirs (first-born sons), students, teachers, criminals, and the physically disabled. Wealthy men could also pay 270 yen for a substitute.[70] Many men fell into the exempted categories, limiting the number of men who could join the newfound national army. Indeed, in the Tokyo and Osaka districts, 85 percent and 70 percent of men respectively failed to meet the physical standards necessary to be conscripted.[71] Other urban districts had similar trouble raising recruits. By 1873 the Japanese could muster only a small force, with 17,900 men under arms. Most of these were the "second and third sons of impoverished farmers" due to the exemptions and exceptions.[72]

Beyond the limitations of conscription, Yamagata and other army nationalists faced another problem: many Japanese opposed conscriptions. The samurai opposed conscription because it undermined their monopoly on the armed services. Peasants disliked conscription because it resembled the Tokugawa Shogunate work levy system.[73] Poorer members of society resented that the wealthy could pay for an exemption. Such widespread resentment of conscription among many elements in society fueled widespread anti-conscription activities. At the local level, responses included men eligible for conscription escaping to the far reaches of Japan or mutilating themselves to avoid duty.[74] Nationally, many within Japan rebelled against conscription ordinance, exemplified by peasant riots in Fukuoka province throughout 1873 and the samurai's Satsuma Rebellion in 1877.[75] While the newfound Japanese army suppressed these uprisings and established the authority of the new state, the opposition to conscription signaled that the populace was not entirely supportive of Meiji Japan's nation-building project. Other measures would need to be taken to win the hearts and minds of the Japanese populace, which still was largely stuck in a pre-Meiji consciousness.

Although conscription represented a great challenge to the Imperial Japanese Army, Japan also needed to modernize other aspects of its armed forces. At first, Japan looked to the French Army as a model for its modernization, largely due to pre-existing ties.[76] These ties led the Japanese government to reach out to the French government in 1872, which dispatched a military mission to Japan that same year. Upon arriving, the French helped the Japanese construct firing ranges, specialized technical schools, and arsenals.[77] The French military mission also helped the IJA establish the Toyama Infantry School in 1873, which taught students small-unit tactics, marksmanship, bayonet practice, and physical education. The new academy not only allowed graduates to specialize in key skills that were necessary for modern warfare but also standardized the small-unit tactical doctrine of the IJA.[78] The Toyama Infantry School was thus a foundational institution for the IJA, which would continue to build its force structure, doctrine, and organization upon the foundation established at the academy.

Non-commissioned officer (NCO) and officer training represented the next major hurdle for the newfound Imperial Japanese Army. After a series of re-locations and false starts, a military academy was formed in Ichigaya, a former *daimyo* stronghold in Tokyo. At Ichigaya, cadets were subjected to either a two-year curriculum (if in the infantry or cavalry branches) or a three-year curriculum (if in the artillery or engineers).[79] Thirteen French instructors taught the cadets French tactics, military organization, debate, and the French language.[80] Ichigaya's curriculum also introduced Japanese officers to Imperial Japanese Army doctrine. The standardization of officer education allowed officers in the IJA to have the same approach to solving tactical problems. The standardized curriculum also provided officers with the ability to operate according to the same principles of command. In this sense, the creation of the Ichigaya military school created a "unitary" officer corps, which in turn facilitated rapid decision-making in the field and coordination between different formations.[81] These advantages would give the IJA significant leadership advantages in conflicts with divided or weak adversaries.

The French military mission to Japan helped the IJA build the foundation for a modern military force. However, the new national army still had several weaknesses. Even though the French taught tactical doctrine and small unit tactics (the battalion level and below), they neglected operational and strategic studies. Consequently, during the 1877 Satsuma Rebellion, operational planning and control above the brigade level was seriously lacking.[82] Furthermore, even though a Staff Bureau existed for the purposes of operational planning, administrators from the Army Ministry were in control of the Staff Bureau, hindering its function as an instrument for operational planning.[83] These deficiencies would be a long-term detriment to the IJA, which needed to master the nuances of controlling large unit operations. To achieve such mastery, a more defined division of labor between the operational and administrative spheres would be necessary.

The solution to the IJA's labor division problem lay in a Prussian innovation: the army general staff. The idea for the formation of a Japanese general staff came from Katsura Taro, who had studied the Prussian Army at Yamagata's behest and noted that it had a General Staff exclusively dedicated to the various tasks of operational and strategic planning. This system allowed staff officers to specialize in certain subjects, and thereby focus their energy on key tasks. Katsura also noted that the Prussian General Staff were only responsible for reporting to the emperor, and not to either the Army Ministry or civilian government.[84] These aspects of the Prussian army were so appealing to Katsura that he recommended that the IJA adopt its system in 1878. Yamagata, who had also studied the Prussian Army, agreed with Katsura's assessment.[85] Therefore, in 1878, the Imperial Japanese Army formed an Army General Staff (AGS), with Yamagata as the first AGS chief of staff. Like the Prussian General Staff, the new AGS was completely independent from the authority of the Army Ministry and civilian government. Instead, it answered only to the "restored" Japanese emperor.

The emperor's role as the commander in chief of the armed forces came to be outlined in the 1882 *Imperial Rescript to Soldiers and Sailors* and the 1889 Meiji Constitution. The "Imperial Rescript" mandated that soldiers must be loyal to the emperor.[86] Article XI of the Meiji Constitution stated that "the Emperor has supreme command of the army and navy," while article XII stated that "the Emperor determines the organization and peace standing of the army and navy."[87] These provisions clearly established the Emperor at the top of the chain of command and cut out the civilian government. But the emperor's control over the armed forces was only theoretical, given that he resided in his palace and was often isolated from the outside world. In practice, the army could simply seize offensive initiative in the field and then secure the emperor's *ex post facto* approval. Therefore, even though the adoption of the Prussian general staff model and chain of command allowed the army to be more efficient in operational planning, it also made the army an active policy-making agent. The army's independent initiative would embroil Japan in wars on the Asian mainland in the future.

The formation of the AGS coincided with the departure of the French military mission to Japan in 1880.[88] In 1884, a Japanese army delegation found a suitable replacement in the form of Prussian staff officer Jacob Meckel, who came to Japan in January 1885. Upon arrival, he helped found the IJA's new staff college where he introduced the

Prussian military education system. This system emphasized the planning and execution of large-scale operations (division level and above) as opposed to exclusively tactical (below brigade level) and technical training.[89] Meckel's influence at the staff college would reform the Japanese army and make it a force capable of projecting power onto continental Asia.

At the staff college, Meckel's revised curriculum included military history, tactics, fortifications, ordnance, gunnery, terrain studies and communication.[90] Of these subjects, military history was among the most useful, as Meckel used historical examples to teach military theory. Japanese officers embraced Meckel's practical approach to teaching, as it made concrete the abstractions of military science and doctrine.[91] Large-scale war games, which Meckel was credited with introducing to Japan, further reinforced the lessons Meckel taught, as did rigorous field exercises. These practical applications of warfare helped drive home Meckel's belief that the quality and quantity of weapons systems was not the main factor in battle. Rather, elan and offensive spirit were of decisive importance. The staff college cadets readily adopted Meckel's outlook into their views on combat, as the teaching reinforced the pre-existing Japanese conception of *seishin*, or fighting spirit.[92] The emphasis on fighting spirit, as opposed to material or numerical superiority, would become a vital part of IJA doctrine down the line.

Changes in the Imperial Japanese Army's doctrine and training accompanied the changes Meckel spearheaded in military education. In 1889, the IJA adopted the Prussian Army's field manual as the basis for its infantry doctrine. This field manual emphasized the importance of the company level in tactical formations and thereby empowered company-level junior officers.[93] These company commanders were also put in charge of training conscripts as part of what was known as the "family training concept." Putting these commanders in charge of training recruits facilitated the standardization of the Imperial Japanese Army's training regimen. With the same people (company commanders) in charge of tactical maneuver and training, the Imperial Japanese Army could rehearse its tactical doctrine through maneuvers and exercises. These reforms to the IJA's training program, combined with the 1887 standardization of regimental training tables, greatly improved the Imperial Japanese Army.[94]

Perhaps the most important change the IJA made in the 1880s was related to the structure of its table of organization and equipment. The domains of Satsuma and Choshu had brought western tables of organization and equipment to Japan in the form of the platoon, company, and battalion. In 1888, the Imperial Japanese Army incorporated a unit several levels above these formations, transitioning from static garrisons to divisions.[95] In terms of raw numbers, these divisions consisted of 18,500 men and 36 guns. In terms of the table of organization and equipment, the new infantry divisions consisted of two infantry brigades, which in turn had two infantry regiments (for a total of four regiments in a division). Each regiment had three battalions, and each battalion had four companies.[96] The new infantry divisions were mobile forces with combined arms capabilities, that included a cavalry regiment (composed of three squadrons), an artillery regiment (with four field and two mountain batteries), an engineer battalion, a logistics battalion (composed of two companies) and two stretcher-bearer companies.[97] The combined arms capabilities of the new infantry divisions allowed them to operate independent of any higher formation, and thereby allowed their commanders to take the initiative in offensives. Yamagata believed that the transition to divisions was a necessary prerequisite to undertaking offensive operations throughout Asia.[98]

IJA TO&E, c.1888	Manpower
IJA Infantry Division (2 brigades)	18,500 (wartime strength)
Infantry Mixed Brigade (2 regiments)	~7,000-8,000 (wartime strength)
Infantry Regiment (3 battalions)	3,280 (wartime strength)
Infantry Battalion (4 companies)	1,088 (wartime strength)
Infantry Company	272 (wartime strength)

Another key development in Japan during the 1880s was the formation of the so-called "Choshu Clique" in the government and army. As one of the key domains that had participated in the "restorationist" side of the Boshin War, Yamaguchi Prefecture (where the domain of Choshu had been located) was a feeder province for government officials, bureaucrats, and army officers in the Meiji state. Commonly bound by their provincial origins, the many figures within the government and army from Choshu coalesced to form a political bloc. The fact that this bloc did not form to exclusively look after the interests of Choshu and Yamaguchi Prefecture is noteworthy; rather, the members of the "Choshu Clique" saw their common provincial origins as a unifying force that created trust.[99] Members of the Choshu Clique also had a common world view that endorsed imperialist expansion on the Asian continent, military independence from the civilian government, and opposed political parties intervening in the army's affairs.[100] These militarist, imperialist views made the Choshu Clique a "hawkish" and aggressive element within Japanese politics that was at the cutting edge of Japanese imperialism during the Meiji era.

Within the Choshu Clique, Yamagata Aritomo was by far the most important and celebrated figure. As one of the founders of the Meiji state and Chief of the AGS, Yamagata used his close connection to the Meiji Emperor to promote the Clique's interests within the IJA.[101] The aforementioned Katsura Taro, who had recommended the general staff system to Yamagata in 1878, became the army minister in 1898, largely thanks to Yamagata's recommendation.[102] Other generals from Choshu such as Kodama Gentaro and Masatake Terauchi also rose to positions of prominence thanks to Yamagata's patronage.[103] Indeed, Yamagata's influence was so great that by 1888, sixteen out of the IJA's forty-two generals were from Choshu.[104] This pattern of promoting men from the Choshu Clique to higher ranks would persist until Yamagata's death in 1922, making the IJA one of the last bastions of *han-batsu* or clan patronage.[105]

Japan Becomes an Empire: Daggers, Lines, and Goodbyes

The growth, expansion, and strengthening of the IJA made it possible for Japan to begin dabbling in imperial expansion in East Asia. Of those who advocated imperial adventures, Jacob Meckel was among the most influential due to his status as the founder of Japan's general staff college. In large part, Meckel's expansionist views stemmed from his assessment that Japan needed to take active defense measures to ensure its sovereignty. Most famously, Meckel noted the Korean Peninsula was a "dagger pointed at the heart of Japan," and that if Japan allowed a foreign power to take control of it, the nation itself would soon be under threat.[106] Consequently, Meckel "spoke in terms of future operations on the Asian continent" when planning exercises and maneuvers of the new Japanese infantry divisions, and emphasized the importance of large-scale offensive operations in his lectures.[107] He also supported the shift from garrisons to divisions in the late 1880s, and vociferously supported the construction of railroads throughout Japan.[108] Meckel's viewpoints on the importance of large-scale offensive operations, active defense, and control over key railroads were to be key themes in Japan's imperial expansion in the years to come.

Even before Meckel's arrival, Japan began its quest for empire. As early as 1873, many within the newfound Imperial Japanese Army wanted to invade Korea to divert the attention of the restive samurai.[109] Only the threat of international alienation and crushing debts forestalled this attack.[110] Soon thereafter, in 1874, Japan dispatched a punitive expedition to Taiwan to punish native tribes for killing shipwrecked Japanese sailors. Two years later, Japan used its newfound naval strength to force the Treaty of Ganghwa on the Joseon Dynasty of Korea; this treaty opened Korea up to trade with Japan and bestowed extraterritorial rights upon Japanese citizens in Korea. By 1879, the Japanese Empire made its first major territorial acquisition outside the four major home islands, the Ryukyu Kingdom. These were Japan's first steps of empire-building.

The Korean Peninsula became the primary site of Japan's strategic efforts in the last quarter of the nineteenth century. In the late 1870s, the Japanese government began the process of establishing the Japanese yen as a reserve currency in Korea when it lent Dai-Ichi Bank money to open its first branch in Pusan in 1878.[111] The Japanese government then offered support for Korean modernizers and reformists, who visited Japan on several occasions to investigate how the growing empire had successfully modernized.[112] Even though the reformists suffered a serious setback when underpaid soldiers of Korea's old army rebelled in the Imo Uprising, Japan was able to take advantage of the situation by imposing a 550,000-yen indemnity for the assassination of Japanese officials and destruction of the Japanese legation during the rebellion. To help the Koreans pay this debt, the Yokohama Specie Bank lent their government 170,000 yen. However, only 50,000 yen went towards paying the indemnity; the remaining 120,000 went towards supporting the Korean reformists.[113] The reformists then launched a coup, with support from the Japanese, against the Korean government in 1884. But the so-called "Gapsin Coup" failed thanks to China's dispatch of 1,500 soldiers under the command of Yuan Shikai. Because of the coup's failure, the Chinese and the conservative Min Faction entrenched their respective positions in Korea.[114] If the Japanese wanted to expand their influence, they would need to change their strategy.

Despite Japan's initial misadventures in Korea, many Japanese saw the acquisition of an empire as necessary for Japan's own national security. Ex-samurai and their descendants were some of the strongest supporters of imperial expansion because they saw it as a cause worth fighting for, in the aftermath of the Satsuma Rebellion. The foremost figure among these ex-samurai was Toyama Mitsuru, who had myriad connections in the financial, industrial, and government worlds. Toyama soon leveraged his high-placed connections and the considerable fortune he gained from investing in coal mining to finance an ultra-nationalist, expansionist organization known as the Genyosha (the Black Ocean society) in 1881.[115]

The stated principles of the Genyosha were to "revere the emperor, love and respect the nation, and to defend the people's rights."[116] But the Genyosha's primary purpose was to promote Japanese ultra-nationalism and the expansion of the Japanese Empire in Asia. To achieve these aims, the Genyosha used a variety of different methods to influence policy—including blackmail, threats, intimidation, and assassination. Perhaps the best example of the Genyosha's use of these tactics was when a member, Kurushima Tsunekichi, attempted to assassinate Foreign Minister Okuma Shigenobu in 1889.[117] While this attempt was unsuccessful, the Genyosha's willingness to use force to influence domestic politics illustrated many of its members' fanaticism.

The Genyosha also established a significant footprint outside of Japan. Starting in the 1880s, they sent agents abroad in order to gain accurate intelligence about various other countries in Asia, including Korea, China, Russia, Indochina, India, Burma, and Thailand. Most of these agents were so-called patriotic adventurers, who donned a wide variety of disguises: teacher, journalist, religious holy man, businessman, tourist, shopkeeper, and professor, among others.[118] Many of these adventurers held pan-Asian views, and sought to help erstwhile Asian revolutionaries (including a young Sun Yat-sen) bring down the old order and help usher in the equivalent of a Meiji Restoration in their own countries.[119] In the minds of these adventurers, there was no contradiction between serving the Japanese Empire and promoting the interests of like-minded Asian revolutionaries, who wanted to establish independent modern countries free of western imperialist domination.

One of the most important "patriotic adventurers" was Arao Kiyoshi. To a great extent, Arao's importance lay in the fact that he was a link between the AGS—he graduated from the War Academy in 1882—and the Genyosha.[120] Upon his graduation from the War Academy, Arao went to China at the behest of the AGS and Genyosha in 1884 where he met a pharmaceutical businessman named Kishida Ginko. With Kishida's help, Arao opened a series of pharmacies, the most notable of which was the so-called the Rakuzendo, or the Hall of Pleasurable Delights. While the pharmacies were nominally popular for the eye ointments they sold, they were in fact a front for Japanese intelligence operations.[121] Arao created a network of hundreds of "patriotic adventurers" who sold drugs, created connections with bandits and brigands, and went on horseback rides throughout Asia to a variety of far regions including Siberia, Xinjiang, Russian Central Asia, British India, and northeastern China. The topics most heavily examined during these rides included economic and financial conditions, individuals likely to support Japan, infrastructure, and the defense capabilities of Asia's various armies.[122] The information gathered during these rides provided the Army General Staff with its first "detailed information" about regions far afield from Japan and significantly boosted their understanding of Japan's two most likely enemies: China and Russia.[123]

While the Genyosha represented a pan-Asian expression of emerging Japanese imperialism, Fukuzawa Yukichi's article "Datsu A-Ron," or "Goodbye Asia," represented a western liberal perspective on the question of empire.[124] In Fukuzawa's piece, he noted how the "wind of Western civilization has blown to the east," and that Japan must "cast their lot in with them" if it wanted to prevent "civilization's" arrival with a "firm resolve." In Fukuzawa's estimation, Japan should treat its neighbors, China and Korea, just as the western countries treated other supposedly "less civilized" nations on account of their backwardness and conservatism. Imperial ventures in Asia would in turn insulate Japan from the designs of the western powers and thereby allow Japan to join the ranks of the west's "civilized" nations.

Yamagata Aritomo gave strategic acumen to Fukuzawa's imperialist screed in an 1890 speech to the diet that advocated for Japan to defend a "line of advantage" outside of the formal borders of Japan (the so-called "line of sovereignty").[125] In Yamagata's view, Japan was being squeezed by strategic pressure from both Russia and the United Kingdom, which were advancing towards Japan's "line of advantage." In the case of Russia, its increased activity in the Russian Far East and Korea potentially foreshadowed an amphibious invasion of Japan, which had been a nightmare scenario of the IJA for decades. The construction of the Trans-Siberian Railway, which began in 1891, only increased suspicion amongst Yamagata and others in the IJA. Further east, the UK was advancing towards northeast Asia through construction of the trans-Canadian railway; meanwhile, Yamagata saw China's influence in the Korean Peninsula as detrimental to Japan's long-term strategic interests.[126] He believed Japan needed to control Korea to prevent foreign powers from using it as a launching pad to attack his empire.[127]

After Yamagata became Japan's Prime Minister in 1889, he began advocating a more aggressive policy in Korea to forestall Chinese influence and Russian aggression. Predicated on the idea of a "line of advantage," this policy would become a defining aspect of Japan's stratagem in the decades ahead.

Chapter 2

Japan's Two Splendid Little Wars

As the 1880s gave way to the 1890s, many in Japan had confidence in the Meiji state's newfound strength. This confidence existed because the empire had built an industrial base and unified its governmental administration under the symbol of Imperial Chrysanthemum. Such political unity and industrial development allowed Japan to modernize its armed forces in terms of equipment, organization, and command. The achievement of the Meiji founders' "rich country, strong army" goal planted the seeds of Japan's imperial expansion, which many in both the government and army started to embrace. This "positive" point of view was especially popular in the IJA which had many young officers that came of age at a time when Japan was ascending and had little memory of Commodore Perry's voyage or the unequal treaties. But others in the government and political parties, including many original founders of the Meiji-state (the so-called *genro*), remained cautious about provoking the ire of the western powers through military adventures abroad. The clash between the "positive" imperialists and the more cautious elements in the Japanese state came to define the next few decades of Japan's foreign policy in Asia.

In the 1890s and 1900s, the advocates of "positive imperialism" got their way, managing to push Japan down the road of war with China in 1894 and Russia in 1904. The First Sino-Japanese War and Russo-Japanese War represented Japan's first significant military expeditions to the Asian mainland in three centuries, cementing in blood the foundations of the Japanese Empire. The First Sino-Japanese War was the *coup de grace* to the ancient Chinese tributary system and made Korea a part of Japan's sphere of influence. The Russo-Japanese War, meanwhile, made Korea a Japanese protectorate and gave Japan control over a strategically vital portion of northeastern China, the Liaodong Peninsula. During both wars, Japan also managed to establish maritime hegemony in the Sea of Japan and the Yellow Sea, allowing it to land forces in Asia at will. Therefore, both of these wars established the Japanese Empire's hegemony in East Asia, opening up the rest of the region to further intrigue in future years.

The Tianjin Convention and China's Influence in Korea

During the 1880s, Japan and China were jockeying for influence in Korea. China temporarily gained the upper hand thanks to the failure of the Gapsin Coup, which it had helped suppress with roughly 1,500 troops under the command of Yuan Shikai. But certain figures in China and Japan, including Li Hongzhang and Ito Hirobumi, wanted to maintain peace and agreed to sign the Tianjin Convention in April 1885. Some excerpts from this treaty are provided below:[1]

> It is hereby agreed that China shall withdraw her troops now stationed in Korea, and that Japan shall withdraw hers stationed therein for the protection of her legation. The specific term for effecting the same shall be four months, commencing from the date of the signing and sealing of this Convention, within which term, they shall respectively accomplish the withdrawal of the whole number of each of their troops, in order to avoid effectively any complications between the respective countries: the Chinese troops shall embark from Masan-po, and the Japanese from the port of Ninsen.
>
> The said respective Powers mutually agree to invite the King of Korea to instruct and drill a sufficient armed force that she may herself assure her public security, and to invite him to engage into his service an officer or officers from amongst those of a third Power, who shall be entrusted with the instruction of the said force. The

> respective Powers also bind themselves, each to the other, henceforth not to send any of their own officers to Korea for the purpose of giving said instruction.
>
> In case any disturbances of a grave nature occurring in Korea, which necessitates the respective countries or either of them to send troops to Korea, it is hereby understood that they shall give, each to the other, previous notice in writing, of their intention so to do, and that after the matter is settled, they shall withdraw their troops and not further station them there.

Of these points, the last was the most important, as it contradicted the previous system of tributary relations that existed in East Asia. Now, instead of sending troops to aid Korea on the basis of a mutually recognized defensive alliance, China would have to notify Japan of its intent to deploy. Given the existing tensions on the Korean Peninsula, this provision was a ticking time bomb, especially after Yuan Shikai coerced Korea into giving the Qing Dynasty control over its foreign policy later in 1885.[2] Thanks to Yuan's actions and the limitations outlined in the Tianjin Convention, East Asia was ready to explode, with only the smallest of sparks needed to light the fuse.

The First Sino-Japanese War

The Donghak Peasant Rebellion of 1894 was the catalyst that sparked war in East Asia. The rebellion began when Korean peasants from Jeolla province rose up due to the oppressive policies of the local government. In response, the Joseon Dynasty of Korea moved to suppress the rebels and asked China for aid. China dispatched 2,500 troops to Korea, in line with the tradition of tributary relations.[3] The Chinese legation to Japan notified the Japanese of their action, stating "it is in harmony with our constant practice to protect our tributary states by sending our troops to assist them… General Yeh, Commander of the troops in Zhili has been ordered to proceed at once to Jeolla and Cheongju in Korea with selected troops and will speedily suppress the disturbance in such manner as he may deem most convenient in order to restore the peace of our tributary state and to dispel the anxiety of the subjects of every nation residing in Korea for commercial purposes, and at the same time the General is commanded to return with the troops as soon as the desired objective is attained."[4] At the end of this message, the Chinese legation also made sure to note that they sent the message to Japan to adhere to the Tianjin Convention.[5] Such a message should have been sufficient for Japan not to intervene, as China had upheld the terms of its written agreement and notified Japan in advance of its expedition to Korea.

The Donghak Peasant Rebellion and China's attempt at suppression occurred at a time when debate over the question of expansion was at its peak in Japan. Despite Yamagata's speech to Japan's legislature, the Diet, about the need for the Empire to pursue a "line of advantage," the Diet had voted to reduce military expenditure in 1892.[6] Further conflicts between the advocates of "positive imperialism" and domestic construction continued to characterize Japanese politics in the following two years. But the Donghak Rebellion represented a great opportunity for the proponents of "positive imperialism" in Asia. As two of its foremost advocates, Foreign Minister Mutsu Munemitsu and Vice Chief of Staff General Kawakami Soroku built consensus in Prime Minister Ito Hirobumi's cabinet for sending troops to Korea, citing a dubious intelligence report from Tianjin that the Chinese had sent 5,000 troops to Korea.[7] These men also concealed the fact that negotiations in Korea were progressing and a path to peace was possible from Ito and his cabinet.[8] Thanks to the efforts of Kawakami and Mutsu, Ito's cabinet was convinced of the need to send forces to Korea in early June. On the basis that China had violated the Ganghwa Treaty by inappropriately referring to Korea as a "tributary state," Japan informed China of its intent to deploy forces to Korea.[9]

Even though the Ito government agreed to send troops to Korea, the questions of how many should be deployed and for how long remained. Knowing that Ito would not want to send too many troops and alienate the western powers, Kawakami and Mutsu conspired to deceive him. Seeking to take advantage of Ito's ignorance about the IJA's table of orga-

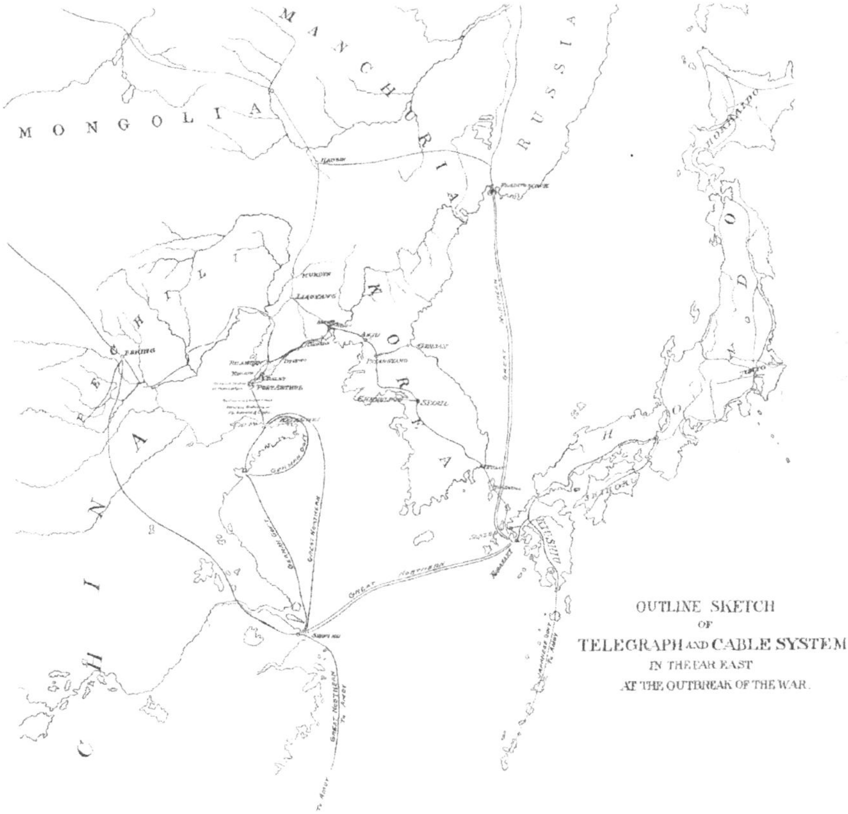

Telegraph Map of China, Korea, and Japan. Note the relative locations of each (War Department Map Collection, RG 77, National Archives at College Park)

nization and equipment, Kawakami told Mutsu of his intent to present the initial force to be sent to Korea as one brigade, which Ito would understand as being equivalent to 2,000 men. But in reality, Kawakami intended to send a mixed brigade consisting of 7,000 men to Korea.[10] In Kawakami's estimation, 7,000 men would be sufficient to overwhelm the Chinese forces in Korea, precipitate further deployments from China, and escalate the conflict.[11] Kawakami went through with his plan in early June 1894 and managed to get Ito to approve the dispatch of a brigade to Korea soon thereafter. Later in June, Japan sent a mixed brigade to the Korean peninsula and threatened China with war if it did not withdraw its forces.[12] Thanks to Kawakami's duplicitous *gekokujo*, Japan was on the war path with China.

In both quantitative and qualitative terms, Japan's threat of war seemed hollow. At the outbreak of the First Sino-Japanese War, the Imperial Japanese Army totaled roughly 68,000 regular troops in seven infantry divisions.[13] This number soon increased to 90,000 with the mobilization of the first reserve, and then 161,000 with the mobilization of the second reserve.[14] While the rapid call-up of reserves was an impressive testament to the Japanese mobilization system, many of the soldiers were of poor quality. Illiteracy was a major problem, with over sixty percent of conscripts either being completely (27 percent) or partially illiterate.[15] Many were also undersized, with the average height of the Azubu catchment area of Tokyo for Tokyo-fu in 1891 being 154 centimeters—roughly five feet high.[16] Nourishment was also insufficient, as IJA rations that consisted of white rice, fish, pickled vegetables, poultry, and tea failed to provide all of the necessary vitamins to soldiers, many of whom contracted beriberi.[17] These and other problems limited the quality of Japanese soldiers on the eve of the First Sino-Japanese War.

Japanese armaments were also a limiting factor for the IJA at the outbreak of the First Sino-Japanese War. The Japanese infantryman's standard issue armament was the 1885 Murata single-shot rifle, at a time when repeating rifles were becoming standard issue in European armies.[18] Only one machine gun unit existed and it was deployed to Taiwan, away from the site of the main fighting.[19] The IJA did have 240 guns, most of which were the Krupp 75mm field gun that could fire at a maximum effective range of roughly five kilometers.[20] However, most of these were bronze instead of steel cannon, which could not use smokeless powder since doing so would foul the inside of the gun barrel. Shell production was also limited due to a lack of capacity, further compounding the shortcomings of Japanese artillery pieces.[21] The IJA was going to have to overcome these constraints with its elan and operational acumen if it were to win the First Sino-Japanese War.

The Chinese army far outnumbered the IJA, with roughly 250,000 men in its eight banner armies and a further 600,000 in the Green Standard Army.[22] The eight-banner army was composed of eight "banners" (corps) and represented the core strength of the Qing Dynasty's army. Each of the banners was composed of three divisions, each division of five battalions, and each battalion five companies of 300 men each.[23] More than half of the men within the eight-banner army were Manchus, and over sixty percent of the Manchus were cavalrymen. Over half of these forces were stationed in northern China. The rest were stationed in northeastern China, Xinjiang, and in city garrisons.[24]

Of the Green Standard Army's 600,000 men, eighty percent were infantry and less than twenty percent were cavalry.[25] Unlike the banners, which were mostly Manchu, the Green Standard Army was an entirely Han Chinese force. The Green Standard Army was also a provincial force, composed of 18 separate provincial armies.[26] The army had two different systems of organization, with some forces being organized into brigades (composed of six to eight battalions) and others into regiments (composed of two battalions).[27] The Green Standard Army's primary function was to garrison interior provinces and secure them against rebellions or internal uprisings. The Green Standard Army could also serve as a pool of reserves in the event of a large-scale conflict.[28]

The Chinese armies possessed a plethora of different weapon systems. Even at the outbreak of war in 1894, many Chinese were still equipped with swords, bows, spears, and matchlock muskets.[29] But thanks to the Qing Dynasty's efforts to modernize the armed forces, a significant minority of Chinese troops were equipped with modern rifles. Many of the Chinese units close to the line of contact with the IJA had breech loading, single-shot Mausers. Others had Enfields, Martini-Henris, Remingtons, Chassepots, Winchesters, Spencers, and Lee-Metfords.[30] In terms of artillery, the Chinese had a significant number of 75mm Krupp mountain and field guns, as well as a series of British guns between 67mm and 76mm in caliber.[31] While these guns represented a formidable element on paper, the wide variety of foreign guns in service complicated logistics and procurement. Moreover, poor deployment, with guns dispersed across positions, limited the effect the new weaponry would have during the course of operations.[32] Such drawbacks would hinder the Chinese forces' application of firepower during the First Sino-Japanese War.

The Imperial Japanese Navy (IJN) was still in its infancy at the beginning of the First Sino-Japanese War. The lack of a maritime threat from China or Korea prior to 1868, combined with minimal ocean-going trade, meant that Japan had little need for a blue water navy.[33] While the arrival of European modernity to Japanese shores changed this reality in the 1850s, the Meiji state did not initially prioritize investing capital in naval construction. Instead, the Imperial Japanese Navy's primary task throughout the 1870s and 1880s was coastal defense.[34] The IJN also did not yet have the necessary technical expertise to build steel-hulled, steam powered warships. The combination of these factors limited the size and technological prowess of the IJN in the 1870s and 1880s.

Despite the IJN's limitations, the IJN began ordering armored, steel hulled ships from British shipyards by the end of the 1870s. These vessels gave the IJN crucial experience in operating and handling the new steel-hulled, steam powered ships. By the middle of the 1880s, Japan built its first naval arsenals at Tsukiji and Yokosuka.[35] In 1882, the IJN stopped building wooden warships and in the mid-1880s, the Yokosuka arsenal began building small iron-hulled, steam-powered ships.[36] The combination of foreign contracts and domestic construction meant that by 1894, the Imperial Japanese Navy was a formidable force that consisted of 1 battleship, 10 cruisers, 8 scout cruisers, 7 gunboats, and 24 torpedo boats.[37]

The Chinese Navy—consisting of three battleships, 10 cruisers, seven scout cruisers, three torpedo gun boats, 36 gunboats, 28 steam-powered gunboats, and 30 torpedo boats—greatly outnumbered the Imperial Japanese Navy.[38] This fleet had been constructed at the behest of Li Hongzhang, who had been at the forefront of upgrading the Chinese navy during the Tongzhi Restoration period. Two battleships, five cruisers, and a significant number of torpedo boats had been built in German shipyards and a further four cruisers in British shipyards thanks to Li's efforts.[39] The Imperial Chinese Navy supplemented these foreign imports with ships built in their native yards at Jiangnan and Fuzhou, which gave China a formidable navy—at least on paper.

Both the Imperial Chinese Army and Navy greatly outnumbered their Japanese equivalents. But Japan had a crucial advantage that China did not: the unity of its government, polity, and armed forces. Japan had one national army and navy with a consistent doctrine, armament, training program, and command structure, epitomized by the formation of the Imperial General Headquarters at the outbreak of hostilities. By contrast, China's "army" was composed of two main forces (the Green Standard and Eight Banner Armies) and a variety of subsidiary forces (including the Huai Army, Manchurian Army, and the Yongying militias), each with its own command structure, table of organization and equipment, and armaments. The Chinese navy was even more divided, with the Beiyang, Nanyang, Fujian, and Guangzhou fleets each having different commanders with different agendas and objectives. Of these fleets, only the Beiyang and Guangzhou fleets even took part in the First Sino-Japanese War, with the Beiyang fleet being the only one to deploy large numbers of vessels. The piecemeal deployment of China's fleets indicated its navy's divided commands and lack of standardization, which in turn hindered the navy's ability to act in unison and coordinate large-scale operations. These would be vital weaknesses against an enemy that had effectively standardized and centralized its systems of administration and command in the previous decades.

Between June 6th and 8th, roughly 2,500 strong Chinese forces arrived at Asan in Korea, soon to be reinforced by 400 more men.[40] However, the Japanese responded by landing 3,000 more men at Incheon on June 12th-13th, followed by the remaining elements of the 9th Infantry Brigade by June 27th.[41] While the Chinese sought to negotiate with the Japanese after they crushed the Donghak Peasant Rebellion, these negotiations broke down, and by July 23rd, the Japanese marched into Seoul and took King Gojong of Korea hostage. The Chinese, forced to retreat north, dug in around Pyongyang with roughly 15,000 men and 32 artillery pieces, hoping these forces could stem the tide of the Japanese advance and allow them to use the city as a launching pad to advance south.[42]

Reinforced Japanese units pursued the Chinese towards Pyongyang, with the Wonsan Detachment landing on the east coast of Korea, and the 3rd and 5th Infantry Divisions landing at Chemulpo to the south.[43]

These forces, along with Major General Oshima Yoshimasa's 9th Mixed Brigade, advanced on Pyongyang from the east and south respectively.[44] Even though the 1st Army's initial assaults south and southwest of Pyongyang were repulsed, the 9th Mixed Brigade continued to attack the Chinese defenders from the south and southeast. Meanwhile, the Wonsan Detachment struck the rear of the Chinese defenders from the northeast. The concentric advance of the Japanese forces nearly encircled Chinese defenders, who were forced to abandon Pyongyang to the Japanese on September 16th.[45]

The Japanese landing at Chemulpo and rapid advance northwards required the Chinese to reinforce their forces in Korea. Instead of sending troops overland through northeastern China, the Chinese sought to speed up their dispatch of forces by sending reinforcements along a more direct route through the Yellow Sea, sending a total of five troop transports, escorted by three battleships, eight cruisers, two gunboats, and two torpedo boats to reinforce the Chinese forces in Korea.[46] As the Imperial Japanese Navy wanted to engage the Chinese fleet in the open, the Chinese played into the IJN's hands in attempting to reinforce Korea by sea.[47] The IJN sent a force centered around eight cruisers and two ironclads to intercept the Chinese convoy and send the Beiyang fleet to the bottom of the ocean.[48]

Soon after the Imperial Chinese Navy met their Japanese counterparts in battle at the mouth of the Yalu River, its lack of standardization and unity began to show. Perhaps no better example exists of the Chinese Navy's lack of unity and standardization than its formational confusion during the opening phases of combat. Even though the Chinese commander ordered his ships to go into line formation, they instead formed a wedge formation due to the differing speeds and signals on each ship.[49] The Japanese, by contrast, were able to easily form columns composed of two elements: a fast "flying" squadron and a main squadron. These forces were able to shape events to their liking, as the "flying" squadron and the main squadron diverged at the center of the Chinese formation. Subsequently, the Japanese attacked the opposite flanks of the Chinese wedge, with the "flying squadron" hitting the right and the main squadron hitting the left before emerging behind them.[50] These forces trapped the Beiyang fleet between them, destroying or damaging eight ships at close range.[51] Japan's victory in this battle established its effective naval superiority over China for decades to come.

Japan's victory in the Battle of the Yalu River, combined with its advances in Korea, meant that it could advance towards northeastern China. On September 25th, the Japanese crossed the Yalu River and entered Chinese territory.[52] Soon thereafter, the Japanese 1st Army diverged onto two axes of advance. One of these axes moved north towards Mukden and the other one moved across the base of the Liaodong Peninsula.[53] The main objectives of the southern force were to seize the harbors of Port Arthur and Weihaiwei. The landing of the 2nd Army's 1st and 12th Brigades on the Liaodong Peninsula supplemented the 1st Army's advance. These forces seized Dalian on November 7th, and then, on November 21st, seized Port Arthur, the two most important port cities in the Liaodong Peninsula.[54]

Further north, the 1st Army continued its advance in northeastern China. On December 13th, the 3rd Infantry Division seized the town of Haicheng with little resistance from the 5,000 strong Chinese garrison. After the seizure of Haicheng, most of the 1st Army stopped advancing due to the winter cold and a lack of transport.[55] Even though the Chinese took advantage of Japan's stalled advance to assemble roughly 20,000 men each at Liaoyang (roughly 60 miles north-east of Haicheng) and Niuzhuang (a few miles northwest of Haicheng) the Japanese repulsed multiple Chinese attempts to retake Haicheng.[56] The failure of these Chinese counterattacks and the warming temperatures meant that the main body of the 1st Army continued its advance in late February and early March, seizing the towns of Anshan, Liaoyang, and Yingkou. The fall of these towns meant there were no more defenders between the Japanese on the Liaodong Peninsula and Shanhaiguan, the primary pass leading towards Beijing.[57]

While the 1st Army advanced north into northeastern China, the Japanese laid siege to Weihaiwei, which hosted the remnants of the Beiyang Fleet. Weihaiwei was defended by roughly 10,600 Chinese, with a further 11-12,000 within 100km of the city.[58] Twenty-one 240mm guns, six 210mm guns, twelve 150mm guns, and fourteen 120mm guns supported

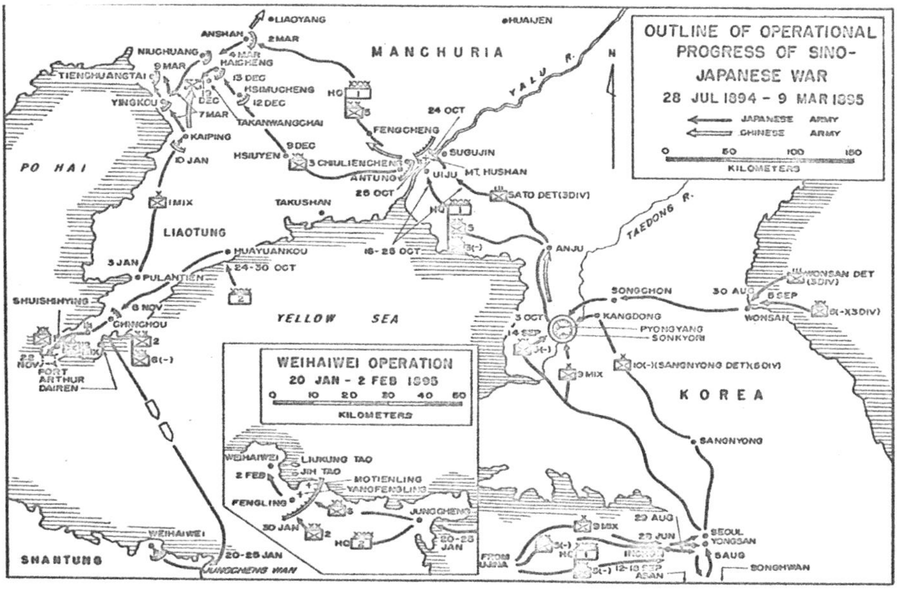

Operational Map of the 1st Sino-Japanese War. Note the location of key places such as Port Arthur (The Imperial Japanese Army in Manchuria, Map 1).

the defenders.[59] The coastal artillery and strong garrison, combined with booms that blocked entrance to the harbor made Weihaiwei a formidable obstacle. Nevertheless, in a joint operation between the Imperial Japanese Army and Imperial Japanese Navy, the Japanese landed on the Shandong Peninsula and laid siege to Weihaiwei for twenty-three days in January and February 1895. The landing on the Shandong Peninsula had the effect of outflanking the Chinese defenders, who were primarily focused on preventing an approach from the sea.[60] The unexpected landing of the Japanese to their rear, combined with effective pressure from the Imperial Japanese Navy meant that the defenders of Weihaiwei could not hold out, and the fortress fell on February 12th, 1895.

The Treaty of Shimonoseki and the Triple Intervention

The Japanese advance in northeastern China, combined with the fall of Weihaiwei, threatened Beijing's security and forced the Chinese to the negotiating table. Japan, the victorious power, dictated the peace terms, which were enumerated in the Treaty of Shimonoseki. The most important terms from the Treaty of Shimonoseki can be seen below:[61]

Article 1

China recognises definitively the full and complete independence and autonomy of Korea, and, in consequence, the payment of tribute and the performance of ceremonies and formalities by Korea to China, in derogation of such independence and autonomy, shall wholly cease for the future.

Article 2

China cedes to Japan in perpetuity and full sovereignty the following territories, together with all fortifications, arsenals, and public property thereon:—(a) The southern portion of the province of Fêngtien within the following boundaries [1]: The line of demarcation begins at the mouth of the River Yalu and ascends that stream to the mouth of the River An-ping, from thence the line runs to Fêng-huang, from thence to Haicheng, from thence to Ying-kow, forming a line which describes the southern portion of the territory. The places above named are included in the ceded territory. When the line reaches the River Liao at Ying-kow, it follows the course of the stream to its mouth, where it terminates. The mid-channel of the River Liao shall be taken as the line of demarcation. This cession also includes all islands appertaining or belonging to the province of Fêngtien situated in the eastern portion of the Bay of Liao-tung and the northern portion of the Yellow Sea. (b) The island of Formosa, together with all islands appertaining or belonging to the said island of Formosa. (c) The Pescadores Group, that is to say, all islands lying between the 119th and 120th degrees of longitude east of Greenwich and the 23rd and 24th degrees of north latitude.

Article 4

China agrees to pay to Japan as a war indemnity the sum of 200,000,000 Kuping taels; the said sum to be paid in eight instalments. The first instalment of 50,000,000 taels to be paid within six months, and the second instalment of 50,000,000 to be paid within twelve months, after the exchange of the ratifications of this Act. The remaining sum to be paid in six equal instalments as follows: the first of such equal annual instalments to be paid within two years, the second within three years, the third within four years, the fourth within five years, the fifth within six years, and the sixth within seven years, after the exchange of the ratifications of this Act. Interest at the rate of 5 per centum per annum shall begin to run on all unpaid portions of the said indemnity from the date the first instalment falls due. China shall, however, have the right to pay by anticipation at any time any or all of the said instalments. In case the whole amount of the said indemnity is paid within three years after the exchange of the ratifications of the present Act all interest shall be waived, and the interest for two years and a half or for any less period, if any already paid, shall be included as part of the principal amount of the indemnity.

According to the terms of the Treaty, China was forced give up its status as the suzerain of Korea, cede the Liaodong Peninsula to Japan, and pay 200,000,000 taels in reparations to the Japanese Empire. These were onerous impositions on a wounded country, as China had already been forced to make severe political and economic concessions to the Western powers in prior decades. That Japan, a relatively small nation, was able to exert such terms on the "Middle Kingdom," illustrates the limitations of Qing China's modernization process relative to Meiji Japan. Thanks to these limitations, Japan ascended to a position of prominence in East Asia and ended the ancient system of tributary relations. Japan's ascendant position was a significant humiliation for China because it long considered Japan to be a lesser state that was on the outskirts of the traditional tributary system. But China would have to adjust to the new reality, as its neighbor now had the ability to project power onto the Asian continent and even into China itself.

Japan's victory over China in the First Sino-Japanese War humbled the once dominant "Middle Kingdom," and brought it the spoils of victory. However, the terms Japan imposed on China in the Treaty of Shimonoseki also brought attention from European powers, who wanted to protect their interests in East Asia. Specifically, Japan's treaty-imposed inheritance of the Liaodong Peninsula was highly controversial because of its warm water ports (Dalian, Port Arthur, and Yingkou) and its proximity to Beijing. German Chancellor von Hohenlohe explained Germany's fears well when he told his boss, Kaiser Wilhelm II that Japanese control over Port Arthur would mean "Japanese dominion over the Gulf of Zhili and therefore a permanent threat to the Chinese capital."[62] The Russians were also greatly concerned by Japan's inheritance of the Liaodong Peninsula, because they coveted Port Arthur as a warm water port with access to the sea.[63] In turn, France wanted to back its ally Russia, with which it recently minted a defense pact.[64] The converging interests of these three countries led them to team up to prevent Japan from gaining control of the Liaodong Peninsula in what became known as the "Triple Intervention."

On April 23rd, 1895, Germany, France, and Russia presented a diplomatic note to the Japanese on the pretext that it was merely "friendly advice." The note indeed read as a relatively benign message, saying "that the possession of the Liaodong Peninsula would be a constant menace to the capital of China, that it would render illusory the independence of Korea and that it would be a perpetual obstacle for pacifying the Far East."[65] But despite the ostensibly peaceful intentions of the note, its broader implication was clear: Japan would be at war with Russia, Germany, and France if it refused. This implication was made more explicit by German representative Baron Alfred von Gutschmid, who used "threatening and peremptory language" to coerce Japan into acquiescing.[66] Faced with the possibility of fighting three of Europe's strongest nations, Japan's *genro* decided to comply with the Europeans' demands and forfeit control over the Liaodong Peninsula. While this decision was controversial with the Japanese public, it was undoubt-

MAP NO. 8

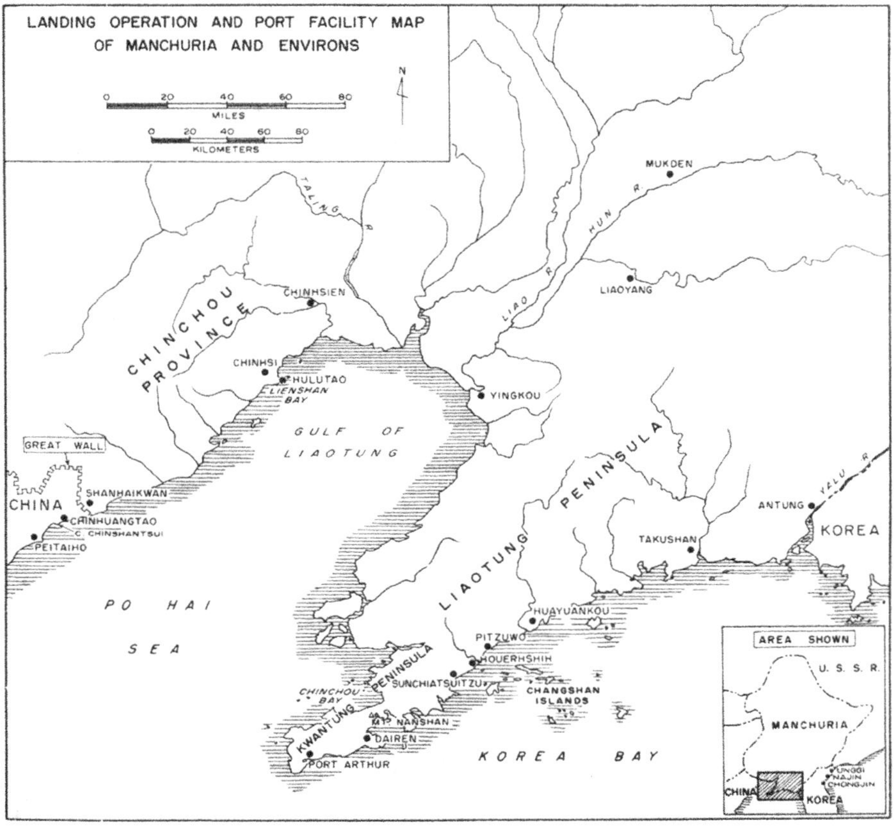

Map of the Liaodong Peninsula. Note the location of Port Arthur and Dalian (Dairen). (Strategic Study of Military Topography and Geography, Map 8).

edly the correct one, as Japan was not prepared to fight such a powerful coalition as Germany, France, and Russia.

Japan's acceptance of the "Triple Intervention" meant that it could not maintain its influence in northeastern China. Consequently, an opportunity opened up for the Russians to fill the vacuum, which they seized in 1896. In May of that year, Russians and Chinese signed the Li-Lobanov Treaty in which China agreed to allow Russian ships to use its ports in the event of emergencies and build an addendum to the Trans-Siberian railway across the provinces of Jilin and Heilongjiang. In exchange, Russia promised to come to China's aid in the event of a Japanese attack.[67] Russia and China also formed a joint bank so that China could get loans to pay off its war indemnities and Russia could fund its new "Chinese Eastern Railway," to Vladivostok.[68] But the most impactful agreement between Russia and China came in 1898, when the latter agreed to lease the Liaodong Peninsula for a period of twenty-five years. The specific provisions of this agreement, which was called the "Convention for the Lease of the Liaodong Peninsula," are excerpted below:[69]

> Art. I.—For the purpose of ensuring that the Russian naval forces shall possess an entirely secure base on the littoral of northern China, H. M. the Emperor of China agrees to place at the disposal of the Russian Government, on lease, the Ports Arthur (Liou-choun-kow) and Ta-lien-wan [Dalian], together with the water areas contiguous to these ports. This act of lease, however, in no way violates the sovereign rights of H. M. the Emperor of China to the above-mentioned territory.
>
> Art. II.—The frontier of the territory leased on the above-specified basis, will extend northwards from the Bay of Ta-lien-wan [Dalian] for such distance as is necessary to secure the proper defence of this area on the land side. The precise line of demarcation and other details respecting the stipulations of the present Convention will be determined by a separate Protocol which shall be concluded at St. Petersburg with the dignitary Siou-tzin-ch'eng immediately after the signature" of the present Convention. Upon the determination of this line of demarcation, the Russian Government will enter into complete and exclusive enjoyment of the whole area of the leased territory together with the water areas contiguous to it.
>
> Art. III.—The term of the lease shall be twenty-five years from the date of the signature of the present agreement and may be prolonged subsequently by mutual consent of both Governments.

> Art. IV.—During the above-specified period, on the territory leased by the Russian Government and its adjacent water area, the entire military command of the land and naval forces and equally the supreme civil administration will be entirely given over to the Russian authorities and will be concentrated in the hands of one person who however shall not have the title of Governor or Governor General. No Chinese military land forces whatsoever will be allowed on the territory specified. Chinese inhabitants retain the right, as they may desire, either to remove beyond the limits of the territory leased by Russia or to remain within such limits without restriction on the part of the Russian authorities. In the event of a Chinese subject committing any crime within the limits of the leased territory, the offender will be handed over to the nearest Chinese authorities for trial and punishment in accordance with Chinese laws, as laid down in Article VIII of the Treaty of Peking of 1860.

The leases of Port Arthur and Dalian gave Russia its coveted warm water ports, while the hinterland reference in Article II gave Russia control of the Liaodong Peninsula. In turn, these ports would be connected to the Russian Empire through the CER and the Trans-Siberian railway. The connection of the Trans-Siberian railway to the newly leased ports of Dalian and Port Arthur meant that Russia could send ground troops to East Asia and threaten Korea and north China. Alternatively, the Russians could conduct an amphibious offensive against Japan from its new bases. These realities meant that Russia occupied a favorable position in northeast Asia, despite its economic and administrative core being located thousands of miles away in European Russia.

Japan saw both Russia's lease of the Liaodong Peninsula and the construction of the Chinese Eastern Railway through northeastern China as a direct threat to its national security.[70] The Kwantung Peninsula gave Russia access to the warm water ports of Dalian and Port Arthur, while the railways across northeastern China allowed the Russians to move troops, equipment and supplies to the Russian Far East. With naval bases in the Liaodong Peninsula and Vladivostok, the Russians could threaten Korea from two sides, and thereby also threaten the security of the Japanese home islands themselves.[71]

Japan's position in Korea had been further undermined in 1895 when King Gojong took refuge in the Russian legation after the murder of his wife, Queen Min, by pro-Japanese factions in Korea.[72] King Gojong's refuge in the Russian legation augured closer ties between the government of Korea and Russia. Soon after Gojong got out of the Russian legation, pro-Japanese Korean officials were dismissed and punished, while Russian influence in the Korean court increased. In turn, Russian officials were appointed, and Russian businessmen were given privileged deals.[73] Perhaps even more worryingly, Korea allowed the Russians to begin surveying the southern part of Korea for prospective naval bases.[74] The increase of Russian influence in Korea threatened to turn it into an outpost for Russia if Japan did not act quickly.

To counter this influence, the Japanese began negotiations with the Russians over both powers' respective interests in Korea and northeastern China. These took the form of sending diplomat Komura Jutaro to Korea and sending Yamagata Aritomo to Russia. The resulting agreements Komura and Yamagata negotiated got King Gojong to return to the Korean palace, favored Korean independence, and sought to equalize the number of Russian and Japanese troops on the Korean Peninsula.[75] However, these rudimentary agreements did little to resolve the underlying tensions between Russia and Japan in northeastern China, which were rooted in the fundamental question of security for the Japanese home islands. Without a settlement that took into account Russian and Japanese interests in both northeastern China and Korea, Japan and Russia would be unable to resolve their respective problems.

For the next six years, the Russians and Japanese attempted to negotiate a settlement over the issue of northeastern China and Korea. Among those who supported negotiations was Ito Hirobumi, the four-time Japanese Prime Minister and one of the founding oligarchs of the Meiji polity. Many members of Ito's cabinet also supported this view.[76] Certain constraints within Russia—like not wanting to alienate Japan over its unilateral acquisition of Port Arthur—played into the hands of Ito and his cabinet, who seized the initiative and managed to negotiate the Nishi-Rosen agreement.[77] This agreement recognized Korea as a sovereign and independent nation while also guaranteeing that Russia would not obstruct the development of Japan's industrial and commercial enterprises in Korea. Even though the new agreement did not explicitly mention northeastern China, the context of the agreement (Russia's acquisition of Port Arthur) could be understood as a de facto *quid pro quo* in which Japan would take the predominant position in Korea while Russia had its way in northeastern China.

Although Ito and his cabinet favored negotiations with Russia over the status of Korea and northeastern China, others in Japan sought a more confrontational approach. Staff officers in the IJA and figures in the Foreign Ministry such as Komura Jutaro and Katsura Taro were the foremost advocates of Japan being more aggressive in East Asia. These men were younger experts that came of age at a time when Japan had narrowed the gap between itself and the West. They believed that Japan should flex its newfound muscle abroad and that a war with Russia was an opportunity to prove Japan's strength. This coalition of IJA staff officers and Foreign Ministry personnel came to be known as the "Kogetsukai."[78]

In the Kogetsukai's view Japan should strike against Russia before it entrenched itself in northeastern China and threatened Japan's interests on the Korean Peninsula.[79] Major General Iguchi Shogo's staff planning corroborated this point of view, which he revealed at an AGS conference in June 1903. In Iguchi's estimation, Japan could concentrate its forces in Korea and northeastern China more rapidly than Russia due to its shorter supply lines. Then, if the Japanese struck the Russian fleet at Port Arthur in a surprise attack, the IJA could defeat the Russian armies separately, or in detail, before its main forces arrived from European Russia. Because Iguchi believed that the Russians could not effectively sustain more than 250,000 men in the Far East, he asserted that they would be unable to overcome Japanese defenses in northeastern China. These handicaps would then force the Russians to arrive at a negotiated settlement after their forces were defeated in the opening phases of the campaign.[80]

Many within the Japanese government, including Ito (now head of the Privy Council) remained cautious about antagonizing Russia. Therefore, the Japanese government had its diplomat in St. Petersburg, Kurino Shincharo, propose the withdrawal of Russian troops from northeastern China in the aftermath of the Boxer Rebellion. But Kurino's diplomatic overtures were fruitless because the Russians simply used negotiations as a screen to reinforce the Russian Far East.[81] Upon realizing the Russians' intent, the Japanese government terminated diplomatic relations with Russia and prepared for war starting in January 1904.

The Russo-Japanese War

Much like the First Sino-Japanese War, Japan's enemy outnumbered it on paper. The Imperial Japanese Army had a total of roughly 850,000 trained men—many of whom were in a tiered reserve system—at the beginning of the Russo-Japanese War.[82] By contrast, the Russian Army had well over a million men at its disposal.[83] From a quantitative perspective, Japan should not have been the aggressor. However, a few key factors were to Japan's advantage; chief among them was space. While large, only 95,000 men from Russia's army were in the Russian Far East, and only a mere 24,000 were in reserve east of Lake Baikal.[84] The lack of manpower required the Russians to move forces east, but the 4,000 mile trip from Moscow generally took at least 15 days.[85] The Russians' disadvantages were exacerbated by the need to traverse Lake Baikal and the Trans-Siberian railway's single track.[86] According to AGS estimates, these hindrances meant that it would take six months to move eight Russian divisions to the Russian Far East.[87]

Japan had 257,000 infantry, 11,000 cavalry, and 894 artillery pieces that could be immediately deployed.[88] Reinforcements and supplies were a short boat ride away from the Japanese home islands, meaning that Japan could more rapidly concentrate its forces in northeastern China in the initial phases of the conflict. The Army General Staff predicated its operational plans on this advantage, with an explicit emphasis on encircling the Russian Siberian Independent Corps and Second Corps near Liaoyang before the Russians could reinforce the Far East.[89] The encirclement of these forces would allow the Japanese to defeat the different Russian forces "in detail," and mitigate Japan's numerical disadvantage.

A second advantage Japan had was its logistical system, which was organized into a series of "lines" that allowed for a layered supply system to take root.[90] These layers included the first line for small units, the second line at the divisional level, and the third line for the corps level and above. In more concrete terms, the first line consisted of frontline soldiers carrying two days of rations and a transport element carrying a further day's worth; the second line, at the divisional level, consisted of a transport column that carried four days of rations; the third line consisted of a "magazine column" that brought supplies from the nearest railhead, port, or depot. The Imperial Japanese Army's organized supply system allowed it to get food, ammunition, fodder, and clothes to its forces in a timely manner.

The Russian logistical system, by contrast, was greatly limited in its scope and capabilities. These limitations were in part the result of great distances (4,000 miles) and the incomplete status of the Trans-Siberian railway south of Lake Baikal in 1904. Even though transport columns carried eight days' worth of biscuits and six days' worth of groats at the regimental and divisional levels, most Russian foodstuffs (including chickens, rice, corn and pigs) had to be purchased or requisitioned from locals in the region.[91] While the Russians set aside cash for regimental commanders to buy food from local inhabitants, these foodstuffs were difficult to obtain in the wide open expanses of northeastern China, where population density was often low.[92] In light of these realities of organization, distance, and circumstance, Russian logistics capabilities lagged behind Japan's.

Japan's third advantage was its forces' table of organization and equipment (TO&E). The wartime strength of a Japanese infantry division in 1904 was roughly 18,500 men, 430 cavalrymen, and thirty-six guns. Of the guns, most were of the Model 31 type, which had a maximum effective range of six kilometers.[93] Russia's armed forces, on the other hand, had two different types of infantry formations: European and Siberian. While the European and Siberian corps each had two divisions, the Siberian divisions were only composed of 3,400 men and

20 guns, with a total of roughly 12,000 men in each Siberian corps. By contrast, the European Corps had 28,000 men and 112 guns at their disposal.[94] Even though Russian guns and howitzers were generally of good quality—the 66mm rapid fire gun could fire to a maximum of 6.4 kilometers at a rate of 15 rounds per minute, for example—the relatively negligible amount of firepower in the Siberian corps limited their ability to suppress or neutralize Japanese formations.[95] Furthermore, the small number of men in Siberian infantry formations meant that it would be impossible to sustain any protracted campaign in the Far East without rapid reinforcement. Given the distances and minimal railway system in the Russian Empire, reinforcement would not be rapid.

The fourth great advantage Japan had lay in the training and education of its officers. Ever since the days of the Iwakura Mission, the IJA had sent students abroad to learn the techniques, technologies, and tactical applications that defined modern weapons systems. In 1882, the Imperial Japanese Army had nine officers in France and three in Germany; in 1898, it had three in France and twelve in Germany.[96] These officers studied the cutting edge of military science that was developing in Europe; upon their return to Japan, they informed their superiors of the latest developments.[97] The education of officers abroad allowed Japan to adapt cutting-edge command, tactical, and strategic principles into its own doctrine. Even though the IJA still utilized some antiquated tactics (e.g. close order formations), it still adopted many of the Prussian Army's command principles and philosophies, which allowed officers in the field to make decisions on their own initiative. In turn, these command principles facilitated rapid decision-making and bold, offensive maneuvers to catch the enemy by surprise.

The Russian Army's command system, by contrast, was defined by passivity, poor military science, and rigid command principles. During the Russo-Japanese War, Russia's initial strategy was primarily defensive due to the lack of forces available in the Far East.[98] Furthermore, when Russian armies did attack, they utilized close order formations and frontal assaults, with few attempts to move concentrically or even outflank enemy positions.[99] That the Russians were pigeonholed into utilizing questionable tactics was in part the result of their command structure, which depended on long, detailed orders that attempted to address every contingency.[100] An overall strategy of passive defense, poor assault tactics, and a rigid command structure meant that the Russians were always reacting to Japanese moves, thereby allowing Japan to shape the course of operations to its liking.[101]

A fifth great advantage the Japanese had was their information and intelligence network, which was greatly boosted by secret societies that had formed in the previous decades. Throughout the 1880s and 1890s, Fukushima Yasumasa of the Genyosha Society used horseback rides across various parts of East Asia to gather intelligence about China's terrain, army, and logistical capabilities.[102] By 1901, as tensions with Russia were increasing, Uchida Ryohei, a former Genyosha member, founded a new secret society called the Kokuryukai, or the Black Dragon Society. Uchida founded the Kokuryukai to counter the Russians, whom he considered the most formidable adversary to Japanese expansion in East Asia.[103] The name of Uchida's society came from the Chinese name for the Amur River that defined the Russian-Chinese border in northeastern China.[104] The name of Uchida's society signaled its ultimate aims, which were conflict with Russia and Japanese primacy in northeastern China.

The specific anti-Russian orientation of the Kokuryukai meant that it dedicated the years between 1901 and 1904 to preparing for war with Russia. To these ends, the Kokuryukai established a Russian and Chinese language school to increase the number of Japanese fluent in both languages.[105] It also sent its members on intelligence missions to China, Russia, and Korea, which allowed the AGS to obtain a significant amount of valuable information with regards to Russian military deployments.[106] But the most important contribution of the Kokuryukai may have been in the field of propaganda, where it made significant headway in promoting war with Russia. The propaganda included a myriad of periodicals in its journal, *The Kokuryu*, and the explicitly anti-Russian writings of its founder, Uchida Ryohei. Uchida also expressed his anti-Russian sentiments to the Japanese public directly through speeches, warning audience members of the looming Russian danger in northeastern China and Korea and urging prompt Japanese action.[107] These speeches contributed to the creation of a jingoistic attitude in certain segments of Japanese society, thereby helping the Kokuryukai achieve its aim of anti-Russian action.

The AGS and the Imperial Japanese Army also had a significant formal network of espionage and intelligence gathering that supplemented the informal networks of the Genyosha and the Kokuryukai. The basis of this intelligence network lay in foreign language acquisition, starting at the staff college where all prospective officers learned French or German for a year.[108] After studying these languages, officers might specialize in others, including Russian and Chinese. The linguistic attainment of Japanese officers facilitated the creation of a strong formal spy network in both Russia and northeastern China, allowing it to survey the terrain, fund Russian dissidents and revolutionaries, and assess the Russian order of battle, troop movements, and logistical system.[109] The real time information about Russian strength that the Japanese spy network provided contrasted starkly with Russia's lack of information about the Japanese, which estimated the IJA's strength to be only a third of its actual level. This underestimation of Japan's strength was one of the foremost contributing factors to Russia's lack of preparedness for war in 1904.[110]

On February 9th, 1904, the Japanese government issued a declaration of war on the Russian government. However, even before this declaration of war, Japan already initiated hostilities against the Russians, as many in Japan believed that declaring war was an "inexplicable European custom" that disadvantaged it as a smaller nation.[111] Taking advantage of the element of surprise, Japan dispatched six battleships, 10 cruisers, 30 destroyers, and four torpedo boats to attack the unsuspecting Russians on February 6th.[112] This surprise attack was a premeditated operation and succeeded in achieving operational surprise against an enemy not prepared for the "proximity of war."[113] Consequently, of seven Russian battleships and six cruisers in Port Arthur's harbor, two battleships and a cruiser were sunk in a nocturnal torpedo attack by ten Japanese destroyers on February 8th.[114] Even though Russian shore batteries prevented the Japanese fleet from engaging the rest of the Russian fleet in the harbor, the Japanese had struck the first blow.

After the surprise attack on Port Arthur, the Japanese quickly mobilized their 1st Army. On February 16th, the 1st Army's 12th Infantry Division landed at Chemulpo. The presence of the 12th Infantry Division played a great role in coercing the Korean government to assent to Japanese advisors handling key administrative tasks in return for paper thin guarantees of sovereignty and independence.[115] The Japanese then strengthened their toehold in Korea by landing the Imperial Guards Division at Chenampo on March 14th. By April 1st, the Japanese 1st Army was able to advance all the way to the Yalu River, marching twenty-five miles a day and encountering little Russian resistance along the way.[116] The Russians withdrew their forces towards the town of Liaoyang, one-third of the way up the southern branch of the CER, and adopted a passive, defensive posture.[117]

The Russians' withdrawal to Liaoyang left the Russian garrison at Port Arthur isolated. Consequently, the Japanese activated the 2nd Army—1st, 3rd, and 4th Infantry Divisions—to land on the Kwantung Peninsula. These forces came ashore at Houershi and established a beachhead, with plans to capture the cities of Jinzhou and Dalian.[118] The Russians, whose forces were stretched thin on the Kwantung Peninsula, could do little to stop the landings, and did not seriously resist them.[119]

Instead of constructing extensive beach defenses along the length of the Kwantung Peninsula's shores, the Russians heavily fortified the narrow strip of land that connects the Kwantung Peninsula with the rest of the Liaodong Peninsula. This strip of land represented a difficult obstacle for the Japanese, as Mount Nan dominated the center of the strip with flat ground only on the flanks of the mountain. Due to the prominence of Mount Nan, the 2nd Army had to conduct a frontal assault on the mountain, and suffered heavy losses in the process.[120] Only after the Japanese tried and failed to take Mount Nan nine times did the 4th Infantry Division on the right finally capture a piece of the Russians' defensive position.[121] The 4th Infantry Division's progress on the right threatened the single Russian regiment defending Mount Nan, which had little margin for error in the cramped space of the Kwantung Peninsula. As General Fock refused to commit any reserves from Port Arthur to counterattack the Japanese, the Russians had to

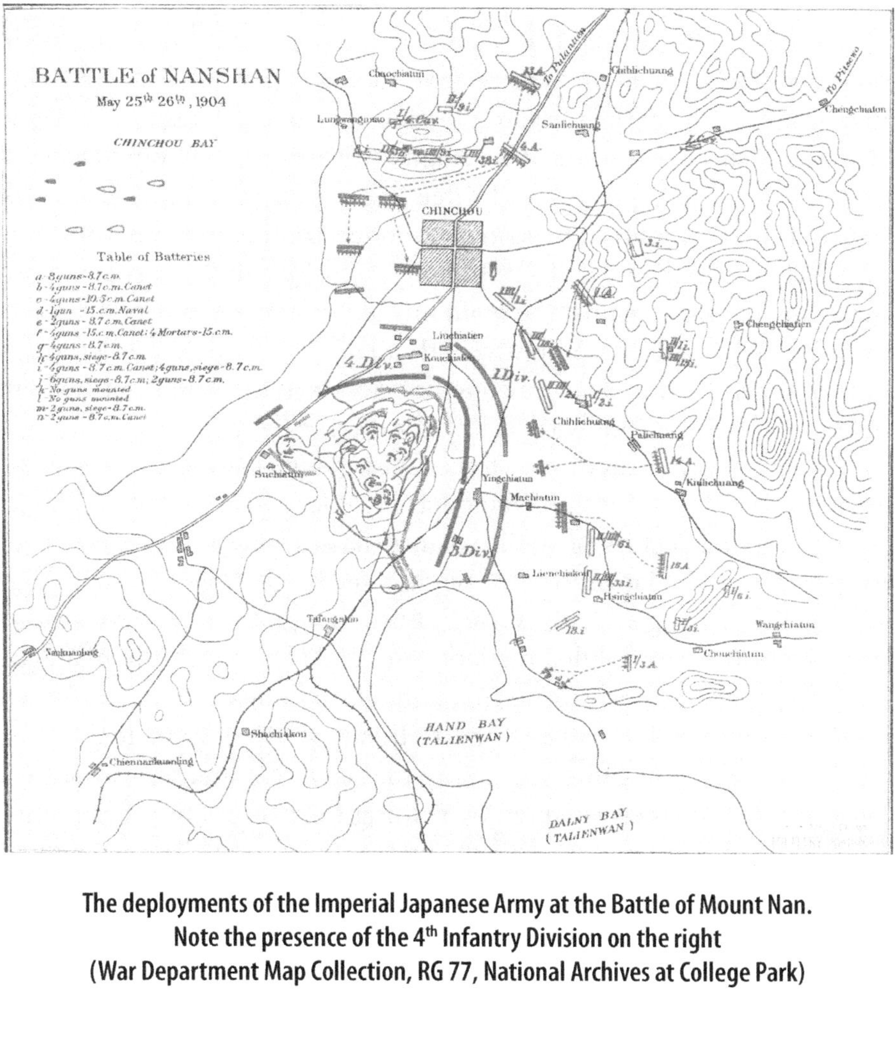

The deployments of the Imperial Japanese Army at the Battle of Mount Nan. Note the presence of the 4th Infantry Division on the right (War Department Map Collection, RG 77, National Archives at College Park)

fall back, lest their forces on the center and right be rapidly encircled in the confined spaces of the Kwantung Peninsula.[122]

The capture of Mount Nan and the establishment of a foothold on the Kwantung Peninsula were the first steps necessary to seize Port Arthur and the Russian fleet embedded there. But to take the heavily fortified city, the Japanese in the Kwantung Peninsula needed new forces, which they received in the form of the 3rd Army.[123] The 3rd Army, which consisted of the 1st, 9th, and 11th Infantry Divisions, had only recently been formed, but was still tasked with assaulting Port Arthur's in-depth, prepared positions. Despite its relative lack of experience, the 3rd Army managed to capture the forward positions of these defenses in July. The 3rd Army's success continued in August with the 1st Infantry Division on the left capturing two out of three lines of Russian trenches on Hill 174. But the 1st Infantry Division could not seize the Russians' third line of trenches, and the 9th and 11th Infantry Divisions in the east could not make any progress against the main line of resistance.[124] Frontal assaults on prepared positions combined with a lack of adequate artillery support doomed further 3rd Army attempts to crack the walnut of Russian defenses at Port Arthur, with its forces suffering heavy losses.[125]

The 3rd Army's heavy losses forced it to change tactics. To narrow the area on which each division would have to concentrate its forces, the 3rd Army assigned sectors of the Russian defenses to each division under its command. The Japanese also sought to advance more gradually, digging in after seizing each objective.[126] Even though the new tactics didn't produce immediate results, the Japanese 3rd Army eventually managed to concentrate its efforts on seizing Hill 203, an objective that had been previously overlooked.[127] Here, the 1st Infantry Division made repeated assaults. The first assaults, like many that came before, were repulsed. However, the Japanese adapted by shifting their forces to other points, which helped them capture Hill 203 by December 5th.[128] The assault had cost the Russians 3,000 men, while the Japanese suffered 8,000 casualties in eight days of assaults on the hill.

The 3rd Army's capture of Hill 203 allowed the Japanese to set up their artillery on the landward side of the hills. Forward observers could guide and correct the fires of Japanese guns from the top. Control of the hills around Port Arthur allowed the Japanese to dominate its harbor and sink four battleships and two cruisers.[129] Inside the city itself, diseases were running rampant and medical supplies were running out. Therefore, even though the Russian garrison still consisted of 25,000 men with sufficient ammunition, the loss of the fleet combined with deteriorating conditions led the Russians to surrender Port Arthur to the Japanese on January 2nd.[130] Port Arthur's surrender marked the end of Russia's dreams of having a warm water port in East Asia.

Defenses of Port Arthur. Note the location of Hill 203 (War Department Map Collection, RG 77, National Archives at College Park).

Further north, the Japanese 1st Army sought to seize the city of Liaoyang, which was on the southern branch of the Chinese Eastern Railway and hung over the Liaodong Peninsula. The Japanese feared that the Russians could use the city as a springboard to launch an offensive that could trap the Japanese 3rd Army. Consequently, the Japanese assigned the 1st, 2nd, and 4th armies to seize Liaoyang and protect the flank of the 3rd Army. These forces would face General Aleksey Kuropatkin's in-depth defenses that had been prepared in the preceding months, with the outermost fortifications being on average roughly 27 kilometers (17 miles) from the city.[131] The depth and strength of the Russian defensive preparations would make Liaoyang a difficult objective for the Japanese to capture.

On the night of August 25th, the Japanese 1st Army on the right attacked the Hongshaling-Hanpoling-Gongchangling line, seizing it with minimal difficulty. Two nights later, the 2nd Army on the left joined in, advancing to a line that connected the towns of Dengapou and Zhongzuotan along the railway line into the city.[132] The Japanese advances between August 25th and 27th forced General Kuropatkin to order a withdrawal from the outermost positions to shorten the Russians' lines. Even though Kuropatkin's withdrawal order made sense, his failure to commit reserves to a counterattack meant that the Japanese advance continued. By the time Kuropatkin finally ordered a counterattack on September 1st, it came too late: his forces had already abandoned the main line of resistance and Japanese artillery was within range of Liaoyang.[133] The failure of the Russian counterattack meant that the Japanese entered Liaoyang by September 4th.

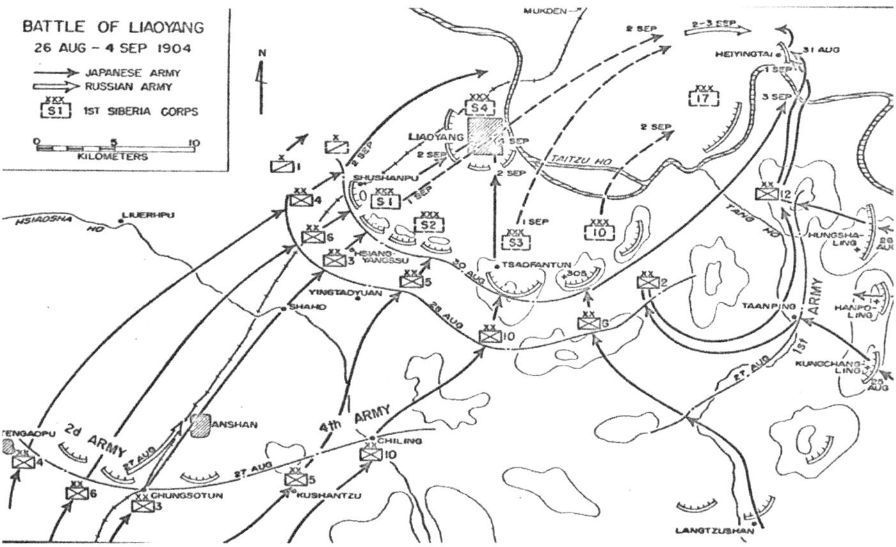

Map of the Battle of Liaoyang (The Imperial Japanese Army in Manchuria, Map 4)

As 1904 gave way to 1905, the Japanese continued their advance into northeastern China. However, time was working against them: Russian reinforcements were continuing to make their way towards the Russian Far East along the Trans-Siberian Railway. Consequently, many Japanese commanders (most notably General Oyama) wanted to initiate a decisive battle at Mukden before these reinforcements could arrive.[134] Here, the Russians were vulnerable with roughly 350,000 men deployed over a front roughly 100 miles long.[135] Such a long defensive line meant that the Russian defenses had little depth, making their forces vulnerable to a concentrated Japanese offensive.

Even though the Japanese only had 300,000 men, they intended to concentrate their forces against the Russians' flanks and encircle the defenders in a pincer movement.[136] To achieve such an encirclement, the Yalu Army on the Japanese right (and Russian left) and the 3rd Army on the left were to attack and penetrate both flanks. The Yalu Army would attack the first, followed by the 3rd Army, which would advance in a wide circle towards the area northwest of Mukden and envelop the Russians.[137] Meanwhile, the 2nd and 4th armies would preoccupy the center of the Russian line in frontal combat.[138] The Japanese hoped that the sequenced deployment of the Yalu and 3rd armies would bait the Russians to deploy reserves to counter the Yalu Army's strike while allowing the 3rd Army to inflict the decisive blow on their right.

The Japanese forces at Mukden began their offensive on February 20th, 1905. The Yalu Army attacked the Russians at Chinghocheng on February 22nd and overran their positions here by February 24th. The rapid fall of Russian positions here led Kuropatkin to reinforce his left (the Japanese right) at the expense of the forces on his right, just as the Japanese intended. Thanks to Kuropatkin's weakening of his right flank, the 3rd Army was able to rapidly advance against relatively little resistance.[139] By early March, the Battle of Mukden was over, with the Russians losing roughly 40,000 to death or capture and a further 49,000 wounded.[140] Japan's victory at Mukden represented its greatest triumph on land during the course of the Russo-Japanese war.

While the Battle of Mukden represented Japan's greatest land battle, a naval threat still existed in the form of the Russian Baltic Fleet, which consisted of 11 battleships, nine cruisers, and nine destroyers. But this fleet had a variety of disadvantages that would handicap it in battle against their Japanese counterparts.[141] First, the speed of its ships was not uniform, forcing it to move at the speed of its slowest

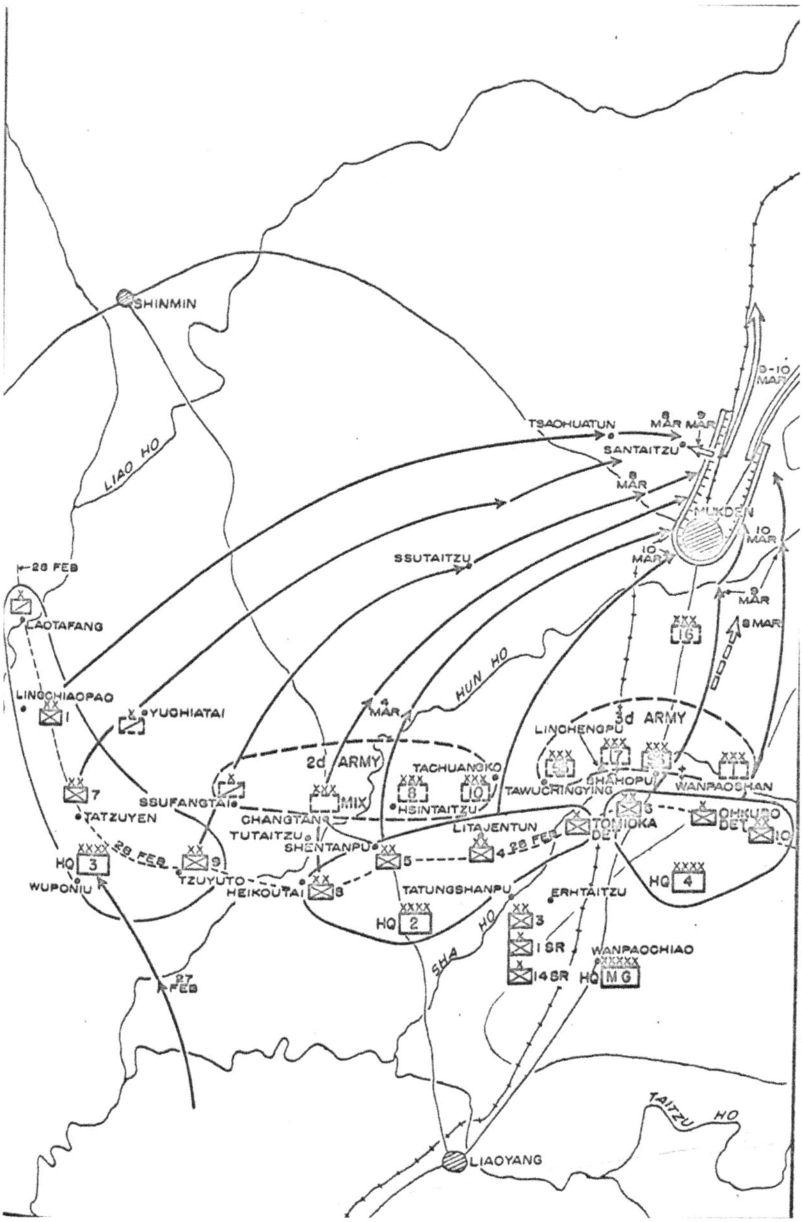

The Japanese scheme of maneuver at the Battle of Mukden (The Imperial Japanese Army in Manchuria, Map 8a).

ship; this reality meant that Japanese vessels moved three to four knots faster than their Russian counterparts. Second, many of the Russian ships attracted barnacles on the long journey from the Baltic, further slowing them down. Third, the Russian fleet was unwieldy and clumsy; it needed to bring colliers for the purposes of re-coaling—they couldn't recoal in neutral ports—as well as vast amounts of provisions. Fourth, Russian sailors were conscripts who were treated vastly different from their officers. Finally, Russian armor-piercing shells had a delayed fuse that often did not explode. By contrast, Japanese vessels utilized shells that could explode on contact and penetrate the non-armored parts of Russian ships.[142]

On the morning of May 27th, 1905, Japanese observers saw the lights of two Russian hospital ships in the Tsushima Strait and reported their presence to Admiral Togo Heihachiro.[143] Soon thereafter, Togo's forces set out from Pusan, Korea, and came into contact with the Russians that afternoon. The Russians were to the north of the Japanese, moving east in two columns. The Japanese were moving south, away from the Russians. At 2:02 p.m., Togo, realizing his forces were moving away from the Russians, opted to make a 180-degree turn from south to north. This decision meant that the smaller ships at the back of Togo's column would not be vulnerable, and also allowed all his ships' broadside guns to engage the Russians' column.[144] By contrast, the Russians could only respond with the forward most gun on their forward most ship. Togo's deft tactical deployment "crossed the T" of the Russians, ultimately allowing the IJN to sink eight battleships. The destruction of these battleships eliminated a significant part of the Russian navy and established Japan as the foremost naval power in East Asia.

The Treaty of Portsmouth and the End of the Russo-Japanese War

The Battle of Mukden and the Battle of Tsushima marked two of the greatest triumphs for both the Imperial Japanese Army and Navy in their respective histories. However, despite these great triumphs, the

Japanese could not sustain the war for much longer. In order to fund the Russo-Japanese War, Japan sold government bonds in Europe and America, which led to a forty-fold increase in its national debt.[145] Furthermore, the Russians continued to reinforce Siberia and the Russian Far East, and their numerical strength was growing every day. Consequently, the Japanese sought a mediator for a peace treaty, which they found in the form of American President Theodore Roosevelt.

Thanks to the U.S. president, Japan and Russia came to terms in the Treaty of Portsmouth, which recognized Japan's claims over Korea, mandated the withdrawal of Russian forces from northeastern China, and gave the Empire of Japan sovereignty over South Sakhalin and the Kurile Islands. Additionally, the Russians also ceded their leases in the Liaodong Peninsula to the Japanese, while maintaining control of the Chinese Eastern Railway. The most relevant terms in the Treaty of Portsmouth relating to Korea and the Liaodong Peninsula are excerpted below:[146]

> **Article 2** The Imperial Russian Government, acknowledged that Japan possesses in Corea paramount political, military and economical interests, engage neither to obstruct nor interfere with the measures of guidance, protection and control which the Imperial Government of Japan may find it necessary to take in Corea.
>
> It is understood that Russian subjects in Corea shall be treated exactly in the same manner as the subjects and citizens of other foreign Powers, that is to say, they shall be placed on the same footing as the subjects or citizens of the most favored nation.
>
> It is also agreed that, in order to avoid all cause of misunderstanding, the two High Contracting Parties will abstain, on the Russian-Corean frontier, from taking any military measure which may menace the security of Russian or Corean territory.
>
> **Article 5** The Imperial Russian Government transfer and assign to the Imperial Government of Japan, with the consent of the Government of China, the lease of Port Arthur, Talien and the adjacent territorial waters and all rights, privileges and concessions connected

> with or forming part of such lease and they also transfer and assign to the Imperial government of Japan all public works and properties in the territory affected by the above-mentioned lease.
>
> The two High Contracting Parties mutually engage to obtain the consent of the Chinese Government mentioned in the foregoing stipulation.
>
> The Imperial Government of Japan on their part undertake that the proprietary rights of Russian subjects in the territory above referred to shall be perfectly respected.

Russia's acknowledgment of Japan's "paramount political, military, and economical interests" in Korea effectively meant that it ceded its influence in the Peninsula, and that Japan would have a free hand to establish control over the country. Additionally, Russia's cessation of its lease over the Liaodong Peninsula and the rights to the southern branch of the Chinese Eastern Railway (soon to be called the South Manchuria Railway) meant that its dreams of obtaining a warm water port with access to the sea were dashed; instead, Japan would have control of this region, which amounted to roughly 3,462 square kilometers of land, along with roughly 1,142.3 square kilometers of railways.[147] China affirmed Japan's new acquisitions in the Treaty of Beijing, and also promised not to build railways "in the neighborhood of and parallel to" Japan's newly acquired South Manchuria Railway (SMR).[148] These treaty terms meant that Japan established a hegemonic position in northeastern China, with a monopoly on transport in the Liaodong Peninsula. But the strategic position of Japan's new territories meant that it could also project power into north and northeastern China, and continue to pursue dreams of imperial expansion. This reality would become more apparent in future decades, as Japan sought to expand its empire and secure its ever more far-flung interests.

Japan's two "splendid little wars" were great successes that established Japan as the most powerful country in East Asia. Not only did Japan's victories signal the success of its modernization program, but they also established Japan's emergence as an Empire in East Asia. The treaties that ended the First Sino-Japanese and Russo-Japanese wars gave

Japan control of Taiwan, Korea, and the Liaodong Peninsula, all of which were ideally situated as base areas for further expansion. Additionally, Japan obtained naval supremacy in the Sea of Japan and the waters of northern China, which would allow it to project power into East Asia with little resistance. In a more macro-historical sense, Japan's defeat of China delivered a death-blow to the tributary system, while its defeat of Russia established it as a formidable regional power, worthy of the west's respect. With these triumphs, Japan had arrived in the world of European modernity as an imperialist power; now it would need to administer, maintain, and defend its newly won territories.

Chapter 3

Japan Consolidates its Empire

In the aftermath of the Russo-Japanese War, Japan stood triumphant: It had defeated Russia's army on the plains of northeastern China, smashed its navy in the Strait of Tsushima, and asserted control over Korea and the Liaodong Peninsula. These achievements helped Japan establish itself as a regional power, alongside the European powers and the US. But this new position remained precarious as Japan had to effectively manage, administer, and extract wealth from its new colonial possessions on the Asian continent. Even more importantly, Japan had to protect its hard-won "line of advantage" in Korea and northeastern China. Fears of a "Russian war of revenge" remained ensconced in the psyche of key Japanese policy makers. After 1911, China became a source of great uncertainty, as a revolution brought down the Qing Dynasty and fragmented the country. Therefore, Japan mobilized significant amounts of manpower, machinery, and funds to defend its empire, including a small new formation called the "Kwantung Garrison" to defend the Liaodong Peninsula. Little did the Garrison's founders realize that this small force, tucked in a remote corner of northeastern China, would go on to change the course of world history.

The Founding of the Kwantung Garrison

In the aftermath of the Russo-Japanese War, the so-called "Kwantung Military Government" administered the Liaodong Peninsula. Under the jurisdiction of General Oshima Yoshimasa, this government ran the day-to-day affairs of Japan's newly won territorial concession, while also preparing to defend the region in the event of a Russian war of revenge.[1] Some of these preparations included so-called "dual use initiatives" to turn the Liaodong Peninsula into a source of supplies for the IJA.[2] But the new military administration was not meant to be a permanent fixture, as evidenced by the fact that the soon to be dissolved IGHQ took command of it in November 1905.[3] In order to defend Japan's interests and assets in the region, a new permanent force would need to be created.

In 1906, Japan formed a garrison force known as the Kwantung Garrison to protect its interests in the Liaodong Peninsula. This force consisted of one infantry division, a heavy artillery battalion, and six independent infantry battalions.[4] The infantry division was garrisoned in the Kwantung Peninsula, while the six independent infantry battalions defended the SMR.[5] Even though the Treaty of Peking nominally allowed the Japanese to deploy fifteen men per kilometer of railway, in practice the number of men deployed was more limited, "having in view the actual requirements."[6] These forces, while small in number, were thought to be a necessary placeholder in the event of another Russo-Japanese war or until China was able to protect foreigners' lives and property.[7] Therefore, the formation of the Kwantung Garrison was a relatively uncontroversial step among Japan's security and foreign policy establishments.[8]

Despite the consensus on the Garrison's formation, controversy soon erupted over its command structure. The Imperial Japanese Army wanted the garrison to be integrated with the Kwantung Leased Territory Administration under the command of the Governor General. This command structure would allow the Governor General—a lieutenant general or general in the IJA—to deploy the Kwantung Garrison to defend key installations (e.g. the SMR line), while also coordinating the mobilization of resources in a war with Russia.[9] But many civilians, especially those in the diplomatic corps, opposed the IJA's proposed command structure. They believed the Governor General's wide-ranging responsibilities would concentrate too much authority in the hands of the IJA.[10] The diplomats' unease with the Governor General's authority was representative of the broader tug-of-war between the security and foreign policy elements of the Japanese system, in which civilians attempted to prevent the military from hijacking policy.

In the debate over the formation of the Governor General position, the IJA had a trump card: its established administration on the Liaodong Peninsula during the Russo-Japanese war.[11] Faced with a *fait accompli*, civilian diplomats scrambled to secure some concessions from the IJA on the issue of administering the Liaodong Peninsula. Their efforts made it so that the new Governor General would have a civilian administrative apparatus under his command, and the Kwantung Garrison would report to the Foreign Minister. Furthermore, the Governor General would also have to report to the government in Tokyo on all subjects unrelated to military matters.[12] But overall, the Governor General still had a great deal of authority, since the Kwantung Garrison was under his command. Moreover, the IJA formed an Army Bureau to help the Governor General handle the Garrison's day-to-day military affairs. This combined authority allowed the Governor General to "take charge of the defense of the territory within the limits of his jurisdiction" and deploy the Garrison in the event of a looming threat to the SMR.[13] While the concentration of power in the Governor General would speed up the decision-making process in the event of war with Russia, it also meant the IJA and the Governor General had disproportionate expertise in and control over Japan's affairs in northeastern China.[14]

The Founding of a Corporation

Even though the formation of the Kwantung Garrison allowed Japan to defend and expand its interests in China's northeast, the question of how to manage the new SMR remained to be answered. In 1905, Count Goto Shimpei and General Kodama Gentaro provided an answer when they proposed the formation of the South Manchuria Railway (SMR) corporation. In Goto and Kodama's view, a corporation separate from the Kwantung Governor General and Kwantung Garrison would streamline tasks related to administration and business in the SMR area and Liaodong Peninsula. The company could also use its administrative and economic role to help the IJA, as Goto later put it, be an instrument of "military preparedness in civil garb."[15] Perhaps most importantly, the company would help Japan extract resources from northeastern China and thereby make inroads into regional markets.[16] For these reasons, Goto and Kodama favored the formation of a new SMR corporation that would allow Japan to have a "proper administration of railroads" befitting its newfound status as an imperial power.[17]

In May 1906, Kodama presented the idea for an SMR corporation to Prime Minister Saionji Kinmochi. His government approved the proposal and established a 77-member special committee that would form the basis of a new "special company" to manage the railway.[18] The Japanese government would have a fifty percent share in the new company, the Imperial household a further one percent, and the public forty-nine percent.[19] This division of shares was born out in the company's initial 200-million-yen capital investment, in which roughly one half came from the Japanese government and the other half from private investors.[20] The Japanese government also maintained the right to appoint the company's executive officers.[21] The Japanese government's control over the SMR meant it was an instrument of Japanese policy rather than an exclusively profit-seeking venture.

The Japanese government appointed Goto Shinpei as the first president of the SMR corporation on November 8th, 1906.[22] Almost immediately, Goto and his staff prioritized policy over profit when they changed the SMR's railway gauge from the narrow-gauge tracks used in Japan to the four foot, eight inch U.S. standard gauge track used in China and Korea.[23] Thanks to the efforts of Goto and the SMR corporation, the process of transitioning the SMR to a standard gauge line was completed within a year.[24] This timely transition allowed Japan to rapidly deploy forces from the Liaodong Peninsula to other places in northeastern China, and made possible the construction of further railroads from Korea. Therefore, changing the SMR's gauge better prepared Japan for Russia's "war of revenge" and exemplified the SMR's role as an instrument of Japanese imperial policy.

The SMR corporation continued to prioritize policy over profit when it took measures to upgrade its railway capacity. Some of the measures necessary to increase capacity included double-tracking, upgrading the railway tracks from 80 pounds per yard to 100 pounds per yard, and buying high speed, high-capacity railway cars.[25] The SMR also negotiated an 80-million-dollar loan from foreign banks to buy 180 locomotives and 2,060 railway cars from the United States.[26] These railway cars were generally high capacity, with no freight cars under 60,000 pounds and many over 100,000 pounds.[27] Due to these adjustments, the SMR would be able to move roughly 12,000 men per day towards Harbin. But the SMR also had the highest daily per mile operating cost for any railroad in the Japanese Empire by 1912.[28] Such high operating costs meant that the SMR would need to find alternative streams of revenue.

To attract business and fulfill the Imperial Japanese Army's "dual use" initiatives, the SMR corporation constructed warehouses and mills, built the port of Dalian to military specifications, founded hospitals, and began selling coal.[29] Nonetheless, soy was by far the most lucrative and important commodity for the SMR by the 1910s. Initially, the railway had a hard time breaking into the soy markets, as the Liao River and Peking-Mukden Railway represented two routes that had both cheaper rates and more established networks of merchants and farmers. But the combination of Dalian's deep-water harbor and the SMR's ability to attract banks, commodity exchanges, and insurance agencies to the city made it a place from which Chinese farmers could securely access capital and international markets.[30] Therefore, the SMR was able to establish itself as the foremost player in northeastern China's soy trade, with forty-seven percent of its revenues coming from soy.[31] Indeed, as early as 1907 the company turned an 800,000 yen profit in six months despite the company's investment in large-scale, expensive railway upgrades.[32] Soy came to occupy a place of such great importance in northeastern China's economy that between 1909 and 1913 soy bean products amounted to 56.6 percent of northeastern China's export trade.[33] In subsequent years and decades, the importance of the soy trade would only continue to increase, as the SMR wrought greater and greater profits from exporting soy products abroad.

SMR policy in the Liaodong Peninsula included another important element: Japanese settlement in northeastern China. Goto Shinpei, the

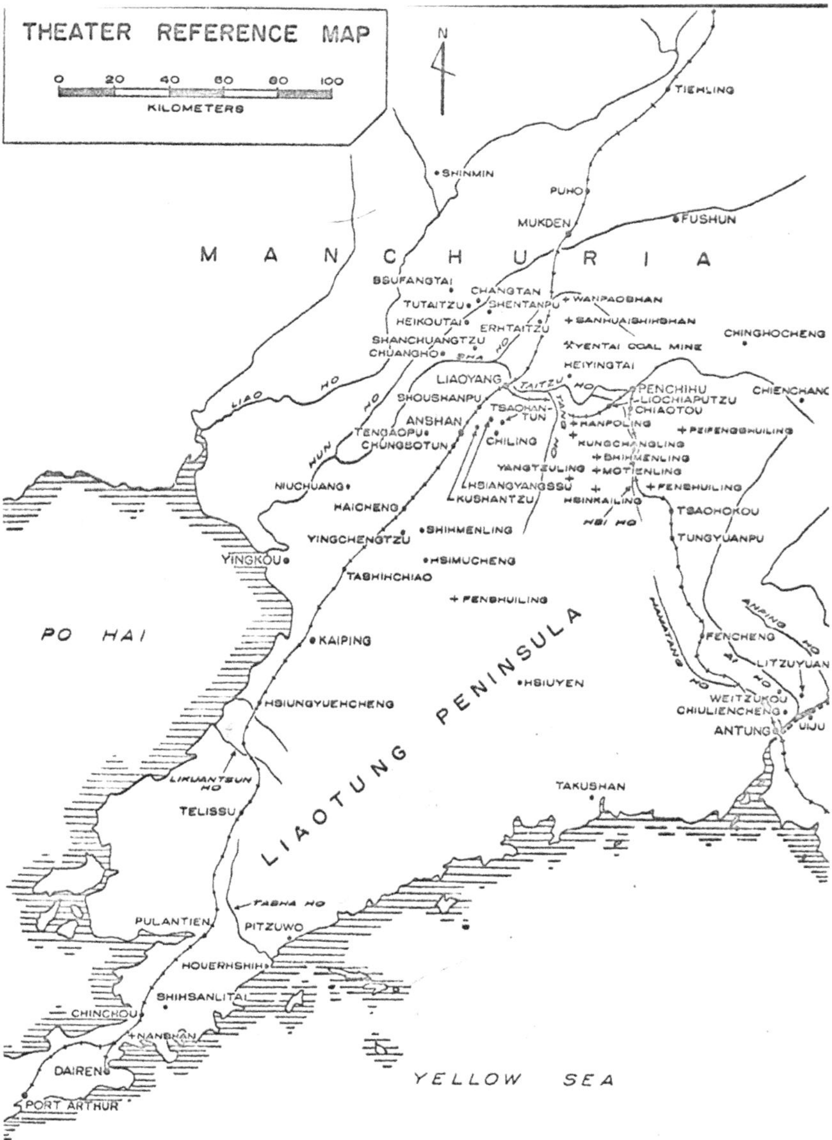

Liaodong Peninsula and the South Manchuria Railway (The Imperial Japanese Army in Manchuria, Map 2).

SMR's first president, was one of the foremost advocates for this policy, which he believed would expand Japan's influence in the region. Goto's views on these matters are elaborated on in the following excerpt, found in Ito Takeo's memoirs:[34]

> The Russo-Japanese war will probably not conclude this situation in Manchuria with a single war. Will a second war actually come at some point in the future? If our prospects for victory are indeed to be anticipated, we must forestall our rival. If the chances for victory are not yet firm, we must devote ourselves to preparation and wait for the right time. Even if we are beaten in the second war, we must reserve sufficient strength for counter-measures. In short, we must occupy a position in Manchuria wherein we are far in the lead, and full of vigor and vitality, confront our rival with fatigue. The necessity of doing things primarily involves the sort of skills needed for running a railroad... Our plan for reaching this end is 1) management of the railroad; 2) opening of the coal mines; 3) bringing over settlers and 4) creating facilities for cattle raising. Among those points, settlers [from Japan] must be seen as the most vital task. We fought a victorious war in vain to gain the superficial appearance today of having sovereignty over Korea, and although our diplomacy met with good results, in fact there is no success with such simple and rapid results. The attainment of sovereignty originated in the earlier emigration of Japanese settlers to Korean soil. The [European] Powers retained a superior position and we proved unable to contest them with words. Furthermore, we can resolve the Manchurian issue by moving people there... If within ten years we are able to induce 5,000 Japanese to emigrate to Manchuria we will not have to commence hostilities recklessly, despite Russian strength. Control over the tempo toward war or peace will fall firmly to our hands. Even were Russia to defeat us in battle, we would still not lose the foundation to recover our success.

Goto's fear of a Russian war of revenge was a relatively common view within Japan's security and foreign policy establishment, but his belief in the necessity of Japanese settler colonialism in northeastern China was more unique. In Goto's estimation, Japanese settlement would start a long-term demographic process that would allow Japan to establish control over northeastern China, regardless of economic or security conditions on the ground.[35] The importance Goto placed on settler colonialism made it a decisive part of his imperial vision, which included all of China's northeast. Others in the Japanese foreign policy establishment, notably Foreign Minister Komura, concurred with Goto's vision, and also favored sending Japanese settlers to northeastern China.[36] Such support for settler colonialism in northeastern China made it an important factor in Japanese policy in the decades ahead.

Starting in the 1910s, the SMR adopted certain measures—financial assistance to purchase land, free use of company machines, and providing convenient services, for example—to encourage settlement in northeastern China.[37] Other measures included granting the families of railway guards roughly 50 acres of farmland within the railway zone once their term of service ended. The Kwantung Government-General also attempted to create experimental farming villages, with specially cultivated strains of rice to survive northeastern China's harsh climate.[38] The SMR and Kwantung Government-General hoped these incentives would attract settlers and thereby help solidify Japan's hold over the Liaodong Peninsula and the railway zone.

Despite the existence of some official support, relatively few Japanese emigrated to the region in the 1910s, with only 37,000 in the Kwantung Leased Territory and a further 25,000 in the SMR zone.[39] Of those that did emigrate, most were "adventurers" that had great difficulty out-competing Chinese farmers in the region.[40] There were two key factors that made the Liaodong Peninsula inhospitable for Japanese settlement at that point. First, there was a lack of arable land available in Japan's colonial territories, which amounted to only 15,000 acres.[41] Second, Japan's modernization process had ushered in a trend of urbanization and opportunities for factory work in cities. In the 1910s, with Japan's industrial economy on the rise, rural peasants still preferred to go to Japanese cities to find work. As Owen Lattimore put it in his book *Manchuria Cradle of Conflict*, "the average peasant would rather move to a town and become a factory worker than move abroad

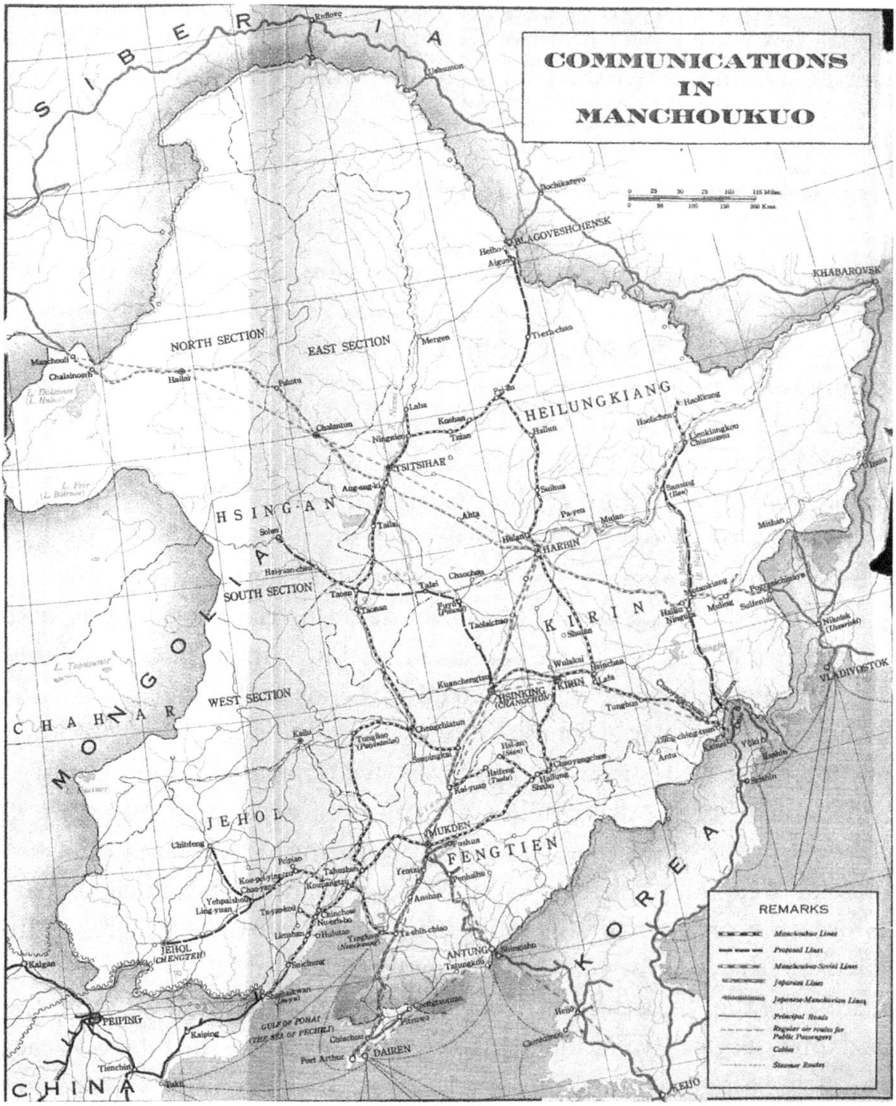

A railway map of northeastern China. Note the location of the South Manchuria Railway and the Andong-Mukden (Andong is spelled "Antung") Railway, denoted in red (Manchukuo Yearbook, 1934)

to take up land. The average townsman will move gladly to a bigger town but not to the country."[42] So long as Japan's economy remained afloat, few settlers would move to northeastern China; only an economic crisis could provide the impetus for a mass emigration.

The Imperial Defense Plan of 1907

Japan's new empire in northeast Asia raised broader strategic questions about how resources should be allocated in the future. This question was the subject of intense debate, with different factions in the armed forces taking different approaches to the question. Yamagata Aritomo represented Choshu Clique traditionalists who placed the utmost importance on preparing for a Russian war of revenge. To counter the Russians, Yamagata proposed the creation of a large fifty-division army, with twenty-five active divisions and twenty-five in reserve.[43] The AGS chief of staff, Kodama, by contrast represented those who disagreed with Yamagata, and believed that Japan did not have the industrial or economic basis to support such a big force. Kodama's faction wanted to create a thirty-eight-division army with nineteen divisions active and nineteen in reserve, with greater emphasis on increasing the new forces' firepower and mobility.[44] The Imperial Japanese Navy represented a force entirely independent of the army and sought to both expand and upgrade its fleet at the army's expense. In the IJN's estimation, it needed to increase its total of capital ships from two to 16 (eight battleships and eight heavy cruisers) in order maintain rough parity with the United States in the Pacific.[45] The complicated relations between the Yamagata and Kodama factions of the IJA, as well as between the IJA and IJN, stoked both intra- and inter-service rivalry that was to continue well into the future, as each competed for government funds and support.

The Imperial Defense Plan of 1907 was an attempt to balance the desires of both the IJA and IJN: The IJN would get its 16 capital ships (eight battleships and eight heavy cruisers) and the army would have a long-term target of fifty divisions (with twenty-five active and twenty-five in reserve).[46] In total, this increase in strength would raise the military budget from 37 million yen in 1906 to 65 million yen in 1908.[47] Russia was designated as the IJA's most likely enemy while the United States was designated as the IJN's most likely enemy, with the army conducting operations on the Asian continent (particularly in northeastern China and the Ussuri river region) and the navy conducting operations in the Pacific.[48] The written intent to achieve the force procurement goals of both the army's Yamagata faction and the IJN indicated a formal parity between the services that had not existed in previous years.

Despite the formal pretense of parity between the IJA and IJN in the Imperial Defense Plan, the IJA still held the upper hand in an informal sense.[49] In part, the IJA's higher status stemmed from the fact that the 1907 Imperial Defense Plan designated Russia, primarily a land power, as Japan's most dangerous threat. The Imperial Defense Plan also noted the importance of defending Japan's mainland Asian colonial territories: Korea and the Liaodong Peninsula. The need to defend these territories is made clear in the preamble of the Imperial Defense Plan, which stated that "having crossed the sea and established claims in Manchuria and Korea, we can no longer consider national defense to be a matter of planning military action within the realm of our island nation. Unless we are prepared to take the offensive against our enemies overseas, our national security cannot be guaranteed."[50] As the IJN had already secured the shipping routes between Japan and Korea, this statement implicitly emphasized the relatively greater importance of the IJA, which would bear the brunt of defending Japan's "line of advantage" in northeastern China. Therefore, even if there was nominal parity between the two branches of the Japanese military, northeastern China's status as the most important operational theater and Russia's status as Japan's most likely enemy made the IJA the first priority of Japanese defense planning.

Operationally, the 1907 Imperial Defense Plan emphasized the need for the IJA to rapidly deploy and concentrate its strength in northeastern China before the Russians could.[51] Some troops would be deployed by sea to the Liaodong Peninsula and move up the SMR; others would sail to Korea and move to Mukden from there. The Kwantung Garrison would be tasked with defending the SMR, and establish a position north of Mukden to cover the deployment of these forces. Once assembled, these forces could advance towards Harbin, dig in and force the Russians to attack along inhospitable avenues of approach. To cover these forces, the Japanese could also launch secondary attacks from northeastern Korea towards Russia's Maritime Province.[52] IJA staff planners thought that if Harbin fell, the Russians would be forced to the negotiating table, as Japan would have an impregnable advantage of interior position. The IJA therefore wanted to build a multitude of railways in northeast Asia to speed up deployment times and supplement the already-existing SMR.

IJA planners assigned the highest priority to reconstructing the Andong-Mukden railway line. While a temporary narrow-gauge railway did exist between Andong and Mukden, the IJA sought the construction of a permanent standard-gauge railway that would make the line between these cities uniform with the other railways in China and Korea.[53] The reconstruction of this railway line would somewhat relieve the IJA's dependence on the IJN to convoy forces to the Liaodong Peninsula, and allow the IJN to instead focus its effort more on the Tsushima Strait.[54] The reconstruction of the Andong-Mukden line would also relieve the bottleneck of the SMR and offer another route to deploy forces toward northeastern China. For these reasons, the IJA supported the reconstruction of the Andong-Mukden line, and regarded this task as vital to the fulfillment of its operational plans and Japan's national security.

Despite the IJA's desire to reconstruct the Andong-Mukden line, the Chinese government stood in the way: While it agreed to allow the Japanese to reconstruct the Andong-Mukden line in the 1905 Treaty of Peking, it did not want Japan to expand its strategic control over its Three Eastern Provinces, Fengtian, Jilin and Heilongjiang.[55] Consequently, the Chinese government "dragged its feet" and put up obstacles to constructing the Andong-Mukden railway line.[56] The Chinese government succeeded in forestalling construction for two years but Japan lost its patience by 1907, taking unilateral measures to prepare the way for constructing the new railway line: purchasing land, deploying guards, and issuing an ultimatum in 1909.[57] Such measures forced the Chinese government's hand; it ultimately removed its opposition to the retracking of the Andong-Mukden line. The only saving grace was that it managed to procure a promise from the Japanese that they would return the Andong-Mukden line to Chinese sovereignty by 1923, the same year the lease on the Liaodong Peninsula expired.[58]

The Chinese government's acquiescence to the re-tracking of the Andong-Mukden railway line was a great victory for the IJA. But another adversary closer to home also opposed the IJA's project: the SMR corporation, which believed that the Andong-Mukden line would be bad for business due to its high projected construction costs (estimated at 37 million yen) and low volume of projected traffic. The SMR instead proposed the construction of a line from Andong to Dashiqiao, which was south of Mukden.[59] However, the IJA insisted on the necessity of building the line to Mukden, since it needed the most direct route to Harbin to fulfill its operational plans.[60] The IJA also reiterated the SMR's primary function was to be an instrument of national policy, not merely a profit-seeking entity.[61] These arguments, combined with the Foreign Ministry's opposition to the SMR's position, forced the SMR to concede and begin the re-constructing the Andong-Mukden line in 1909. Re-construction was complete by 1911, and the IJA had its second railway line to Mukden.[62]

Between 1905 and 1911, Japan consolidated its control over the Liaodong Peninsula and SMR line by establishing the Kwantung Garrison and SMR corporation. These measures made it possible for Japan to defend, exploit, and administer its new territories in northeastern China. But in the process, the Japanese government placed a significant amount of influence in the hands of the Kwantung Garrison and the Imperial Japanese Army. In the coming years, these entities would leverage their proximity to events in the colonies to influence the direction of Japan's China policy. More often than not, this influence brought Japan closer to war on the Asian continent.

Open Door, Boxer Rebellion, and Revolution in China

Japan's two wars netted it a significant sphere of influence in northeastern China. But Japan was not the only imperialist power trying to carve up China: the United States, United Kingdom, Germany, France, Russia, and Italy all sought to get their "slice of the Chinese melon" between 1850 and 1900. During this period, all of these imperialist powers expanded their spheres of influence using aggression and intimidation, to the point where it seemed a partition of China would be likely in the near future.[63] Such a partition would likely lead to each power expanding their spheres of influence at the expense of the others, who would not be allowed to engage in commercial activities outside their respective spheres.[64]

In 1899, United States Secretary of State John Hay announced his "Open Door" policy, which sought to establish the principles of free trade and equal opportunity for all countries seeking to do business in China and also nominally sought to uphold China's sovereignty and territorial integrity.[65] The Open Door policy was designed to prevent the denial of access to China's vast market through the formation of exclusive "spheres of influence." Hay's policy should therefore be understood as being designed for the benefit of the United States, which had fewer concessions in China and hence sought an arrangement where all powers were allowed to freely engage in unfettered commercial activity in the country.[66] Even though the other imperialist powers were initially somewhat hesitant to accept Hay's new policy they eventually capitulated.[67] Consequently, the Open Door Policy became an integral part of America's China strategy, serving as the framework through which future US administrations would approach Sino-US relations.

Despite the differing interests of the imperialist powers in China, they were forced to unite in 1900 to suppress the so-called Boxer Rebellion. This rebellion was started by a Chinese secret society known as the "Society of Righteous and Harmonious Fists" which sought the removal of foreigners by force.[68] While the "Boxers," had some initial success in cutting the Beijing-Tianjin railway and besieging the foreign legation in Beijing, the imperialist powers' maritime supremacy allowed them to land an expeditionary force at the vital Dagu Forts, which had coastal artillery that defended the approaches to Tianjin and Beijing.[69] After capturing the Dagu Forts, troops of the so-called "eight nation alliance" (United States, United Kingdom, Japan, Russia, France, Germany, Italy, and Austria-Hungary) took Tianjin and eventually Beijing, putting down the Boxer Rebellion and forcing the Qing government to come to terms.[70] In the Treaty of Beijing, the great powers entrenched their domination over China, imposing roughly 450,000,000 taels in reparations on the Qing government and forcing China to permit foreign troops to be garrisoned in Beijing's vicinity.[71] The foreign powers' coordinated invasion of China, combined with the excessive reparations, meant that in many regards the Treaty of Beijing marked the low point of China's "Century of Humiliation."

To pay the "Boxer indemnity," the Qing government nationalized multiple railways (most notably the Szechuan-Hankou and Hankou-Guangdong railways) with the intent of selling their rights to western banks in exchange for loans.[72] Many Chinese landlords, gentry, and merchants who held stock in these railroads vehemently opposed the Qing government's "nationalization" project, as did many politically-conscious students, workers, and peasants. These forces together formed the "Railway Protection Association" in Szechuan province in 1911, with the objective to "protect the railway and break the [foreign-loan] treaty."[73] This movement eventually formed the basis for a broader "rights recovery movement" in the whole of China to take root.

At first, the rights recovery movement and the "Railway Protection Movement" merely protested the Qing government's subservience to foreign powers. However, on October 10, 1911, elements of the New Army revolted against the Qing Dynasty in what came to be known as the Wuchang Uprising.[74] The leaders of the New Army's revolt were ideologically influenced by the "Tongmenghui," a secret society of Chinese nationalists led by Sun Yat-sen and Song Jiaoren.[75] Despite the "New Army's" relative youth, its revolt succeeded in establishing a new military government in Hubei province. This success allowed eighteen other provinces in south and central China to secede from the moribund Qing Dynasty.[76]

The success of the Wuchang Uprising surprised Sun, Song, and the other leaders of the Tongmenghui, who had been forced into exile in earlier years. Sun, who was in America at the time appealing to overseas Chinese for financial support, returned to China in December 1911.[77] Upon his return, Sun found a new ally in the form of Yuan Shikai and his Beiyang Army, which had entered negotiations with the Republican rebels in November 1911.[78] In the course of the negotiations, Yuan (who had been ordered by the Qing court to suppress the Republican rebellion) betrayed the Qing Dynasty, forcing three-year-old Emperor Puyi to abdicate.[79] The Qing Dynasty subsequently fell, and China began its belated entry into the world of western modernity with Sun becoming the president of the new Republic of China on January 1st, 1912

The new Republic of China's elections went ahead in 1912, with Sun Yat-sen's newly formed Kuomintang (KMT) winning the overwhelming majority of the seats in the National Assembly.[80] But Yuan Shikai had little intention of allowing the establishment of a "Chinese Republic" and representative government. Therefore, in 1913, Yuan assassinated Song Jiaoren, one of the two most important leaders of the KMT.[81] Soon thereafter, the newly formed KMT rebelled against Yuan's rule, but did not have the necessary force to overthrow Yuan and his Beiyang Army. Instead, Yuan dissolved the recently elected parliament, and the KMT was once again forced to go into exile.[82]

Yuan Shikai's pole position in Chinese politics allowed him to take out a 25,000,000-pound loan from a consortium of foreign banks.[83] Moreover, he also anointed himself Emperor in 1915.[84] But many Chinese were not keen on paying back foreign creditors or having another emperor. Consequently, Yuan soon found himself facing a rebellion in Yunnan province, where a military officer named Cai E formed a "National Protection Army" to take up arms against Yuan's newly formed government.[85] At Cai's behest, Yunnan declared independence from Yuan's new government in December 1915.[86] Cai's declaration of independence from Yuan's new government led to a chain reaction of governors and military officers in other provinces, including Guangxi, Guangdong, Zhejiang, Szechuan and Shaanxi declaring independence from Yuan Shikai's government.[87] While Yuan desperately tried to assert his authority over the rest of the country, he was forced to give up his claim to the title of emperor in 1916.[88] Shortly thereafter, Yuan died, and China's edifice of unity passed for the foreseeable future.

Japan's Activities in Northeastern China

In Japan, there was a lot of debate about how to handle China's dissent into revolution and fragmentation. Some supported an activist stance in northeastern China, while others sought to avoid conflict. Of those who supported a more active stance, most were in the Imperial Japanese Army. Many army officers saw the chaos in China as an opportunity to expand Japan's rights and concessions or even partition the country into a series of different smaller polities.[89] Of these figures, Yamagata Aritomo was the most prominent due to his status as a *genro* and influence over the IJA and emperor.[90] In policy terms, Yamagata and his acolytes in the "Choshu Clique" wanted to send two infantry divisions to the Three Eastern Provinces, which dovetailed with their desire to continue the expansion of the IJA outlined in the 1907 Imperial Defense Plan.[91] These measures would strengthen the IJA's position vis-à-vis the IJN and the government, while also prioritizing the IJA's desired outpost of northeastern China.

While Yamagata was the most prominent supporter of intervention in northeastern China, other members of the security establishment similarly sought to take advantage of China's disarray to advance Japan's interests. Many of these figures saw China as weak and vulnerable given its revolutionary upheaval. Their point of view is reflected in the passage below, which originates from a December 1911 report about China from the Military Affairs Bureau and Police Bureau:[92]

> The current Chinese policies and internal disorder have already reached a critical point. This war is expected to have long-lasting consequences. Reports from Japanese consular and military officials indicate that tensions are rising across various regions and that firm countermeasures may become necessary.

> A few days ago, a figure from the former cabinet confided to a friend in private, saying, "China's current arrogance is an affront to Japan's national dignity. To yield to such unreasonable claims would be intolerable.
>
> I cannot imagine this country continuing to exist for long. It is unbearable to accept humiliation from a nation that recently lost a war.

Even though these figures advocated "firm countermeasures" and an intervention in China's internal affairs, it is notable that the authors believed that the Japanese public was still not ready for firm measures and a war that might result. Such a war might have disastrous political consequences, especially in light of the fact that Japan just got done with the costly Russo-Japanese War a mere six years earlier. Nevertheless, they still advocated intervention in China on the basis of defending Japan's honor, which they believed was under threat from Chinese nationalists. In the opinion of the report writers, Japan's victory in the First Sino-Japanese War meant that it should be able to enjoy its hegemonic position in northeastern China without resistance. Such imperious attitudes meant that many Chinese saw Japan as their most rapacious adversary despite its relatively late entry into the imperialist "great game."

The IJA's belligerent China policy encountered significant opposition from a coalition of different actors including civilian politicians, the Imperial Japanese Navy, and bureaucrats in Prime Minister Saionji Kinmochi's government.[93] All of these groupings had different reasons for opposing the IJA's aggressive China policy. Prime Minister Saionji's Cabinet opposed Yamagata's policies because it did not want to fund the two-division expansion of the IJA. The Seiyukai political party, under the leadership of Hara Kei, wanted to create a government controlled by the political parties and believed aggression in northeastern China would only strengthen the IJA's position. The IJN wanted to pursue its dreadnought construction agenda at all costs, and was therefore a competitor for funding from the IJA. Many Japanese bureaucrats were from Satsuma and greatly resented the Choshu Clique's attempts to establish a stranglehold on politics. The united bloc of these disparate interests all made it exceedingly difficult for the IJA to form a coalition that could see to its interests and desires.[94]

With regards to the Chinese Revolution, the Saionji government's official policy was to avoid supporting any of the combatants.[95] While some in the government supported using the chaos to gain further concessions in the Kwantung Peninsula and the Three Eastern Provinces, they stopped short of seeking to actively undermine China's sovereignty over its northeast.[96] But despite the Saionji government's official policy, many within the IJA still sought to implement aggressive policies in northeastern China. They found useful allies outside the government in the form of the Kokuryukai, which had spent much of the previous decade operating behind the scenes to facilitate the expansion of the Japanese Empire in northeastern China.[97] The Kokuryukai's efforts to create discontent in China would be a point of collaboration for the AGS, the Kwantung Garrison, and the Kokuryukai's "patriotic adventurers" who all sought to establish Japanese control over the region. As their actions contradicted the Saionji government's official policies, they fell under the broad umbrella of *gekokujo*, and helped establish the precedent for middle-ranking military and intelligence officials to formulate government policy.

The most notable and effective Kokuryukai operative was Kawashima Naniwa, who first went to China during the Boxer Rebellion as a translator for the IJA. After the Rebellion, he was the principal of a police school and used his position to establish a series of connections that would help him infiltrate Chinese politics.[98] The most important connection Kawashima formed was with Prince Su, a member of the House of Aisin-Gioro (the Qing Dynasty) and China's future interior minister.[99] Kawashima also made strong connections among Mongolian separatists in northeastern China, securing the support of eight different Mongolian princes for a project of Inner Mongolian separatism.[100] Kawashima's influence was so extensive that he even had agents advising Sun Yat-sen and Yuan Shikai, the two most important figures in post-revolution China. The extent of Japanese infiltration of Yuan's inner circle is evidenced by the fact that twenty Japanese officers served as his advisors by the time of the Russo-Japanese War in 1905.[101]

Kawashima used his connections to the Kokuryukai and the AGS to lobby for separating the Three Eastern Provinces from the rest of China.[102] Kawashima had a more than willing audience in the form of the Utsunomiya Taro and the AGS's Intelligence Division, which had formed in 1908 for the purposes of centralizing Japan's intelligence collection. The Intelligence Division was home to many China hawks who sought a more aggressive China policy; indeed, the division had used the Kokuryukai's networks in China to probe the feasibility of expansionist plans in the Three Eastern Provinces.[103] Kawashima's established connections and plans made his gambit attractive to the AGS, which could maintain the veneer of not contradicting official Japanese government policy while maneuvering to support Kawashima's vision behind the scenes. Kawashima was therefore the perfect vessel for the AGS because he could initiate and drive the action without overt AGS direction.

Kawashima, the Kokuryukai, and the AGS laid the foundation for the Kwantung Garrison to trigger a "Manchurian" separatist movement. The Kwantung Garrison provided the catalyst for action when it demolished a railway bridge near Shanhaiguan (a strategically important mountain pass that separated northeastern China from the rest of the country) in February 1912.[104] Soon thereafter, Governor General Oshima (who also was commander of the Kwantung Garrison) urged the central government in Tokyo to send two divisions to augment his force. Governor General Oshima's ostensible intention was to deploy these divisions in a punitive campaign against the supposed Chinese perpetrators.

In reality, Oshima's request was part of a larger scheme that included the Intelligence Division, AGS, and Kawashima Naniwa. Prior to the demolition of the bridge at Shanhaiguan, Kawashima had built a Mongol-Manchu army. He had made arrangements to arm and train these forces, smuggling weapons through Haku Masumi's joint Chinese-Japanese gang, and also found a political figurehead in the form of Prince Su, whom he smuggled out of Beijing to Port Arthur to lead the new army and state in late January 1912.[105] Moreover, Kawashima used his contacts in the AGS and Intelligence Division to procure funds, weapons, and advisors for his new separatist army.[106] Kawashima's influence among the

AGS was so great that he got it to approve his signing of a contract on Japan's behalf pledging support for the separatist cause.[107] This contract was in brazen contradiction to the Japanese government's official policy, and therefore represented a severe act of *gekokujo* on Kawashima's part.

Unfortunately for Kawashima, the AGS, and the Kwantung Garrison, others outside of their orbit realized something was amiss. One of the first to realize the Japanese plot was the Chinese viceroy of the Three Eastern Provinces, Zhao Erxun, who complained of Japanese arms smuggling in the region.[108] Soon thereafter, Ijuin Hikokichi, the Japanese envoy in Beijing, heard of Kawashima's training and armaments program for Mongol separatists. His counterpart at the Mukden consulate, Ochiai Kentaro, learned of the "Manchurian" separatist plans that same month from Colonel Takayama Kimimichi, and promptly informed the Foreign Ministry. The fact that figures who opposed the plot learned of its existence indicates the plotters' failure to maintain the standards of operational security necessary to keep their plans secret. This lack of operational security jeopardized the plot's viability, which was at the mercy of government officials in Japan and China.

In Japan, Foreign Minister Uchida Kosai had been aware of Kawashima's plot prior to being informed by Consul General Ochiai Kentaro. Upon hearing the news, he vehemently opposed it, as did Prime Minister Saionji, who did not want to bog Japan down in a costly war in northeastern China.[109] To clarify their stance on the Chinese Revolution, Saionji's cabinet decided to reiterate its position of neutrality on February 20th. Their public statement was an implicit warning to Kawashima and his fellow plotters that the government would not support their efforts to foment Mongol-Manchu separatism. Foreign Minister Uchida and Prime Minister Saionji drove this point home in conversations with AGS chief of staff Fukushima Yasumasa and Kwantung Territory Governor General Oshima, who they told to get their subordinates in order and put an end to the plans for an armed separatist uprising.[110] Therefore, Saionji and Uchida placed their faith in the integrity of the chain of command and the obedience of their subordinates.

Fortunately for the Saionji government, Fukushima began taking steps to rein in the plotters in northeastern China. In February 1912, he asked that Kawashima send Prince Su away from Port Arthur. When Kawashima responded that Prince Su could not leave, Fukushima, Kawashima, and Oshima brokered a deal in which Prince Su could stay in Port Arthur as long as he did not participate in any independence movement.[111] In March, Fukushima wrote to Major Taga Mukeyuki that it was not the correct time to foment a Manchu-Mongol separatist movement and that Japan must wait for the right circumstances to arise before it stirred the Manchurian pot.[112] While Major Taga continued to arm and train the Mongol separatists, the April recall of Kawashima to Tokyo and Fukushima's re-posting from AGS Chief of Staff to Kwantung Territory Governor General deprived the Mongolian independence plan of two of its strongest backers. Kawashima's recall and Fukushima's removal as the AGS Chief of Staff effectively ended the IJA's first attempt at fomenting "Manchurian" independence.[113]

The Saionji Kinmochi government's position on expansion in northeastern China combined with its decision to veto the IJA's two division expansion ruffled feathers in the IJA. Consequently, Yamagata and the Imperial Japanese Army refused to nominate an Army Minister to Saionji's Cabinet in the wake of Army Minister Uehara Yusaku's resignation, which they engineered.[114] The IJA's ability to make such a move stemmed from the so-called "active-duty rule" of 1900, which mandated that an active duty army officer—who was still institutionally liable to his colleagues in the IJA and could be pressured on policy matters—had to serve in the post of Army Minister. The resulting chaos led to Saionji and his government stepping down.[115] However, when the IJA-favored Katsura Taro attempted to form his new Cabinet later in 1912, the Imperial Japanese Navy also refused to nominate a Navy Minister, leveraging the same 1900 active-duty rule the army had. Katsura was forced to appeal to Japan's recently anointed Emperor, Taisho, to grant him a rescript that would force the IJN to nominate a Navy Minister.[116]

Katsura's government was short-lived, in no small part because of the controversy that ensued after he obtained the Emperor's rescript.[117] The Imperial Japanese Navy opposed Katsura because of his association with the IJA and because he called on the emperor to intervene in the IJA-IJN political dispute. Katsura's formation of a new political party, the Doshikai, alienated the Seiyukai and Hara Kei, who led an effort to issue a vote of no confidence in him in the Diet.[118] The press also vehemently opposed Katsura, as many journalists were liberal-minded people that wanted separation between the throne and the government. Many strong newspaper articles against Katsura's rescript maneuver emerged, and these in turn incited many protests, some of which turned violent.[119] Indeed, even Katsura's former friends in the IJA were dissatisfied with him; Yamagata did not like his lack of political astuteness, and his failure to approve the addition of two more divisions into the IJA meant he was not privileging the IJA's agenda. These factors, combined with Katsura's poor health, forced him to resign in February 1913.

After Katsura resigned, the government of Admiral Yamamoto Gonnhyoe ascended to power. Yamamoto's government collaborated with Hara Kei and the Seiyukai to end the "active-duty rule" and reduce the IJA's power. The new administration also overtly favored the IJN, promoting naval expansion while opposing IJA expansion. But despite the Yamamoto cabinet's anti-IJA bias, it still supported measures to expand the empire in northeastern China.[120] These included negotiations around iron mining concessions in Anshan, fostering Manchurian-Korean-Japanese economic integration through the reduction of tariffs on the Sino-Korean border, and re-negotiating Japanese rights to railway branch lines that fed into the SMR. The new administration also countered Chinese railway expansion in the Three Eastern Provinces by proposing the construction of five Japanese funded railways.[121] Even though the Chinese did not want to accept these terms, the Foreign Ministry's threat of "serious repercussions" meant that they signed an agreement to construct the five railway lines in principle. This agreement came to be known as the Manchuria-Mongolia Five Railways Agreement.[122]

The concessions the Yamamoto government won in northeastern China still did not placate the Imperial Japanese Army or the patriotic adventurers. In large part, these elements' dissatisfaction with Yamamoto and his government stemmed from their unwillingness to support China's anti-Yuan Shikai uprising in the summer of 1913. Despite

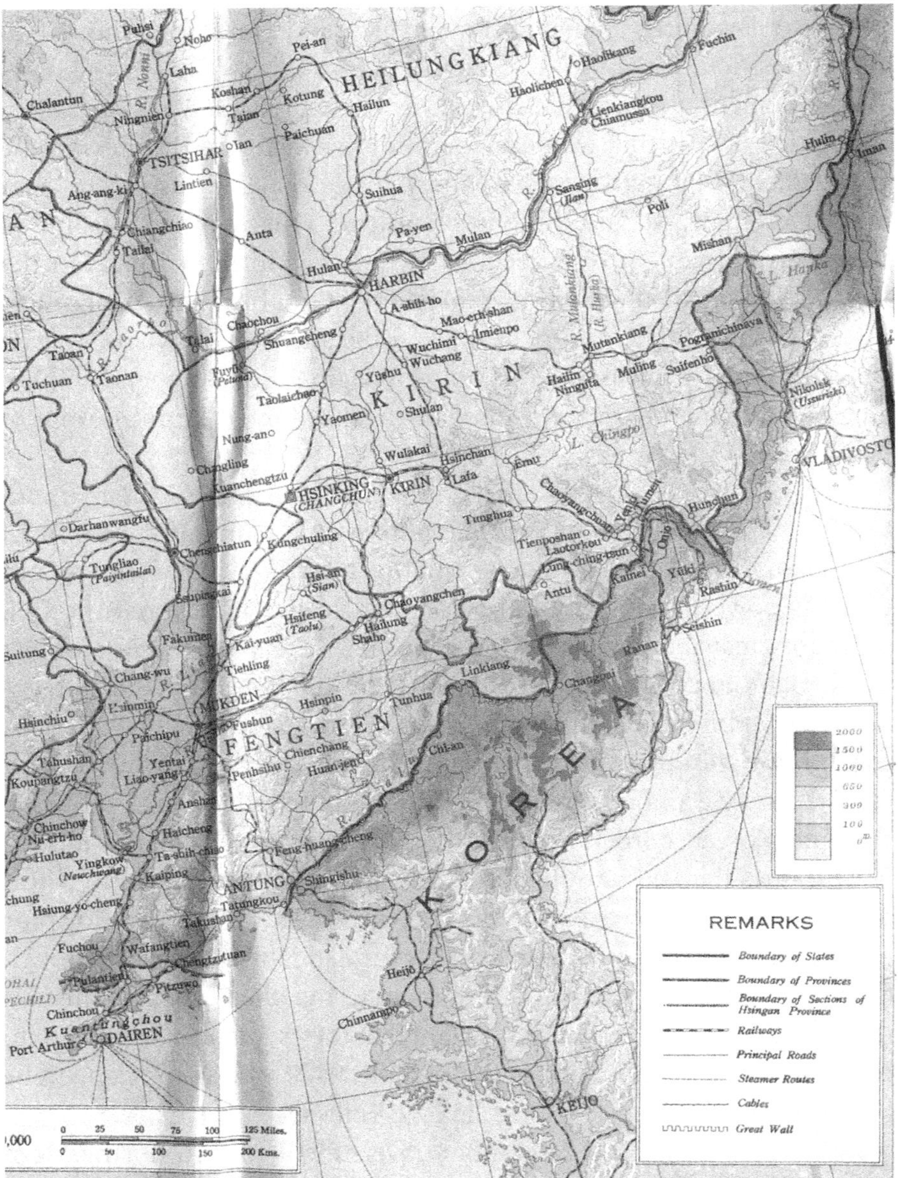

Railway map of northeastern China. Note the locations of the Jilin (Kirin)-Mukden Railway and the Jilin-Taonan railway (Manchukuo Yearbook, 1934)

the Yamamoto government's official position of neutrality with regards to Yuan Shikai, many of the patriotic adventurers in northeastern China provoked skirmishes with Chinese police and attempted to procure weapons for the anti-Yuan rebels. These actions forced the Yamamoto government to send an envoy to China to prevent the adventurers from sending weapons to the anti-Yuan forces and suggest that Intelligence Division China experts be replaced with more reliable figures.[123] While Army Minister Kusunose rejected Yamamoto's suggestion regarding the AGS's China experts, Foreign Minister Abe Moritaro made an even more radical suggestion: the Foreign Ministry should have total control of Japan's foreign policy and all the independent activities of the patriotic adventurers, AGS, and officers in China be curtailed. Unfortunately for Abe, these views directly led to him being stabbed by two IJA assassins, which led to his death in September 1913.[124] Abe's murder illustrated the willingness of army radicals to use assassination as a tool to remove threats to imperial expansion on the Asian mainland.

After the Imperial Japanese Navy became embroiled in a corruption scandal in August 1913, the pro-IJN Yamamoto cabinet was forced to resign. Okuma Shigenobu took Yamamoto's place as Prime Minister. Unlike the Saionji and Yamamoto administrations, Okuma and his cabinet sought to pursue an aggressive policy in China.[125] The posture of Okuma and his cabinet was a great boon to the China hawks in Tokyo and in northeastern China, who wanted to balkanize the country while it was engulfed by the chaos of revolution.

Kawashima took advantage of the new government's policy by arranging the sale of weapons and ammunition to the anti-Yuan Prince Su and Prince Babojab (a Mongol separatist) in the summer of 1915. Kawashima then created a detailed plan of action for Prince Su and Prince Babojab, which included procurement of food and ammunition, movement timetables, reconnaissance and intelligence, and perhaps most importantly the recruitment of soldiers.[126] He hoped to use these forces to help restore the Qing emperor to the throne. General Tanaka Giichi, the vice chief of the General Staff, supported Kawashima's idea of "Manchurian" separatism, sending three China experts to aid Kawashima's anti-Yuan Shikai rebels in fomenting an uprising in

Mukden.[127] Tanaka hoped to pressure Yuan into accepting more Japanese concessions in northeastern China.

Contrary to Tanaka's expectations, Yuan moved to suppress the rebels, which caused the AGS's China experts to encourage Tanaka to support the overthrow of Yuan Shikai. Tanaka agreed with his subordinates' assessment, and began lobbying the Japanese government to support a full-fledged "Manchurian" independence movement and the overthrow of Yuan Shikai in February 1916. By March 7th, the cabinet came around to Tanaka's point of view and supported overthrowing Yuan.[128] Soon thereafter, Tanaka and the AGS sent more advisors to northeastern China with arms, supplies, and money. Additionally, Kwantung Governor General Lieutenant General Nakamura Akira—nominally responsible to the Foreign Ministry—ordered all diplomats in northeastern China to not interfere with the efforts of Kawashima and the other patriotic adventurers to overthrow Yuan's government and establish an independent state in the Three Eastern Provinces.[129] In this manner, the Kwantung Garrison and its commander abetted the Tanaka-Kawashima plan to overthrow Yuan Shikai.

Despite the gung ho attitude of many Japanese towards overthrowing Yuan Shikai, many others opposed such a course of action. Among those who opposed the overthrow of Yuan were officers within the Kwantung Garrison, including 17th Infantry Division commander Hongo Fusataro and Independent Garrison Commander Major General Fujii.[130] More importantly, two Foreign Ministry consuls in Mukden, Yoshida Shigeru and Yada Shichitaro also opposed the Tanaka-Nakamura-Kawashima plan and swayed Japanese Foreign Minister Ishii over to their point of view.[131] The Foreign Ministry then pressured Tanaka and the AGS to back off their support from the Qing-restorationist variant of "Manchurian" separatism supported by Kawashima and the Kwantung Garrison.

Instead of supporting Qing-restorationism, the Foreign Ministry supported Zhang Zuolin, an ethnic Han Chinese warlord, to be the face of "Manchurian" separatism. Zhang was a vastly different figure from the Qing court associates who got the support of Kawashima and the Kwantung Garrison. With roots as a bandit in the countryside, Zhang had worked his way into a position of prominence in northeastern China by the twilight of the Qing Dynasty, in large part due to the collapse of Qing control over its Three Eastern Provinces.[132] By 1907, Zhang had 3,500 troops under his command in seven battalions, with five battalions commanded by either family members or bandit associates.[133] Zhang's elevation of friends and family to high places facilitated loyalty among his subordinates, which in turn fostered his rise in the political landscape in northeastern China.

After the 1911 Revolution, Zhang's forces helped put down republican attempts to establish control in northeastern China. Initially a staunch monarchist, Zhang still harbored dreams of restoring the Qing Dynasty in 1912. He had even contacted Kawashima that year about restoring the Qing Dynasty and called for a punitive expedition against Yuan Shikai after he came to an agreement with the KMT.[134] But reality soon forced Zhang to drop all pretense to his monarchist ideals, and instead cultivate his power base in Fengtian, where he established a strong network of local connections. These connections, which outstripped those of Fengtian Governor General Duan Zhigui, facilitated his rise to be the most powerful figure in the province, despite only being the second in command to the Governor General.[135] Zhang's position therefore made him the perfect candidate to spearhead a "Manchurian" separatist regime that would safeguard Japanese interests.

Elements within the Foreign Ministry, particularly Yada and Yoshida in Mukden, understood the reality of Zhang's strong position in northeastern China, and wanted to co-opt his power for Japan's purposes.[136] They opposed the romantic idea of Qing restoration, which was impractical and anathema to many Chinese. Fortunately for the Foreign Ministry, it succeeded in getting through to Tanaka, who telegraphed Kwantung Army Chief of Staff Nishikawa Torajiro not to go through with its plans to restore the Qing.[137] Instead, the Foreign Ministry would implement its plan to support Zhang as its puppet.

Yada told Zhang that if he were to declare "Manchurian" independence, the Japanese would give him arms, supplies, and money. However, Kawashima and many within the Kwantung Garrison did not believe that Zhang would secure Japanese interests in the region, and opposed the Foreign Ministry's attempts to collaborate with him to achieve its aims.[138] Zhang reciprocated these feelings towards Kwantung Garrison officers, 2nd Department agitators, and patriotic adventurers who he saw as troublemakers and criminals. Indeed, the antipathy between them became so strong that 25-year-old second lieutenant Mitsumura Yutaka placed a bomb under his coat and threw himself at Zhang's carriage in an attempt to assassinate him, likely at Kawashima's behest.[139] While this assassination attempt failed, Zhang realized that he would have to curtail the patriotic adventurers and the various freelance IJA officers in northeastern China if he were to survive. Zhang therefore made demands to the Japanese Foreign Ministry to get these rogue elements under control if they wanted him to collaborate with them. Zhang's demand, combined with the Foreign Ministry's influence over the AGS and the Kwantung Garrison, led to yet another order from Tanaka to Colonel Doi Ichinoshin to postpone Kawashima's separatist movement.[140]

The death of Yuan Shikai in June 1916 marked a temporary end of Japan's support for Zhang Zuolin. Instead, Japan shifted its support to China's new president, Li Yuanhong. But Yuan's death further accelerated disorder in China more broadly in the years ahead, largely due to the rise of warlordism. Divided between warring factions with no coherent vision for unity or restoring national strength, China found itself the object of great power schemes and proxy wars. Of the great powers, Japan represented the most salient threat to Chinese sovereignty largely due to its desire to expand its economic and security interests in north and northeastern China. But Zhang's collaboration would be necessary for advancing Japan's interests in these regions, largely due to his strong position in the Three Eastern Provinces. Therefore, even though Japan temporarily scorned Zhang in 1916, the Foreign Ministry's attempt to foment "Manchurian" separatism would not be the last time Zhang and the Japanese would work together.

Chapter 4

World War I, Japan's Imperial Opportunism, and the Rise of the Liberal International Order

World War I broke out in Europe in the summer of 1914, as the Allied powers of Britain, France, and Russia went to war with Germany, Austria-Hungary and the Ottoman Empire. Even though there was little fighting in East Asia, the war significantly impacted the region. With the Western powers bogged down in Europe, Japan was able to expand its imperial footprint in Asia economically and geopolitically. In 1915, Japan made the so-called Twenty-One Demands, which expanded its imperial footprint in northeastern China and Shandong province. Soon thereafter, it made the so-called "Nishihara loans" in a bid to support warlord Duan Qirui and his Anhui Clique. These moves greatly increased Japan's influence in China at minimal cost. But at the end of World War I, US President Woodrow Wilson ushered in a new paradigm of international relations: a liberal international order that on paper prioritized the principles of sovereignty, free trade, and the self-determination of nations. This new liberal democratic order sought to do away with the "old order" of the past, defined by brute force, secret alliances, and spheres of influence. For Japan, Wilson's new order clashed with the gains it made during World War I, and threatened to undermine its burgeoning empire in East Asia. The clash between Wilson's universal liberalism and Japan's distinctly national interests would continue in the coming decades, before ultimately culminating in World War II.

The First World War and the Twenty-One Demands

Due to the demands of war production, the European powers mobilized their economies for war, leaving little room for commercial ventures in East Asia. Moreover, due to Britain's naval blockade on Germany and unrestricted submarine warfare, what little trade could occur was severely hampered. Between 1913 and 1918, British imports to China declined from 96 million to 48 million, French imports from 5.2 to 1.5 million, and German imports from 28 million to 0.[1] The sudden disappearance of European competition allowed Japanese companies to fill the vacuum of exports to China. Japan's economy greatly benefited from expanded market share in China and import substitution at home, with exports increasing threefold in money value, and manufacturing increasing fifty-four percent between 1914 and 1918.[2] These economic realities made the First World War a splendid time for Japanese capitalism, which achieved unprecedented heights during the period.

On August 23rd, 1914, Japan declared war on Germany. A few weeks later on September 2nd, the IJA's 18th Infantry Division landed in Shandong province and laid siege to the German concession at Qingdao. After two months of siege, the encircled German garrison was forced to surrender in November due to a lack of ammunition and incessant attacks from the Japanese.[3] Meanwhile, the Imperial Japanese Navy also seized the Germans' Pacific Islands possessions including the Marshalls, Marianas, and Carolines. These campaigns represented the extent of the fighting in the Far East during World War I, as Germany had no more territorial possessions in the region. But Japan emerged from the fray with a plethora of valuable territories, most notably on the Shandong Peninsula. Here, the German concession at Qingdao would become the subject of a great struggle between the rising tide of Chinese nationalism and Japan's burgeoning empire.

Japan had a significant upper hand during World War I in its relationship with China. The European powers, distracted by operations in Europe, could not give much thought to events in Asia.[4] For its part, China was still divided and in chaos after its revolution in 1911. The confluence of these two realities opened up opportunities for Japan in China, which it did not hesitate to capitalize on.[5] While some in Tokyo advocated caution in handling Sino-Japanese relations, the dominant trend among Japanese officials was to promote aggressive policies in China. In their minds, such an opportunity to expand Japan's influence may never come again, which meant Japan had to take advantage of World War I to incorporate more of China into its sphere of influence.[6]

Japan's first attempt to fill the power vacuum in China was its issuance of the "Twenty-One Demands" in 1915. If accepted, the demands would have represented a great blow to Chinese sovereignty.[7] They included the permanent transfer to Japan of previously German concessions in Shandong province, Chinese government employment of "influential Japanese as political, financial, and military advisers," a 99-year extension to the already existing 25-year lease of the Liaodong Peninsula, and further economic and extraterritorial privileges for Japanese nationals in northeastern China.[8]

In China, the informed public reacted negatively to Japan's issuance of the Twenty-One Demands. Forty thousand people protested in Shanghai against the demands, while other cities, such as Wuhan, Beijing, and Mukden, also hosted significant demonstrations.[9] Chinese merchant associations organized their opposition to Japan's demands through a nationwide boycott.[10] Chinese students overseas also joined the nationalist chorus, encouraging Yuan Shikai to reject Japan's onerous terms. As an overseas Chinese student group at Columbia University put it, Yuan must reject Japan's demands and be prepared to fight if necessary because it was better to "live as fighters than die as slaves."[11] The strong backlash against the Twenty-One Demands indicated an ever-rising nationalist, anti-imperialist consciousness among a vanguard of China's intellectual and political elites. Yuan and his government would have to contend with these elements if it wished to stay in power.

Japan's issuance of the Twenty-One Demands put the Yuan Shikai government in an unenviable position. If Yuan accepted Japan's demands, he would be seen as a traitor to the country by the KMT and its pro-Republican allies.[12] But if he did not accept, he would likely find himself embroiled in a challenging war. The Japanese drove this latter point home in an ultimatum issued on May 7th, 1915, stating that if Yuan and his government did not give a "satisfactory reply" by May 9th, the Imperial Government would "take any steps she may deem necessary."[13] Such coercion from the Japanese government induced Yuan to accept most of the demands, including all of those pertaining to the extraterritorial rights of Japanese citizens in northeastern China and the 99-year extension of the leases on the Liaodong Peninsula and SMR. Instead of fighting to the death, Yuan surrendered large parts of the country to Japan; China would now have to deal with the fallout of signing such an onerous treaty.

The Warlord Era and the Nishihara Loans

Yuan Shikai died in 1916, marking the end of his reign as China's head of state. In his place, Li Yuanhong ascended to the presidency and anointed Duan Qirui, a subordinate of Yuan's in the Beiyang Army, as his premier.[14] Despite being appointed by Li, Duan effectively controlled his government, refusing to recognize the 1912 Constitution crafted by Sun and the KMT.[15] Duan's refusal to recognize the constitution sparked outrage among republican-minded officers and could have led to an immediate civil war had not another ex-subordinate of Yuan's, Feng Guozhang, advised Duan to restore the charter. Following Feng's advice, Duan restored the 1912 Constitution and allowed the KMT to return to political relevancy. These moves reunified China after Yuan's death and seemingly restored order to the country.[16]

The restoration of parliament and the KMT was short-lived. In 1917, another Beiyang army officer named Zhang Xun attempted to restore Qing emperor Puyi to the throne.[17] While Zhang's attempt at Qing restorationism failed, his attempt to re-install the Qing government exposed the contradictions between Sun Yat-sen's KMT and the Beiyang Army. The Beiyang Army opted to set up a "Provisional National Council" with elections set for 1918.[18] The KMT, realizing that it would not be a significant factor in this new legislative assembly, relocated to Guangzhou, where it set up an alternative government on August 31st, 1917. This second split between the Beiyang Army and the KMT led to a short conflict between the two in the autumn of 1917, which ended inconclusively insofar as neither side could claim a clear victory.

The end of the KMT-Beiyang Army conflict in November 1917 precipitated Duan Qirui's resignation as Premier of the new government. Questions arose afterwards regarding whether China should enter World War I and how to go about reunifying the country.[19] On one side, Beiyang Army officer Feng Guozhang and the so-called "Zhili Clique" preferred to stay out of World War I and reunify China peacefully. Duan and his "Anhui Clique" took the opposite stance, favoring entry into World War I and reunifying China by force. To a great extent, Duan favored entry into World War I because Japan promised to send him military and economic aid if he declared war on Germany. Indeed, the Japanese preemptively sent Duan 17-million-yen worth of weapons systems to entice him to enter the war.[20] But even though these weapons were ostensibly sent to equip Chinese units that might participate in World War I, in reality Japan sent them to Duan to better equip him for the coming civil war against the Zhili Clique and the KMT. These weapons shipments set the stage for a clash between the Anhui and Zhili Cliques in the near future.

Japan's February 1918 weapons shipments to Duan Qirui never reached his forces because Zhang Zuolin intercepted them in the Three Eastern Provinces.[21] Zhang, who had quietly amassed power in northeastern China throughout the 1910s, had emerged as a serious force to be reckoned with by 1918. He demonstrated his newfound strength in March of that year, sending 50,000 men to march on Beijing with the stated goal of reunifying China.[22] Zhang made a series of demands, including the return of Duan as Premier, recognition of his position in northeastern China through the creation of an "inspector general" for the "Three Eastern Provinces," and expanding the armed forces. Feng

Guozhang decided to honor Zhang's demands, reinstating Duan as premier, and recognizing Zhang as the "inspector general" of the "Three Eastern Provinces in Manchuria." After a short campaign in Hunan, Zhang's forces withdrew back to northeastern China in December 1918.[23] Nonetheless, they had still demonstrated their potentially decisive role in determining the fate of which warlord clique would come out ahead in the coming conflicts.

China's collapse into warlord chaos also facilitated Japanese financial imperialism. One of the most important financial imperialists was Nishihara Kamezo, who advocated the creation of a unified Japan-Korea-Manchuria trading bloc where the Japanese paper yen would be used as the common currency. In Nishihara's conception, this "yen bloc" could expand into other areas of China including Rehe, Shandong, and the Yellow River area.[24] To achieve Nishihara's vision, he recommended a series of policies in his "Outline for Opportune Economic Facilities Vis a Vis China," including the establishment of "business investment in China," the creation of provincial banks in Dongshan and Rehe to issue paper gold currency, and the installment of Japanese financial advisors in these provincial banks.[25] Nishihara hoped that the formation of such a "yen bloc" would allow Japan to become more self-sufficient, with greater access to Chinese raw materials.[26]

While Nishihara's recommended policies were not implemented outright, he was able to move towards his goal of the yen bloc by offering a series of loans to Duan Qirui's Anhui Clique in 1917 and 1918. These loans, which came to be known as the "Nishihara loans," were essentially a means of smuggling money and arms to Duan in return for his willingness to secure Japan's rights and interests in China.[27] Because Japan was a member of a multi-national China consortium, there were many types of loans it could not make unilaterally; hence, the Nishihara loans carried the veneer of being "industrial loans," which Japan could make without other powers.[28] Some of the "industrial loans" Nishihara negotiated with Duan are listed below:[29]

1. Jilin-Huining Railway Primary loan—10 million yen
2. Mine and Forestry Loan—30 million yen

3. Manchuria-Mongolia Four Way Railway loan—20 million yen
4. Shandong Two-Way Preliminary Loan—20 million yen
5. War Participation Loan (for China's entry into World War I)—20 million yen

On the surface, these loans would create infrastructure and develop China's resources on the grounds of "mutual benefit" for both Japan and China. But the loans "more than doubled the outstanding debt of the Chinese government to Japan," with little tangible benefit to China's national interest.[30] Moreover, in return for these loans, the Japanese secured confirmation of several key points listed in the Twenty-One Demands, including the recognition of its claims in Shandong, control of the railways in Shandong, and the expansion of its extraterritorial rights in northeastern China.[31] The loans also helped arm Duan's forces to fight his domestic enemies on behalf of Japan. Therefore, despite the veneer of industrial development and mutual benefit, the Nishihara loans were in fact quite one-sided, with Japan as the primary beneficiary.

Doctrinal Adjustments and Autarkic Empire

While World War I changed Japan's geopolitical position in East Asia, it also greatly affected the Imperial Japanese Army's doctrinal philosophy. At the tactical and operational level, the war demonstrated the absolute superiority of firepower on the battlefield, with weapons systems such as bolt-action rifles, machine guns, and rifled artillery establishing the superiority of defense and undermining close-order assault formations that had defined warfare from the beginning of time. Many young officers within the IJA took note of these trends, and sought to incorporate the lessons of World War I into the IJA's doctrine and policy. In the view of these reformers, the IJA needed to shift from a force built on the spirit of close-order formations and offensive élan towards a force built on defensive firepower that had greater access to mechanized systems.

Despite the expertise and observations of many young reformers, the "old guard" of the army, largely embodied in Yamagata's "Choshu Clique" represented a significant hurdle for them.[32] Of the older officers, AGS Chief of Staff Uehara Yusaku was their most formidable opponent.[33] He, along with many other older officers, came up in an era when the full effects of the evolution in firearms had not been revealed, and consequently still venerated élan, hand-to-hand combat, and abstract moral principles such as physical courage or morale. Under Uehara's direction, Japanese infantry doctrine continued to emphasize the importance of hand-to-hand combat and offensive spirit.[34] The continued usage of close-order formations, poor application of artillery, and the lack of machine guns, tanks, aircraft, and automobiles also caught the eye of foreign observers at the Grand Maneuvers in 1918. One western intelligence officer went so far as to say that "the mechanical mind has not yet been revolutionized in Japan, and in case of a modern war of any duration, it would probably be through this failing that it would lose."[35]

Japan's unpreparedness for a protracted war also concerned many within the IJA. IJA leaders began to study the European Theater of World War I in 1915, with the formation of the Army Ministry's "Provisional Military Research Committee."[36] This committee, which studied everything from grand strategy to platoon-level tactics, quickly concluded that total, protracted wars would define future conflicts. To emerge victorious, Japan would need to mobilize its entire economy, society, and population; the committee also concluded that Japan would need to change its national defense policy to reflect this reality. Therefore, the AGS created a "military research committee" in 1916 to change Japan's national strategy, and made Vice Chief of Staff Lieutenant General Tanaka Giichi the head of this committee.[37] This committee's task was to research the issue of "total war," and the necessary policies needed to implement total war measures in Japan.

Tanaka came to the conclusion that autarky, or economic self-sufficiency, was absolutely necessary to effectively pursue "total war." In general, five criteria were necessary for a state to be completely autarkic: 1) an obedient and motivated population 2) reliable and secure source of natural resources, 3) a comprehensive industrial base that could manufacture these raw materials into a variety of complex weapons systems 4) a strong research and development apparatus to make technological improvements, and 5) a detailed mobilization plan to rapidly bring the nation's strength to bear.[38] Of these criteria, Japan only fulfilled the first point; obtaining points two and three would be a difficult endeavor, largely due to a national lack of resources and industries.[39] But Tanaka saw a solution in China, whose abundant land and resources could make Japan self-sufficient if it could secure control over them.[40] To investigate his proposed solution more thoroughly, Tanaka toured China for two months and wrote down his thoughts in a piece titled "The Exploitation of China's Resources."[41] This excursion further entrenched Tanaka's view that exploiting Chinese resources would allow Japan to fight a protracted war and survive any blockade that might otherwise severely hinder Japan's efforts in a later conflict. Thereafter, Tanaka supported a more active policy in northeastern China, with security considerations no doubt playing a key role in his position.

Tanaka's conclusions were corroborated by other IJA officers who analyzed the First World War in Europe. One of these officers, Chiga Tsurutaro, wrote an article in 1917 called "Japan's Position Toward the European War," to make his case. He believed that Japan needed to form a union with China and create a self-sufficient economic and political bloc. Chiga's article is excerpted below:[42]

> The greatest lesson to be learned in the present war is that with population and wealth alone attainment of ultimate victory remains in doubt. A nation must be independent not only in machinery but also in materials as well and must be able to acquire such goods on its own terrain before it can be said to have gained self-sufficiency in war supplies. Japan will not avoid being completely inept in a future war if it continues presently as is. First, Japan lacks steel and its coal is not plentiful. Without steel, it cannot sustain a future. Japan's pressing needs are thus to acquire the necessary steel and coal and to create institutions and economic organizations to make it independent in military supplies.

Chiga clearly emphasized the importance of coal and steel to Japan's future survival. Without these resources, production of key weapons systems and munitions was impossible. Because northeastern China's plentiful coal and iron mines offered critical natural resources it represented a logical place to expand the empire. From there, Japan could extend its empire to the rest of China, and form the bloc Chiga desired. Chiga acknowledged that forming a new China-Japan bloc may require the use of force, but justified the proposed aggression on the basis of Japan's self-defense. In Chiga's words, "it is necessary for self-defense that we must merge even if we have to compel China to accept it by sheer force."[43]

Colonel Kuniaki Koiso also corroborated Tanaka and Chiga's conclusions in a 1917 document titled "The Empire's National Defense Resources." Koiso noted how the First World War caused a "global [economic] transformation", because countries that once depended on international trade "began retreating inward, focusing on national survival and domestic supply chains."[44] To prepare for a future war along the lines of World War I, Koiso advocated that Japan "must therefore construct an empire that prepares for wartime economic conditions."[45] An empire in China would help Japan circumvent a blockade from foreign adversaries, and thereby avoid the fate of Germany during World War I, whose civilian economy completely collapsed under the weight of the British naval blockade. Koiso believed this empire to be absolutely necessary for Japan's future survival, and advocated taking policy measures to form it. As he put it, "to neglect Chinese resources is, in fact to be unconscious of the ruin of divine Japan."[46]

Koiso's document went on to provide a detailed analysis of the raw materials and industries necessary for Japan to be self-sufficient in a future conflict. In Koiso's assessment, Japan in 1917 was too dependent on imports of rice, industrial raw materials, and manufactured products to emerge victorious in a "total war" à la the one in Europe.[47] To remedy these problems, Koiso and other officers like him advocated two solutions: the increase of industrial and agricultural productivity in metropolitan Japan, and the subjugation of China.[48] Koiso believed that China could provide sufficient amounts of rice, wheat, beans, meat, cotton, coal, iron, tin and other ores to allow Japan to survive a protracted "total war." Northeastern China was especially important, as it not only had many of these key resources but also occupied a key geo-strategic position between Japan, the rest of mainland China, and the Soviet Union.[49] The combination of resources and geography meant that Northeastern China was Japan's "lifeline" in Koiso's opinion, and therefore needed to be under Japan's effective control.

Siberian Intervention

Japan's relations with Russia also underwent great change as a result of the First World War. This change came about thanks to the October Revolution, which saw the Bolsheviks come to power and establish communist rule. But this new communist order existed on unstable ground, as the anti-Communist "White Army" took up arms against the Bolsheviks. The fighting between communist "red" and anti-communist "white" Russians resulted in the Russian Civil War, which divided and greatly weakened Russia. The world's various capitalist and imperialist powers, including the US, UK, France, Germany, Austria-Hungary, and Japan saw great opportunity in the chaos of the Russian Civil War, seeking to either expand their empires at Russia's expense and/or prevent the establishment of a communist Russia. These realities led all of these nations to dispatch expeditionary forces to Russia in support of the whites and to secure their own interests, even while World War I was ongoing.

Many in Japan saw the chaos in Russia as an opportunity to expand the empire. Of these figures, Vice Chief of Staff Tanaka was the most prominent, as he saw a prospective Siberian expedition as a means of creating a buffer zone east of Lake Baikal.[50] Others, most notably Chief of Staff Uehara, believed that sending an expedition to Siberia would be a surefire way of eliminating Japan's traditional Russian nemesis, and thereby securing its interests in northeastern China.[51] Still others worried about the implications of the Bolsheviks seizing power in Russia, and the potential for communism to spread in East Asia. The spread of communism could potentially foment instability in Japan's colonies of Korea and the Liaodong Peninsula, and thereby weaken Japan's geopolitical position in East Asia.[52] For these reasons, many Japanese officials embraced sending troops to Siberia.

Immediately after the Bolshevik Revolution, the AGS drafted a document titled "Plan for Sending Troops to the Russian Far East to Protect Foreign Residents," emphasizing the need to send a preliminary expedition to the Russian Far East to protect Japanese nationals and their property.[53] Soon thereafter, the IJA and IJN formed a "Joint Committee on Military Matters," headed by Vice Chief of Staff Tanaka Giichi. By March 1918, the Joint Committee began drawing up plans to send two armies to Siberia. The 1st Army (two divisions) would occupy Vladivostok, Khabarovsk, Nikolsk, and Ussurisk in the east and the 2nd Army (five divisions) would advance to the city Chita and the Trans-Baikal region to the west.[54]

While the IJA made plans to send an expedition to Siberia, other Japanese—most notably Prime Minister Terauchi Masatake, head of the Seiyukai Hara Kei, and Yamagata Aritomo—opposed sending forces to Siberia. These figures feared that Japan was not strong enough to send forces to Siberia, and that such a venture would be costly, both in terms of blood and treasure.[55] They also feared receiving a bad reaction from the western powers, especially the Americans.[56] As Yamagata himself put it in a March 15th memorandum, "the present time is not opportune for sending our armed forces to Siberia. If we dare to send them to the Russian territory without any request on the part of the Russian people, it will cause suspicion among the British, and especially among the American people, and we will not be able to get their help."[57] This sentiment roughly summarized the views of Terauchi and other figures in his government, and led them to not dispatch forces to Siberia in March 1918.

Even though the Terauchi government initially opted to not send forces to Siberia, it decided to send a small force to Vladivostok in the name of protecting Japanese lives and property in April 1918. Soon thereafter, US President Woodrow Wilson proposed sending 7,000 troops to Siberia in a combined US-Japanese expedition in July 1918, ostensibly to secure allied supplies in the port of Vladivostok and rescue the stranded Czechoslovak legion.[58] After a concerted effort by the AGS, the Terauchi government agreed to send forces to Siberia, their doubts somewhat allayed by the fact the US had proposed the expedition. But instead of sending a few thousand men to Siberia as the US had suggested, the Imperial Japanese Army sent ten times the number of men the US had requested, and occupied large swaths of Siberia, including key points along the Trans-Siberian railway.[59] The IJA also negotiated free passage of Japanese troops through Zhang Zuolin's territory to easily reinforce Siberia.[60] These moves effectively established a sustained Japanese presence in Russia's Far East and the northern part of the Three Eastern Provinces for the foreseeable future.

Although the Siberian Expedition represented a great commitment of force on the Asian continent, great changes were occurring at home. In September 1918, the Japanese population, fed up with high rice prices, took to the streets and rioted against the Terauchi government.[61] In response, the Terauchi government attempted to suppress the rioters with the IJA. While the IJA exercised force to suppress the riots, many within the IJA privately opposed the use of violence to suppress civilian demonstrations. Such measures undermined the hard-won support the army had cultivated among the populace in the years after the 1877 Satsuma Rebellion. This tactical error cost Terauchi's moribund government its support within the army and ultimately led to its dissolution.[62] In its place, a government headed by Hara Kei of the Seiyukai came to power in September 1918. Hara's government represented the first headed by the elected chief of a political party, and inaugurated the era of "Taisho democracy" in Japan.

The End of World War I, the Treaty of Versailles, and the Liberal International Order

The "war to end all wars," concluded in November 1918 with a new arrangement of the international order. To create the new order, the victorious powers—including the United States, United Kingdom,

France, Italy, and Japan—gathered at the Versailles palace in Paris to discuss peace terms. The United States was the most dominant power at the conference, as it had not spent as much blood and treasure on the fighting and maintained a great industrial economy. President Wilson sought to use US influence to establish a new liberal international order at the Versailles Conference and do away with the pre-war international system that had facilitated the descent of the world into war. His "Fourteen Points," included limitations on "private international understandings," freedom of the seas, the abolition of "economic barriers," guarantees for arms reduction to the bare minimum necessary for self-defense, references to the self-determination of nations, and most importantly, the creation of a "general association of nations," that would safeguard the "political independence and territorial integrity of great and small states alike."[63] Of these points, the last one was the most important, as Wilson hoped that the formation of a "League of Nations" would provide a medium for dispute settlement to avoid war in the future. In Wilson's imagination, this institution would forever change the landscape of international relations, leading to the creation of a "New Diplomacy" predicated on principles of trust, openness, and peaceful negotiation.[64]

In East Asia, Wilson's liberal internationalist vision took the form of abolishing the imperial powers' spheres of influence in China, enforcing the Open Door Policy, doing away with the unequal treaties, and using the League of Nations to safeguard China's sovereignty and territorial integrity.[65] For the United States and United Kingdom, these measures were also convenient, as the two largest imperialist powers could preserve their interests in China by preventing Japan from expanding in the name of Chinese sovereignty. But for Japan, which had a much smaller empire and more acute national security interests, Wilson's liberal internationalist vision directly undermined its position in East Asia. Therefore, many Japanese opposed and resented Wilson's liberal internationalist vision, and sought to assert Japan's hard-won rights in Shandong and northeastern China at the Versailles Peace Conference.[66]

Soon after it became clear that Wilson's Fourteen Points would be the touchstone of a post-war peace conference, Japan's Advisory Council on Foreign Relations met and agreed that 1) the Japanese government would only return sovereignty over Jiaozhou Bay to China after the German government surrendered the rights to its leased territory in Shandong and 2) Japan would insist on the right to settle the issue of Shandong bilaterally with China, outside of the peace conference.[67] The Council also discussed Wilson's proposed League of Nations, and advocated the postponement of its formation due to the continued existence of racism among the Western powers, which the council feared would negatively impact Japan. On the issue of the League of Nations, the Council therefore decided to 1) resist its formation and 2) insist on a racial equality clause to limit racial discrimination in the event that it was formed.[68] Because the Council contradicted Wilson's vision for a liberal international system, a clash at Versailles over the terms of peace between Japan and the US seemed likely.

Another Wilsonian proposal also ruffled the feathers of the Japanese prior to the Versailles Conference: the formation of a multi-national banking consortium to control the lending of money to China. This new system was designed to prevent any single power from weaponizing credit for its own purposes when lending to China, especially with regards to railways.[69] The new consortium would do away with the loophole of "industrial loans" that Japan had leveraged to lend money to Duan Qirui in the Nishihara loans, while also getting rid of the system of "spheres of influence" for the great powers.[70] The United States hoped that these measures would better secure a market for American capital in China relative to its imperial competitors such as Japan and the United Kingdom.[71]

At the Versailles peace conference, the question of the multilateral consortium was discussed informally between the representatives of the United States, United Kingdom, France, and Japan. Initially, Japan opposed the consortium, as the elimination of unilateral loans and "spheres of influence" would severely hinder its interests in China. Moreover, the talks also included the proposal of "pooling" each state's concessions into the consortium, depriving each individual state of its own holdings.[72] The idea of surrendering hard-won concessions in China to a multi-lateral consortium was distinctly unattractive to the Japanese, who attempted to negotiate a new agreement regarding a multi-lateral consortium. These negotiations eventually resulted in an acknowledgment of Japan's special interests in the Liaodong Peninsula and Mongolia, with a status outside of the consortium's jurisdiction. Thanks to this guarantee, Japan acceded to the US's proposed multi-lateral consortium on September 1st, 1919.[73]

The League of Nations also was a source of controversy for the Japanese at the Versailles Conference. Even though Japan wanted to oppose the formation of the League, Wilson's favor forced Japan to adopt its back-up plan: a proposed anti-racial discrimination amendment to the League Covenant in the spring of 1919. But the proposal encountered much opposition from then Australian Prime Minister Billy Hughes, who was an advocate of the "White Australia" immigration policy that totally prevented non-white immigration.[74] As Australia was a Dominion of the British Empire, the United Kingdom would not be able to accept this clause. The United States would also have difficulty accepting this clause due to its own anti-Asian immigration policies and sentiments, especially on the west coast. Wilson also did not want the debate surrounding racial equality to undermine the formation of the League of Nations. As the American president himself put it, "My own interest is to quiet discussion that raises national differences and racial prejudices. I would wish them particularly at this juncture in the history of relations with one another, to be forced as much as possible into the background."[75]

On April 11th, a final vote was held on Japan's racial equality clause. Eleven out of seventeen delegates voted in favor of Japan's proposal, including Japan itself, France, Italy, Serbia, Czechoslovakia, Greece, China, and Brazil.[76] Of those that abstained, the United States and United Kingdom were the most prominent. Despite the abstention of the US and UK, a majority had voted in favor of Japan's proposed racial equality clause. But owing to the sensitivity of the issue and Wilson's desire to form the League of Nations, he declared that a unanimous vote would be necessary on the issue of racial equality.[77] Given the opposition of Australia and therefore the United Kingdom to Japan's proposal, a unanimous vote in favor of racial equality would have been impossible. Therefore, Wilson's declaration effectively ended Japan's attempt to get a racial equality clause included in the League of Nations covenant, and undermined one of Japan's key objectives at the Versailles conference.

For Japan, the question of Shandong was the most important issue to be discussed at the Versailles Conference. Japan's arguments for its right to control Shandong hinged on its contributions to the allied effort in World War I and the treaties it signed recognizing its jurisdiction over the region, including the Twenty-One Demands as well as bilateral agreements with the UK, France, and Italy.[78] But Wellington Koo, one of the Chinese representatives to the League, rebutted Japan's arguments, saying that the Twenty-One Demands were signed under threat of war and that they could not be seen as legitimate under such circumstances. He also argued that Shandong rightfully belonged to China under Wilson's principle of self-determination.[79] Wilson and the Americans were sympathetic to the plight of Koo and China, in large part because Wilson saw the Shandong question as vital to the issue of self-determination.[80] Wilson had even told Koo of his intention to support China's claim to Shandong in January 1919, and requested that Koo attend the Versailles peace conference in the first place.[81] The combination of Chinese opposition and American support made Japan's desire to get its jurisdiction over Shandong recognized a difficult endeavor.

Unfortunately for China, the rest of the powers, including the UK, France, and Italy, did not recognize its rights in Shandong, instead referring to their own bilateral agreements with Japan. The Japanese delegation also stated that its orders from the government mandated they not sign the treaty "unless they were placed in a position to carry out Japan's obligations to China."[82] This threat forced Wilson to reconsider his previous position on Shandong, because if the Japanese withdrew, they would not become part of his coveted League of Nations. Consequently, Wilson decided to betray his previous promises to Koo and opted to broker a compromise on the question of Shandong. In this compromise he recognized Japan's economic rights and the right to an extraterritorial settlement in Shandong in return for giving sovereignty over the rest of the province to China.[83] In brokering this compromise, Wilson severely undermined the principles that supposedly foregrounded his liberal international order, setting the stage for future great power competition in China.

In China, many liberal intellectuals and students who had placed their hopes and dreams in Wilson's promise of self-determination felt betrayed by his settlement at Versailles.[84] In response to this betrayal, these activists sparked the so-called "May 4 Movement," which protested Wilson's decision to hand over economic rights in Shandong to the Japanese and called for China's self-determination. The protestors denounced Wilson both for lying to Koo and for his hypocritical promotion of self-determination. They also denounced their own government for its feudal, reactionary, and warlord tendencies that had made China weak and vulnerable in the face of the Western powers and Japan.[85] In place of the old order, the protesters wanted an infusion of modernity into China, perhaps best embodied in their calls for "Mr. Science" and "Mr. Democracy."[86] The protesters hoped these concepts could make China a strong, independent nation immune to being humiliated on the international stage ever again. The desires of these students and intellectuals for a strong, modern China sparked a renaissance of anti-imperialist, anti-feudal, and nationalist consciousness in China, which would come to define its position in 20th-century world affairs.

Wilson's Fourteen Points and League of Nations created a bedrock for later post-war international initiatives of disarmament, dispute settlement, and respect for sovereignty in the 1920s. But the cases of the racial equality clause and the Shandong Compromise clearly exhibited the limitations of this system. As events at Versailles clearly illustrated, Wilson's calls for "self-determination" primarily applied to Europeans. All non-western countries, regardless of their status and power, were not truly considered part of the "family of nations." For China, the inadequacies of Wilson's system meant that it would continue to languish in a semi-feudal, semi-colonial state for the foreseeable future. Its dissent to this reality is perhaps best encapsulated in the unwillingness of its delegates to sign the Treaty of Versailles. For Japan, its eventual dissatisfaction with its role in Wilson's liberal order, symbolically foreshadowed in the western powers' rejection of its proposed racial equality clause, would eventually put it on a collision course with China, the British Empire, and the United States. Therefore, even though Wilson's liberal international order was designed to prevent the outbreak of another World War, it in fact only added a thin veneer of formal legitimacy to a global system defined by the large-scale distribution and application of force. In East Asia, the grim reality of force as the main determiner of political outcomes would be realized in northeastern China, where the forces of Chinese nationalism and Japanese imperial aggression would collide. This clash would come to define East Asia's geopolitical reality in the twentieth century.

Chapter 5

The Foundation of an Army

In the late 1910s and 1920s, Japan sought to find its way in the liberal international order. The old *genro* grip on power and decision-making was loosening due to the effects of time and old-age. The IJA's authority had been discredited somewhat by the various governmental controversies of the 1910s and the unpopular Siberian Expedition. While Japan's technocracy was starting to come into being, it was still at a relatively young stage and needed time to mature before it became a consequential political force. Consequently, the liberal government of Hara Kei had great control over Japanese politics in the late 1910s and early 1920s, which enabled it to integrate Japan into the liberal internationalist system.

Japan's integration into the broader Wilsonian system was a great victory for Japanese liberals who wanted closer relations with the United States and United Kingdom. But their vision for harmony between Japan and the Anglo-Americans was undermined by a variety of factors, both at home and abroad: the ongoing Chinese Civil War, the status of Japan's special rights and interests in northeastern China, the post-World War I decline of Japan's economic position in East Asia, and the Imperial Japanese Army's staunch opposition to budget cuts and disarmament. The conflicting interests that arose from these factors eventually resulted in the decline of liberal parliamentarism in Japan, and the rise of a new entity that would become a chief driver of policy in northeast Asia—the Kwantung Army. The Kwantung Army's actions would in turn facilitate the downfall of a variety of liberal parliamentary governments, before eventually ushering in the ultimate end of parliamentary democracy in Japan in 1931.

The Formation of the Kwantung Army

Ironically, the Kwantung Army's roots lay in the very desire of Hara's liberal government to assert greater civilian authority over the military. Prior to 1919, an IJA Governor General ran the military and civilian affairs of the Liaodong Peninsula. As the force designated to defend the peninsula, the Kwantung Garrison was at the Governor General's command; this structure was designed for the period following the Russo-Japanese war. If a Russian "war of revenge" were to erupt, centralized command would be needed to facilitate rapid-decision making. But with the descent of Russia into civil war, the national government viewed this arrangement as concentrating too much power in the hands of the IJA. Hara Kei sought to overturn the command structure that had existed since 1906 and replace it with one more easily controlled by the powers to be in Tokyo.

At Hara's behest, the position of Governor General and the so-called "Army Bureau" were abolished in 1919. A civilian governor appointed by the emperor would instead take over the civilian functions of the Governor General, and a new Kwantung Army Headquarters would take over the military functions.[1] This new headquarters would report to the Army General Staff and the Army Ministry in Tokyo (as opposed to the Governor General).[2] The creation of two separate entities for civilian and military functions in the Liaodong Peninsula decentralized the previously existing command structure, which had been centered around the Governor General. Hara hoped that this new arrangement would give his civilian-government greater control over the Kwantung Army and reduce the influence of the IJA in the Liaodong Peninsula.[3]

Despite Hara's best intentions, these structural reforms had the opposite effect in northeastern China. Hara's separation of military and civilian spheres left the civilian governor with no control over the Kwantung Army. Freed from its previous responsibility to the Governor General and Foreign Ministry, the Kwantung Army now felt it was only responsible to its superior agencies in the IJA, the Army Ministry and AGS.[4] These organs in turn only answered to the Emperor, and were not subject to the authority of the Cabinet, the Diet, or the political parties, as per the Japanese Constitution. As the Emperor technically had "Supreme Command" over the armed forces, he was responsible for the functions of declaring war and overseeing the activities of the armed forces. In practice, however, both Emperor Yoshihito and later Emperor Hirohito lacked knowledge of the inner-workings of the armed forces and the increasingly radical views of its officers.[5] The Imperial Palace's lack of control of armed forces effectively made the AGS and Army Ministry responsible for the actions of the newly formed Kwantung Army.

Adding to the oversight problem, the AGS and Army Ministry would have great trouble exercising control over the newly formed Kwantung Army. The civil war in China and unstable situation in the Three Eastern Provinces tantalized Kwantung Army staff officers who sought to capitalize on China's division to expand Japan's influence in the region. Simultaneously, many Kwantung Army officers perceived the rise of Chinese nationalism and the Soviet Union to be severe threats that necessitated militant action to be addressed, even without Tokyo's knowledge.[6] The combination of these factors led the Kwantung Army to take a series of unilateral actions in the 1920s, often at the behest of staff officers. Acting in the name of the "Supreme Command Prerogative," the Kwantung Army was able to circumvent scrutiny and accountability for its actions despite the fact these actions were oftentimes in contradiction of formal Japanese policy, not to mention the post-war liberal international order.[7] Removed from events on the ground in Tokyo, the AGS and Army Ministry had trouble keeping up with the Kwantung Army's schemes. While these organs were sometimes able to assert their authority over their subordinates and temporarily moderate the worst excesses, supervisory officials all too often acquiesced to Kwantung Army staff officers' schemes. The unstable situation in the Three Eastern Provinces, combined with the lack of oversight for the Kwantung Army, turned it into a hotbed of *gekokujo* activity in the 1920s.[8]

Transition in the Imperial Japanese Army: The Demise of the Choshu Clique and the Rise of a New Generation of Officers

The 1920s ushered in a decade of transition and adjustment for the Imperial Japanese Army. In light of World War I, which was defined by the total mobilization of national resources and the primacy of firepower, many younger officers realized the need to reform the IJA. These young officers represented a generation of new technically minded reformists, by and large outside the Army's traditional "Choshu Clique." But amid the overall climate of universal liberalism and disarmament that defined the early 1920s, the IJA faced budget cuts and difficult decisions. These cuts sparked internal debate about the army's priorities. Simultaneously, civil war in China, the rising tide of Chinese nationalism and the establishment of the USSR all threatened the Empire's future interests in north and northeastern China, and provided counter pressures to disarmament and budgetary retrenchment. These serious external threats led many in the Kwantung Army to advocate taking preemptive action in these places. As these threats to Japan's empire grew more severe, the new generation of officers sought increasingly radical solutions, sometimes backed by superiors that sought budgetary expansion and strategic action against perceived adversaries. The interplay between these various trends in the IJA came to define it as an institution and set the stage for the radicalism of the 1930s.

The decade of the 1920s also saw an acceleration of the split between the "modernizers" and "traditionalists" in the IJA. Within the Choshu Clique, a few key "modernizers," including General Tanaka Giichi and his protégé Ugaki Kazushige, advocated a series of measures to promote modernization. These reforms included the design and mass production of new weapons, the provision of greater firepower for infantry divisions, and a reduction of existing "square divisions" (composed of two brigades and four regiments) to "triangular divisions" (composed of one brigade and three regiments) to increase divisional mobility.[9] They also advocated reducing army personnel to pay for greater technical modernization. Tanaka and Ugaki believed that these reforms would better position the IJA to win a protracted war of total mobilization where the superiority of firepower had an outsize weight in determining the outcome of conflict.[10]

The majority of the Choshu Clique opposed Tanaka and Ugaki's proposed modernization reforms.[11] Led by General Uehara and his AGS, many IJA "traditionalists" believed Japan's lack of natural resources and industrial base totally precluded any attempts to prepare for a protracted "total war." Instead, Uehara and his cohort believed Japan should predicate its war-fighting philosophy on the principles of élan and fighting spirit to achieve rapid victory in "battles of annihilation."[12] To win such battles quickly and avoid the pitfalls of protracted war, Uehara and his clique focused on manpower over modernization, believing that Japan needed to maintain a large standing army capable of rapidly mobilizing, concentrating, and overwhelming any prospective enemy. Uehara wanted to use the Army's limited budget to maintain a force structure of forty divisions, even at the expense of modernizing the army.[13]

Outside the Choshu Clique, many young officers that had attended the IJA's staff college supported modernization. In 1921, three of these figures—Major Nagata Tetsuzan, Major Obata Toshiiro, and Major Okamura Yasuji—met in Baden-Baden, Germany. All three were members of the sixteenth graduating class of the IJA's Military Academy.[14] All three had also graduated the IJA's Staff College with honors before ascending to positions of importance in the AGS.[15] None were from Satsuma and Choshu, and therefore had limited pathways to make it into the IJA's upper echelons, which were still dominated by the Satsuma and Choshu Cliques.[16] These officers met in Baden-Baden to find common ground on key issues, and establish a network of "outsiders" that could surmount the obstacles presented by the aforementioned cliques.

Thanks to their technical training at the Staff College and their studies of the First World War, Nagata, Obata, and Okamura all understood that the Imperial Japanese Army's armaments and procurements systems were out of date and in grave need of modernization. They also agreed with German General Erich von Ludendorff's view of war as a totalizing exercise of national mobilization that required the efforts of the whole national economy and society.[17] To the frustration of these officers, Uehara's Choshu Clique faction stood in the way of their "total war" modernization efforts.[18] Uehara and his fellow Choshu Clique officers remained ascendant thanks to the existing network of patronage sponsored by Yamagata Aritomo.[19] Nagata, Obata, and Okamura, all resolved to oppose clique politics, and agreed to support reform of the IJA's personnel policy at their meeting in Baden-Baden.[20] They also agreed to military organization and general mobilization reforms, in line with Erich von Ludendorff's theories of total war.[21] The three officers hoped that these reforms would allow them to advance their own careers, which would in turn make it possible for them to put the IJA on the path of military modernization.[22]

Soon after the meeting at Baden-Baden, Okamura began recruiting other like-minded officers to build a united front against the Choshu Clique. Of these officers, a certain major named Tojo Hideki would go on to be the most prominent. With officers such as Tojo on board, the Baden-Baden group was able to form the first Issekikai, an officer discussion group dedicated to bringing down the Choshu Clique in 1921.[23] This initial group paved the way for other discussion groups with similar aims to the Issekikai such as the Dojinkai (the Universal Benevolence Association) soon thereafter. In 1923, the Dojinkai changed its name to the Futabakai after the French restaurant the group met at in Tokyo in 1923. This group, composed of roughly 18 to 20 members, opposed the Choshu Clique's conservatism, and resolved to implement the necessary reforms to modernize the IJA.[24]

The Futabakai's members were not alone in their desires to destroy the Choshu Clique's influence and pursue a program of modernization. By the end of the First World War, more than half of the staff college's graduates were from provinces other than Satsuma and Choshu, indicating a broader trend away from provincial patronage in the army.[25] The growth of non-Satsuma and Choshu staff college graduates suggested that others in the IJA agreed with the Futabakai's viewpoints on patronage and military modernization. New associations also formed to combat the clique's hegemony as a result. The most important of these associations, the so-called "Saga outsiders," was organized at the behest of Taro Utsunomiya, a major from Saga prefecture. Many figures who would later reach positions of prominence in the IJA were members of Taro's group, including Muto Nobuyoshi, Muraoka Choatro, Masaki Jinzaburo, Araki Sadao, and Yamashita Tomoyuki.[26] While Taro died in February 1922, his legacy lived on through the younger officers in his organization, who would see through Taro's desired plan of personnel reform and territorial expansion.[27]

The formation of groups such as the Issekikai, Futabakai, and Saga outsiders was indicative of a larger trend of non-Choshu officers ascending in the ranks of the army. As time passed, greater numbers of young officers joined the ranks of these officer discussion groups and secret societies.[28] While Nagata, Obata, and Okamura did not give birth to the ultra-nationalist and expansionist young officers' movement that would rise later in the 1920s, the rise of these new discussion groups set the precedent of young officers undermining the IJA's existing institutions. The *gekokujo* that Nagata, Obata, and Okamura initiated at Baden-Baden came to define a generation of officers, who sought to impose their own ideas and views on what they saw as conservative and out-of-touch superiors.[29] This trend would come to a head later in the decade as Japan was forced to reckon with economic and geopolitical threats.

The various anti-Choshu elements within the IJA also began a bottom-up process of re-engineering the IJA's officer corps. This anti-Choshu program sought to prevent Yamaguchi prefecture (where Choshu was located) applicants from entering the IJA's staff college, an important stepping stone to advancing up the army ranks. Because anti-Choshu Clique elements had infiltrated the staff college, they were able to ensure that none of the Yamaguchi prefecture candidates passed their oral exams.[30] Tojo Hideki was instrumental in instigating this scheme in his capacity as head of the staff college between 1922 and 1926. During this period, few Choshu men successfully entered the staff college.[31] Thanks to the countermovement's efforts, the steady pipeline of patronage that had maintained the Choshu Clique since the late 1870s was cut off at the source, never to re-emerge.[32]

Up until 1922, the Choshu Clique was ascendant in the IJA, with outsize influence in the AGS. The overall conservatism of the Choshu

Clique hindered the modernization process that many felt necessary after World War I. Yamagata's position as a *genro* had ensured the Choshu Clique's hegemony in army politics, and thereby kept Uehara's conservatives in power. But Yamagata's death in 1922 opened the door to many figures outside the Choshu Clique to advance up the ranks of the IJA and challenge the hegemony of the old establishment. The rise of these figures signaled the decline of the *genro's* influence in the Imperial Japanese Army and gave rise to new, increasingly radical perspectives and worldviews.

Rumblings in the Three Eastern Provinces

The emergence of the Kwantung Army and the IJA's radical young officers overlapped with a period of greater instability in China and the rise of Chinese nationalism. The latter trend threatened Japan's "special rights and interests" in northeastern China. The Japanese government, IJA, and the Kwantung Army turned to warlord cliques to help safeguard Japan's hegemonic position in the region and prevent the rise of anti-imperialist Chinese nationalism. Japan's foreign policy during the 1920s was defined by the contradictions between its meddling in China's internal affairs to support these figures and its commitments to the liberal international order, which on paper swore off such meddling. The contradictions between these two realities deepened as the decade wore on, before eventually culminating in 1931.

In 1919, Hara Kei's Cabinet and the *genro* agreed to secretly provide even more direct support to Duan and his Anhui Clique, equivalent to roughly five to ten million yen of weapons and supplies per year.[33] The Cabinet's decision was controversial, because Japan had made previous announcements promising neutrality and non-intervention in China's internal affairs. Hara's Cabinet had even put an arms embargo on sending forces to China.[34] But the opportunity to support Duan, who had received Japanese support in years past to advance their joint interests in China, was too good to pass up. Hara allocated more aid in the form of equipment and advisors to Duan's Frontier Defense Army, originally a formation in the so-called "War Participation Army" during the First World War.[35] By the summer of 1920, the Frontier Defense Army was composed of four brigades under the command of Xu Shuzheng, and was highly regarded.[36] The Frontier Defense Army's apparent strength made Duan Qirui's army a force to be reckoned with in central China.[37]

Japan's support for Duan's forces represented a direct threat to Duan's adversaries in the Zhili Clique. Despite Duan's previous pledges to not utilize these forces in domestic disputes, his subsequent aggression against the KMT in southern China and his advocacy for reunifying China by force made him untrustworthy in the eyes of many Chinese warlords. Sensing this political shift, a variety of Zhili Clique warlords including Feng Guozhang, Cao Kun, and Wu Peifu, rallied against Duan and the Anhui Clique in the summer of 1920.[38] Zhang Zuolin also opposed his former ally Duan, who had recently advanced into Mongolia and threatened Zhang's sphere of influence. The shared interests between Zhang and the Zhili Clique led them to broker a secret anti-Duan alliance in the summer of 1920 in Beijing.[39] Soon thereafter, on July 13th, Zhang ordered the dispatch of forces south to Beijing in preparation for war against Duan and the Anhui Clique.[40]

Much to the surprise of many observers, Duan's army collapsed under the weight of a two-front conflict with the Fengtian and Zhili Cliques.[41] Unable to fight a two-front war and suffering from morale problems, Duan's forces collapsed in a mere four days.[42] Having committed to a policy of non-intervention in China, the Hara Cabinet was left to scramble for answers in the wake of the collapse of its foremost partner on the Chinese mainland. Japan would need a new partner in China to fulfill its strategic objectives and safeguard its interests.

Of the various warlords jockeying for power in warlord-era China, Zhang Zuolin was the best positioned for the role of safeguarding Japan's interests. Even before the Anhui-Zhili War, elements within the Kwantung Army had supported increased cooperation with Zhang due to his strong position in northeastern China. For his part, Zhang had obliged the Kwantung Army's desires by purging an anti-Japanese Jilin governor and suggesting that the Kwantung Army could expand Japan's influence in the far north of his domain in the event of a war.[43]

Zhang had also allowed the Japanese to use local railroad infrastructure to reinforce their forces in Siberia. However to the Hara Cabinet, Zhang's sudden betrayal of Duan made him an untrustworthy figure.[44] Some within the Kwantung Army also recommended caution in dealing with Zhang, largely due to perceptions that he sought to expand his reach beyond his domain in northeastern China, south of the Great Wall.[45] Zhang's contradictory actions colored broader Japanese opinion of him in the immediate aftermath of the Anhui-Zhili war, making the prospects of collaboration between them murky.

Nonetheless, Zhang sought Japan's support to equip and supply his forces. To court the Japanese, he told Major General Minami Jiro of the North China Garrison Army his belief that Japan and the Fengtian Clique had a confluence of interests against the Anglo-American supported Zhili Clique. Similarly, when Zhang met Kwantung Army intelligence officer Major General Kishi Yajiro, he again stressed the need to join forces against Anglo-American hegemony in China and the Zhili Clique. Zhang also communicated that he would allow Japanese forces into Jiandao to suppress Korean nationalists and independence activists. Zhang further let it be known that he sought Japanese assistance to prevent Soviet control over the Chinese Eastern Railway (CER). Zhang hoped that these appeals would go up the chain of command and win him the support of key figures in Tokyo, thereby securing his desired partnership with Japan.[46]

Word of Zhang's willingness to work with Japan got back to Hara and his government in Tokyo. Upon hearing of Zhang's appeals, Zhang sent Major General Sato Yasunosuke to meet with Zhang.[47] Even though Sato did not totally trust Zhang, he still recommended a policy in which Japan would help manage his fief in northeastern China, while declining to assist Zhang in conquering the rest of China. Army Minister Tanaka Giichi seconded Sato's recommendation. These figures' endorsement of Zhang influenced Hara, who realized Zhang had common interests with the Japanese Empire.[48] The Hara Cabinet's receptiveness to working with Zhang increased after the latter sent his secretary, Yu Chonghan, to Japan in November 1920. After Yu's visit, the Hara Cabinet gave its verbal approval to working with Zhang, and Hara stated his opinion that

Map of Northeast China and its two most important railroads. Harbin lays at the intersection of the Chinese Eastern Railway and the South Manchuria Railway, making it strategically important (Japanese Intelligence Planning Against the USSR, Map 1).

Japan needed to work through a strong figure such as Zhang to protect its interests in the Three Eastern Provinces.[49] With Hara's endorsement, Zhang would be Japan's confederate in China.

In May 1921, the Hara Cabinet assembled civil and military officials at the so-called "First Eastern Regions Conference" to discuss Japan's foreign policy in East Asia. A variety of issues were discussed, including the status of the Siberian Expedition, the question of Japan's rights in Shandong, and the situation in northeastern China. Regarding these issues, the Cabinet approved a policy statement that advocated three points: 1) negotiations with the pro-Bolshevik "Far Eastern Republic" that could eventually lead to Japan withdrawing its forces from the Russian Far East, 2) Japan handing over the Shandong railway to Chinese railway guards in preparation for Sino-Japanese talks on Shandong, and 3) the necessity of Japan asserting greater authority in northeastern China due to its military and economic importance to the Empire's security.[50] Soon thereafter, Hara made clear his agreement with the Cabinet's resolutions, stating his desire to end Japan's costly and prolonged presence in Shandong and Siberia despite objections from Kwantung Army and Siberian Army commanders.[51] But on the question of northeastern China, Hara made clear his desire to maintain Japan's special rights and interests, principally by working through Zhang Zuolin. Consul General Akatsuka and Kwantung Army commander Misao Kawai concurred with Hara on the need to work with Zhang. The consensus of these political and military figures at the First Eastern Regions Conference cemented Japan's policy in northeastern China, and ensured its support for Zhang for much of the rest of the decade.[52]

Washington Naval Conference and Nine-Power Treaty

The years following World War I represented a golden age for liberal internationalism in the Far East. In Japan, the Hara Cabinet's rise to power demonstrated the ascendance of liberal democracy. In China, the US realized its "Open Door Policy" through the formation of a four-power banking consortium for lending money and preventing unilateral concession hunting. In Russia, the Siberian Expedition brought the various liberal powers of the world into a coalition dedicated to suppressing Soviet communism in its infancy. Perhaps most importantly, the formation of the League of Nations theoretically created new structures to settle disputes and avoid conflict. These all represented great victories for liberal internationalism. These advances also served to harmonize relations between the various imperialist powers—most notably the US, UK, and Japan—around foreign policy in the Far East.

Despite the formation of a liberal consensus among the imperialist powers from 1918 to 1921, many contradictions and potential points of conflict still remained in the Asia-Pacific region. These included the questions of naval disarmament, the situations in Shandong and Siberia, and broader questions of sovereignty in China. While Wilson's Fourteen Points had advocated measures such as the end of secret treaties, global disarmament, national self-determination and respect for sovereignty, geopolitical realities had prevented the inclusion of all these principles in the Treaty of Versailles. To implement these Wilsonian liberal principles and address inter-imperial contradictions in the Asia-Pacific additional agreements would be necessary.

The Washington Naval Conference, held from November 1921 to February 1922, was the medium through which American liberals attempted to address the various inter-imperialist contradictions in the Asia-Pacific region. The original *raison d'etre* for the conference lay in the question of naval disarmament, as the United States and Japan were accelerating the construction of warships despite the end of World War I. At the conference, the main imperialist powers in the Asia-Pacific Region—the United Kingdom, United States, and Japan—agreed on a 5:5:3 ratio of battleships, with battleships being defined as those ships over 10,000 tons and/or possessing guns eight inches or greater.[53] According to the 5:5:3 ratio, the United States and United Kingdom permitted battleship tonnage was 525,850 while Japan's was 315,000 tons.[54] No individual battleship was permitted to exceed 35,000 tons.[55] Even though the 5:5:3 ratio seemed to be a bad deal for Japan on the surface, the US and UK had multiple oceans to control, whereas Japan could concentrate its naval resources on the Pacific. Therefore, despite appearing unbalanced, the 5:5:3 ratio succeeded in creating a roughly proportionate balance of power in the Pacific.

On the questions of Shandong, Siberia, and northeastern China, Japan did not want to discuss them as a matter of sovereignty. Indeed, Japan had been hesitant to even accept the US invitation to the Washington Naval Conference because it wanted to avoid broaching these topics.[56] Despite such hesitancy, Shidehara Kijuro, the ambassador to the United States and lead negotiator of the Japanese delegation, recommended that Japan "make the first move in setting the agenda for discussion," due to US Secretary of State Hughes' ignorance of the Asia-Pacific region.[57] Thanks to Shidehara's encouragement, the Japanese government agreed to attend, but also informed the US of its desire to avoid discussion over Shandong, Siberia, and northeastern China, all of which were regarded as the "sole concern" of Japan.[58] But Hughes was intent on discussing all of these matters at the conference, informing Japan that by accepting the invitation to the conference, it agreed to discuss all issues.[59] Hughes's steadfastness ensured that these questions were all on the table at the Washington Conference.

As per the resolutions of the First Eastern Regions Conference, Japan agreed to make great concessions on the issues of Siberia and Shandong. On the question of Siberia, the Japanese publicly stated their intention to withdraw from the Russian Far East.[60] With regards to Shandong, negotiations were more complex, but by February 1922 Japan and China finally arrived at an agreement. Japan, for its part, agreed to surrender its claim to the Jiaozhou Bay area, and sell the Jinan railway back to China. China agreed to allow a Japanese manager and chief accounting officer for the Jinan railway until payment was complete.[61] Shandong's coal mines would also be under joint Sino-Japanese management.[62] In this settlement, Japan surrendered the sanctity of its far-reaching claims in the "Twenty-One Demands," limiting its rights to a few areas. By surrendering its far-reaching claims, Japan allayed a key sore spot for the new liberal international order, as China largely recovered its sovereign authority over Shandong and therefore exercised its right to self-determination.

While the Washington Naval Conference put to rest the issue of Shandong, the contradiction between the US's Open Door Policy for China and Japan's "special rights and interests" in northeastern China still loomed large. For Japan, northeastern China was the most important issue at the conference, as the army believed this region was necessary for the Empire's economic survival and national security.[63] It wanted the US to recognize its "special rights and interests" in northeastern China at the conference. Japan's emphasis on the importance of northeastern China was aptly summarized by Masanao Hanihara, a delegate to the Washington Naval Conference:[64]

> As to that territory (northeastern China) the Japanese delegates desire to make it clear that Japan has no intention at present to relinquish the important rights she has lawfully acquired and at no small sacrifice. The territory in question forms a part of Manchuria—a region where, by reason of its close proximity to Chinese territory, more than anything else she has vital interests in that which relates to her economic life and national safety… the commercial and industrial interests they have established there are of such importance and magnitude to Japan that they are regarded as an essential part of her economic life.

Hanihara's statement made clear Japan's position on the issue of northeastern China, which it saw as fundamental to its economic and security interests. Secretary of State Hughes, who confirmed Japan's designs on northeastern China through the US diplomatic code-breaking apparatus, understood Japan's position, and opted not to ask Japan to forfeit its rights in the region.[65] But Hughes and the Americans did coax Japan into signing the so-called "Nine-Power Treaty," which upheld the territorial integrity and sovereignty of China and prevented further attempts to expand one nation's "sphere of influence" at the expense of others. These principles were enumerated in the most relevant parts of the Treaty, Articles I and III:[66]

Nine-Power Power Treaty:

Article I

The Contracting Powers, other than China, agree:
(1) To respect the sovereignty, the independence, and the territorial and administrative integrity of China;
(2) To provide the fullest and most unembarrassed opportunity to China to develop and maintain for herself an effective and stable government;
(3) To use their influence for the purpose of effectually establishing and maintaining the principle of equal opportunity for the commerce and industry of all nations throughout the territory of China;
(4) To refrain from taking advantage of conditions in China in order to seek special rights or privileges which would abridge the rights of subjects or citizens of friendly States,[2] and from countenancing action inimical to the security of such States.

Article III

With a view to applying more effectually the principles of the Open Door or equality of opportunity in China for the trade and industry of all nations, the Contracting Powers, other than China, agree that they will not seek, nor support their respective nationals in seeking-
(a) any arrangement which might purport to establish in favour of their interests any general superiority of rights with respect to commercial or economic development in any designated region of China;
(b) any such monopoly or preference as would deprive the nationals of any other Power of the right of undertaking any legitimate trade or industry in China, or of participating with the Chinese Government, or with any local authority, in any category of public enterprise, or which by reason of its scope, duration or geographical extent is calculated to frustrate the practical application of the principle of equal opportunity.

> It is understood that the foregoing stipulations of this Article are not to be so construed as to prohibit the acquisition of such properties or rights as may be necessary to the conduct of a particular commercial, industrial, or financial undertaking or to the encouragement of invention and research.
>
> China undertakes to be guided by the principles stated in the foregoing stipulations of this Article in dealing with applications for economic rights and privileges from Governments and nationals of all foreign countries, whether parties to the present Treaty or not.

The Nine-Power Treaty's explicit reference to "more effectually applying the principles of the Open Door Policy," and against "taking advantage of conditions in China in order to seek special rights or privileges which would abridge the rights of subjects or citizens of friendly States," clearly reflected America's desire to do away with the old "sphere of influence" imperialism in China that advantaged Japan. The new treaty created an alternate framework where all the imperialist powers could do business unimpeded. The clauses referring to the necessity of respecting the "sovereignty, the independence, and the territorial and administrative integrity" and providing "the fullest and most unembarrassed opportunity to China to develop and maintain for herself an effective and stable government" were designed to prevent any outcome that might constrain the ability of other powers to engage in trade and economic relations with China. These guarantees to protect China's sovereignty echoed many of Wilson's Fourteen Points, which advocated the self-determination of nations, free trade, and gradual de-colonization in the name of world peace. By putting these principles into writing in the Nine-Power Treaty, the US sought to deploy Wilson's vision of universal liberalism in East Asia to counteract Japan's influence in China.

The Nine-Power Treaty put in writing the key principles of Wilsonian liberalism in East Asia. But one great exception to the principles in the Nine-Power Treaty still remained: the Three Eastern Provinces. Because many in Japan understood the Liaodong Peninsula and the rest of China's northeast to be an indispensable "lifeline," Japan would not recognize China's sovereignty in the region. While the US and UK were content to allow such a state of affairs many Chinese were not, and were becoming increasingly militant in their nationalism. With Chinese nationalism on the rise, and elements within Japan seeking to expand their influence in northeastern China, it was only a matter of time before these forces clashed. The clash between Chinese nationalism and Japanese expansionism would come to define the decade of the 1920s.

Chapter 6

Zhang's Adventurous Wars

In China, the defeat of Duan Qirui in the Anhui-Zhili war left Feng Guozhang's Zhili Clique and Zhang Zuolin's Fengtian Clique as the main contenders for power. Both received backing from foreign powers, with Feng and the Zhili Clique receiving aid from the UK and US while Zhang received support from Japan.[1] With arms and support from imperialist powers, both Feng and Zhang spent 1921 nervously eyeing the moves of the other, seeking to gain an advantage. Zhang would ultimately initiate action in 1922, venturing south of the Great Wall to attack the Zhili Clique. Zhang's attempt to reunify China would lead to war and instability in the country throughout the 1920s, as he and the Zhili Clique fought over who would claim domain over China. Because of Zhang's boldness, his Japanese backers would be forced to choose between protecting their interests in the Three Eastern Provinces and keeping their powder dry, in terms of both hard and soft power. Their decisions would shape China's geopolitical landscape and set precedents for their successors.

The First Fengtian-Zhili War

While Japan had committed itself to a policy of non-intervention in China, many Japanese saw the northeast as an exception to this policy. Japan supported Zhang Zuolin's Fengtian Clique to help maintain the Empire's special rights and interests in the Three Eastern Provinces. From the perspective of policy makers in Tokyo, stability in northeastern China was vital; for this stability to exist Zhang and the forces under his command needed to remain in the Three Eastern Provinces, north of the Great Wall.[2] However, Zhang had other ideas and sought to venture south of the Great Wall to vanquish his previous allies, the Zhili Clique. In part, Zhang's desire to destroy the Zhili Clique stemmed from his fear of their growing power, which threatened his fiefdom in China's northeast.[3] He also wanted to be the one to reunify China and succeed where many others had failed.[4] For these reasons, Zhang began making moves to prepare for war with the Zhili Clique.

In November 1921, the Chinese central government was facing a financial crisis so acute that police and soldiers near Beijing had been unpaid for several weeks, leading to a simultaneous run on two of China's central banks. The run on these banks caused Liang Shiyi, the director of the Bank of Communications, to invite Zhang to intervene.[5] Zhang's solution to the crisis was to force the resignation of Premier Jin Yunpeng's Cabinet under the pretense of establishing a new, unified government.[6] In their place, Zhang selected Liang to be Premier and proposed the formation of a new Constitution.[7] He ostensibly believed that the new Liang government would be more suited to govern China in a unified manner than Jin's government had been.[8] But Zhang likely lied about his views, as both he and the Zhili Clique had previously endorsed Jin's government.[9] Instead, Zhang's motives for intervening were to take control of Chinese financial institutions (the Treasury, Bank of China, and Bank of Communications) and gain greater legitimacy in the eyes of the world before his showdown with the Zhili Clique.[10] Zhang's quiet achievement of both his objectives put him in a good position to fight the Zhili Clique.

The Zhili Clique was not content to stand idly by and let Zhang dictate the central government. The new government drew their ire when it failed to deliver on a commitment from the previous government to fund Zhili Clique member Wu Peifu's forces.[11] Wu, alongside a variety of other Zhili Clique members, denounced the new government and demanded Liang's resignation.[12] Forced into a difficult position, Liang took sick leave in January 1922, leaving the position of premier in the hands of his Foreign Minister. But even this measure was did not satisfy Wu, who denounced the new Cabinet's Minister of Finance in February for embezzlement.[13] These sudden and repeated denunciations of Zhang's handpicked Cabinet signaled that the Zhili Clique was not willing to go along with his attempts to establish control over China's central government. Therefore, Zhang would have to begin the quest for allies and support to subdue the Zhili Clique by force.

In February 1922, Zhang formed an alliance with the KMT and remnants of the Anhui Clique against the Zhili Clique.[14] While Zhang's new alliance pressured the Zhili Clique on multiple fronts, Zhang needed Japanese support both to equip and supply his army and to provide security in his rear in the event his campaigns in the south went poorly. Working through his Japanese advisor Machino Takema and Yu Chonghan, Zhang put out feelers to the Japanese Minister in Beijing and the Kwantung Army for weapons and support. He also sought security assurances from General Kishi regarding the Kwantung Army's willingness to secure his rear in the event of conflict with the Zhili Clique.[15] Zhang hoped that his overtures to the Japanese would allow him to gain their support for his venture south of the Great Wall, even in spite of Japan's official stance.

The Kwantung Army largely supported Zhang's planned offensive south of the Great Wall, and advocated giving him arms and supplies. Kishi Yajiro, an army special service agent in Mukden, supported arming Zhang because doing so would "have an important bearing on the existence of the empire."[16] Kwantung Army Chief of Staff Fukuhora Yoshiya argued that Japan had to support Zhang's efforts to overthrow the Zhili Clique to counter the US and UK, who were covertly supporting this clique to advance their own interests in China. The Japanese military attaché in Beijing, Azuma Otohiko, shared similar views as his Kwantung Army colleagues.[17] The consensus of these leaders contradicted the Hara Cabinet's decision at the First Eastern Regions Conference to avoid supporting any of Zhang's ventures south of the Great Wall. Despite the government's established policy, these figures still believed it necessary to support Zhang to counter British and American influence in China, which might eventually threaten Japan's special rights and interests in the northeast. Tokyo would have to formulate a policy in the heat of the moment in order to properly manage the situation.

Few in Tokyo were willing to support Zhang's venture south of the Great Wall. The Foreign Ministry cited the international arms embargo on China and the Nine-Power Treaty as reasons not to equip Zhang's forces.[18] Foreign Minister Uchida also sent a long telegram to the Kwantung Army explaining the importance of upholding global public opinion and Japan's international treaty commitments, both of which would be undermined in the event Japan sent weapons to Zhang.[19] While Akatsuka and some Kwantung Army officers sent more messages reiterating their view that Japan needed to support Zhang, Tokyo did not budge, withdrawing its advisors to Zhang, Machino and Honjo Shigeru, for the rest of the crisis. The capital also withdrew Kishi Yajiro from his post at the Mukden Special Service Agency.[20] These decisions were carried out because the Japanese government wanted to adhere to the "spirit of the nine-power treaty" and demonstrate a "line of impartial neutrality and fairness" towards China's internal affairs.[21]

When war between Zhang and the Zhili Clique broke out in April 1922, Zhang once again asked the Japanese for weapons and supplies. While some in the Kwantung Army lobbied Tokyo to fulfill his request, Tokyo held fast to its previous decision, opting not to support Zhang in his fight south of the Great Wall.[22] Despite this setback, Zhang still had a good chance to defeat the Zhili Clique on paper, especially when one considers his alliances with the KMT and remnants of the Anhui Clique. But Sun's KMT was unable to take the offensive against Zhili Clique forces in the south thanks to the presence of southern warlords, meaning that remnants of the Anhui Clique in Zhejiang also could not attack since their territory was surrounded on three sides by Zhili-aligned forces.[23] The inability of Zhang's allies to take the offensive further south meant that Zhang was left to fight the Zhili Clique alone in north China, at numerical parity. Such a fight would be a challenge for Zhang's under-equipped and supplied forces.

The first skirmishes between the Fengtian and Zhili forces occurred on April 28th, 1922, along the Beijing-Mukden and Tianjin-Pukou Railway lines. These skirmishes were inconclusive, but did result in the deployment of both forces along a wide area, which linearized the conflict.[24] But Zhang's forces' front line collapsed soon thereafter on May

4th. Multiple explanations exist for this sudden demise:[25] First, Zhang's forces collapsed because Feng Yuxiang's 11th Infantry Division outflanked them on their right on the night of May 3rd, leading to a hasty withdrawal. Another explanation asserts that Zhang Jinghui, a Fengtian Clique commander responsible for its western front, conspired with the Zhili Clique to undermine the defense of North China. The final explanation contends that Zhang Jinghui heard rumors that three supposedly neutral divisions in the vicinity of Beijing were in contact with Wu and would attack Fengtian's rear, which produced a panicked withdrawal.[26] Whatever the cause of the Fengtian forces' collapse, the result was the same: Zhang and his forces were thrown out of the provinces of Rehe, Chahar, and Suyiuan, forced to retreat back to Zhang's home base in the Three Eastern Provinces.[27] While the presence of the Kwantung Army saved Zhang's forces from total annihilation, defeat in North China meant that Zhang could not lay claim to the rest of China as he had intended.[28]

Zhang's defeat in the Fengtian-Zhili war meant that he was dismissed from his official position as governor of the Three Eastern Provinces (Fengtian, Jilin, and Heilongjiang) on May 10th, 1922. Zhang's dismissal from office facilitated a radical step: his declaration of independence for the Three Eastern Provinces.[29] His declaration essentially separated these provinces from the rest of China, and therefore served as a great boon to Japanese policymakers in Tokyo.[30] But Zhang's declaration of independence belied his desire to reunify China and expand his sphere of influence south of the Great Wall.[31] Zhang's ambitions south of the Great Wall meant that Tokyo would be forced to try and control his actions, along with those of their subordinates in the Kwantung Army. The Kwantung Army in turn would have to balance Zhang's independent pursuits north and south of the Great Wall with their own objective of expanding Japanese hegemony in northeastern China. The push-pull between these elements would come to define events in the coming years, as each of these contradictions sharpened and eventually came to a head.

Zhang's desire to reunify China was roughly concurrent with an upsurge in Chinese nationalism. The Fengtian Clique, seeking to capitalize on the upsurge in nationalism that occurred after the May 4th Movement, began taking a more nationalist political stance in the 1920s. Soon after the end of the First Fengtian-Zhili War, a Chinese nationalist organization, the "Society for Promotion of Civil Government in the Three Eastern Provinces" was formed, putting forth a nationalist, anti-imperialist line, overseen by a chief secretary, Zhao Zhafei, with ties to the KMT.[32] Soon after the formation of this organization in October 1922, Zhang's Zuolin's son, Zhang Xueliang provided a 20,000-yuan subsidy for the "Three Eastern Provinces People's Paper" to form.[33] This paper, which openly advocated the KMT's line, was the first pro-KMT paper to form in northeastern China, as well as the first to explicitly promote Chinese nationalism. The formation of other nationalist papers and organizations soon followed as Chinese nationalism became more popular. The increasing popularity of Chinese nationalism in the Three Eastern Provinces put pressure on Zhang Zuolin to embrace the new movement despite previous opposition.[34]

Starting in 1923, Zhang Zuolin began to embrace Chinese economic nationalism. Early that year, the civil governor of Fengtian Province named Wang Yongjiang came to Zhang and informed him of his intention to construct a railway between Mukden, Hailung, and Jilin with the help of private Chinese investors.[35] This rail line, which would also be linked to the port of Huludao, would rival the SMR's line and thereby threaten its treaty-protected monopoly in the Liaodong Peninsula. To ensure the construction of the new railway, Wang appealed to the SMR to secure funding and assure the Japanese that the construction of his railway line was not intended to undermine the SMR's monopoly. In return, he also tried to entice the Japanese by offering to build an extension of the Beijing-Mukden Railway line between Taonan and Qiqihar, which Japan long desired. Wang and Zhang hoped this deal would allow local Chinese businessmen in northeastern China to establish their own strength and challenge the SMR's monopoly in the region.[36]

Wang's railway proposal offered different advantages and disadvantages to the Japanese. For the Imperial Japanese Army, Wang's proposal was quite convenient, as the construction of the Taonan-Qiqihar railway would allow the IJA to preempt the deployment of Soviet forces along the CER west of Harbin.[37] The SMR also had some interest in the project, as the Taonan-Qiqihar railway offered to give it a leg up in its competition with the CER.[38] But two main drawbacks also existed: 1) the railway's construction would allow the international consortium to meddle in northeastern China and 2) the railway's construction would establish the precedent of a Chinese-owned railway near the SMR. Ultimately, the Japanese Foreign Ministry decided to approve Wang's scheme and agreed to build the Taonan-Qiqihar and Mukden-Hailung-Jilin railways upon the recommendation of Matsuoka Yosuke, a member of the SMR's board of directors.[39] But Zhang's assertion of Chinese economic self-interest reinforced the idea that he was an unreliable partner in certain Japanese circles, thereby establishing the basis for an eventual rift between himself and his Japanese sponsors.[40]

Zhang's economic challenge to Japan expanded beyond railways and into the field of industrial enterprises. In January 1924, he called a conference on the issue of establishing a series of industries, where he pledged to raise twenty million yuan for the establishment of Chinese-owned factories in three months. He also pledged a further 2-million-yuan increase in capitalization for Chinese-owned sugar, cotton, timber, and coal industries in his domain.[41] Zhang even sponsored the expansion of a Chinese-owned electric power grid, establishing twenty new companies between 1923 and 1927. These electric power companies competed with the SMR corporation to provide electricity to cities such as Changchun, Harbin, Andong, Mukden, and Sipingjie.[42] The formation of these Chinese-owned industries undermined the SMR's monopoly in its non-railway industries and further threatened Japan's hegemonic position in northeastern China's economy.

The growth of Chinese nationalism and anti-imperialism was accelerated by both the Chinese government's repudiation of the Twenty-One Demands and calls to restore sovereignty over the Kwantung Leased Territory, SMR, and Andong-Mukden Railway to China. These possessions, which should have been open to purchase for China in 1923, 1928, and 1933 respectively, remained part of the Japanese Empire.[43] When Japan cited the Twenty-One Demands as grounds to continue holding them, Chinese nationalists triggered a nationwide boycott of Japanese goods, signaling the rise of China's national consciousness.[44] The formation of more anti-imperialist organizations—such as the Jilin Society for the Support of Diplomacy, the Society of Fellow Provincials of the Three Eastern Provinces Resident in Beijing, and a Society for Support of the Return of Port Arthur and Dalian—further illustrated the rise of this national consciousness. These kinds of nationalist groups sponsored many protests that demanded the return of Japan's concessions in the Three Eastern Provinces to Chinese sovereignty.[45] As many of these groups had gained a foothold among the populace, Japan's grip over northeastern China was loosening, which greatly worried many Japanese policymakers. New solutions would be needed to combat the rising tide of Chinese nationalism and maintain Japan's hegemonic position.

The SMR as an Instrument of Imperial Policy: The Monopolization of Northeastern China's Economy

Japan's counter to Zhang's economic challenge in northeastern China was the growth and expansion of the SMR Corporation, which continued to increase its profits. The company's revenue on the railways it controlled between 1907 and 1920 increased by an average of 8.3 percent, while between 1920 and 1933 it increased by 7 percent.[46] These increases resulted in the SMR's income tripling by 1913-14, subsequently doubling in 1917-18, and doubling again by 1920-21.[47] In total, company revenue totaled roughly 185 million yen in 1923, with total assets exceeding 1 billion yen by 1930. In comparison, Korea's entire colonial government had a budget of only 101 million yen, and Taiwan's government budget only 100 million yen.[48]

Of the SMR's main sources of revenue, soy continued to be the most important. The Three Eastern Province's production of soybeans doubled to 221,000,000 bushels or 5,300,000 tons annually between 1915 and 1929.[49] This amount of soy was equivalent to roughly two-thirds of the world's total.[50] Of the 5.3 million tons of soy Japan produced, roughly 4.7 was exported from the ports of Dalian, Niuzhuang, and Yingkou to other destinations, with the rest of China receiving 13 percent, Japan receiving 41 percent, and Europe receiving 44 percent.[51] Because the SMR's main branch line terminated at Dalian and the harbor there could handle 2,500 ships a year (roughly 2.8 million tons of cargo), Dalian was by far the most important embarkation port for Japanese exports.[52] The SMR's monopoly access to the port meant that it also received 62 percent of the Liaodong Leased Territory's imports from Japan. The SMR's control over the port facilities in Dalian allowed it to exert monopoly control over the soy trade in northeastern China and charge high rates for its exports abroad. These factors were key to the SMR's profitability, which continued to increase throughout the 1920s.[53]

While the SMR profited greatly from monopolizing the soybean trade, local Chinese suffered due to the SMR's monopoly.[54] The cultivation of soybeans came at the expense of traditional subsistence crops like gaoliang, millet, and corn, which declined from sixty percent of cultivated land to forty-five percent between 1907 and 1927. Overreliance on soy also made local farmers vulnerable to fickle global demand and price fluctuations. To even get their crops to market, Chinese peasants had to compensate multiple layers of comprador middle-men to ship their goods to the SMR, which ate further into their profit margins. Furthermore, the SMR's transportation monopoly in the Liaodong Peninsula meant that it could charge higher prices to export the peasants' agricultural goods than it did to transport imported manufactures into the Liaodong Leased Territory. These factors all degraded the ability of Chinese peasants to make a decent living when global demand dried up, as evidenced by the fact that between 1927 and 1931, peasant family incomes in northeastern China more than halved, from 170 to 81 yuan.[55] The economic hardship that Japan's imperialist system created inevitably led to banditry and unrest in northeastern China, with the number of bandits tripling from 20,000 in 1920 to 60,000 in 1931.[56] Banditry would continue to grow in future years, especially as Japan began to intervene more directly in northeastern China's economic and political landscape.

Due to the predominance of soy in northeastern China, agriculture remained the most important sector of the economy, registering at 45.7 percent of GDP in 1929.[57] But despite importance of soy, private Japanese investors already began creating the basis for the formation of industrial enterprises in the Liaodong Peninsula. The process of building these enterprises in the Three Eastern Provinces began with textile factories, seventy-two of which were built by 1913.[58] After the Twenty-One Demands in 1915, Japanese private investment in the Liaodong Leased Territory continued to expand, with the number of companies in the region increasing from 124 in 1915 to 914 in 1925, with a total investment of 536 million yen.[59] The number of factories also rose, nearly tripling from 152 in 1909 to 450 in 1919.[60] While large Japanese monopolies like Mitsui controlled most of the initial output in the region, the expansion of Japan's industrial capital in these years laid the foundation for future efforts.[61]

The private sector was the primary source of Japan's first industrial ventures in northeastern China. But starting in the 1920s, many in policy-making circles began to advocate state-led efforts to build industries in the Liaodong Peninsula.[62] The SMR's revenues made an expansion of its industrial footprint possible and even desirable, as the company sought to expand into other ventures to achieve vertical integration. Consequently, the SMR founded a variety of different spinoff companies, including Dalian Oil and Fat, the South Manchuria Glass Company, Anshan Iron and Steel Works, and Dalian Ceramics.[63] The efforts of the SMR greatly contributed to the growth of industrial production in northeastern China, which increased 3.1% per year on average between 1924 and 1929.[64] In more concrete terms, the number of factories increased from 450 in 1919 to 789 in 1929.[65]

Of the various spinoff companies the SMR created, one of the most successful in terms of output was the Anshan Iron and Steel Works. In part, the output of this iron and steel works is attributable to the abundance of raw materials to be found in the Liaodong Peninsula. The coal mines at Fushun were vital to the function of the Anshan Iron and Steel Works, as these mines produced roughly 2.5 million tons of coal in 1918.[66] Iron ore deposits at Anshan were also significant, registering approximately 100,000 tons in 1918.[67] The voluminous production of iron and coal in and around Anshan made it a perfect site to set up a steel production facility, which the Japanese built in 1919. The new facility, consisting of two blast furnaces, received 45,900,000 yen in investment capital between 1919 and 1926.[68] These investments bore fruit by the middle of the 1920s, producing eighteen percent of pig iron used for steel production in Japan and a further thirteen percent of pig iron used for casting.[69] The total annual production at the Anshan Iron Works increased from 31,620 tons in 1919 to 210,443 tons in 1929.[70] Anshan's importance to Japan's domestic steel industry, and the increased output of the firm itself, made it a useful enterprise for Japanese industry in China's northeast.

In the 1920s, a series of other sponsored industries also sprung up in the Liaodong Peninsula and the SMR zone. One of these industries was the bean oil industry, where mills produced bean oil for cooking and lighting, as well as bean cakes for fertilizer and cattle feed. By 1929, there were 465 mills which produced 2,200,000 tons of bean oil and bean cakes.[71] The textile and fabrics industry also continued to expand, producing 124,050,000 pounds of fabric per year valued at 46,540,000 yen.[72] Even the ceramics industry became significant, with over forty-six factories operating by 1929.[73] These industries brought in significant profits for the SMR and other Japanese corporations, and helped the SMR in particular make up for losses incurred in certain sectors (e.g. administration, iron-steel production, hotels). The spinoff companies also helped the SMR establish vertical integration over key sectors, particularly the soybean trade. The profitability and productivity of these spinoff corporations meant that they could largely be considered successful ventures.[74]

The SMR's enterprises greatly benefited from its research and development program, which was rooted in the Central Experimental Institute. One of the most important innovations that the Central Experimental Institute pioneered was the creation of special earthenware and porcelain containers for storing soybeans, which improved the process of soybean fermentation. In turn, oil could be extracted from these fermented soybeans and used for a variety of different purposes.[75] The

Central Experimental Institute also invested a significant amount of time in shale-oil processing and created products from the "dry distillation of oil shale."[76] In 1928, the Institute further augmented its oil extraction from shale with the invention of a combustion process to achieve the aforementioned dry distillation. The "electrically operated" Fushun colliery was the means by which the coal shale was extracted.[77] The Fushun colliery also had state of the art technology in the form of mechanical excavators that could dig coal up to 750 feet below ground and transfer it to wagons above ground.[78] Such state-of-the-art technology made coal one of the most profitable sectors for the SMR corporation, despite high upfront costs.

Japanese settlers also started arriving in the Liaodong Peninsula in greater numbers as the 1910s gave way to the 1920s. In 1915, the Japanese population in the combined territory of the Liaodong Leased Territory and the SMR zone was 85,000; by 1920, this figure had grown to 137,000.[79] Ten years later, by 1929, the Japanese population had grown again to 203,022.[80] Many of the new migrants came with their families, which fulfilled the objectives of Goto and Nakamura for long term Japanese settlement in the region. The SMR was a key employer for new Japanese transplants with 21,824 employees, primarily in white-collar positions.[81] The SMR's influence over Japanese life in the Leased Territory and SMR zone was omnipresent, with the company providing important amenities such as electricity, gas, and water to residents in the Liaodong Leased Territory and the SMR zone. Ito Takeo detailed his impression of the SMR's influence in the Liaodong Peninsula while on a tour of the region:[82]

> One could neither see nor hear a single thing that did not have some relation, directly or indirectly, with the SMR's activities in the lives of Japanese living along its railways beginning at Dalian. I was astonished to discover this. We spent several days sight-seeing in Dalian, staying at a dormitory of the Dalian School of Engineering, which had been founded by the SMR. The factories we observed were of course managed by the SMR, as was the restaurant within the park where we escaped from our chaperones. Electricity, gas, water works—all had

> the SMR's markings on them. In our youthful minds, the SMR was merely a railway company and we misunderstood its real character.

Ito's impression of the SMR's far-reaching influence is a testament to its wide-reaching dominance in a variety of different fields. The SMR's control over a diverse array of economic sectors, combined with its utilization of state-of the art technology, allowed it to claim impressive profits, amounting to an average yearly revenue of 218 million yen throughout the 1920s.[83] But the company still primarily existed to fulfill Japan's policy objectives in the region, which were growing ever more ambitious. In future years, the SMR would need to balance the pursuit of profit with the maintenance of Japan's economic and political hegemony in the region. The Empire's future in northeast Asia was linked to the SMR's efficacy in achieving both of these objectives.

The Rise of Shidehara Kijuro's "Friendship" Diplomacy, the Second Fengtian-Zhili War, and the Guo Songling Affair

Despite Japan's meddling in northeastern China, on paper it remained committed to the liberal international system that defined global politics in the decade after World War I. The foremost defenders of the liberal system in Japan were the political parties and elements within the Foreign Ministry. Of this cohort, Shidehara Kijuro, Japan's Foreign Minister between 1924 and 1927, was by far the most prominent; he was a key architect of Japan's adherence to the liberal international order, having been the chief representative of the Japanese delegation at the Washington Conference. Shidehara truly believed in the new international order he had helped create, famously saying that "the age of Machiavellian aggressive state policy is over" and that "foreign policy must be conducted in accordance with the great principles of peace and justice."[84] Shidehara's benign views on statecraft led him to take relatively friendly stances on issues pertaining to China during his tenure as Foreign Minister.

Shidehara began his tenure as Japan's Foreign Minister in 1924, serving in the Cabinet of Japanese Prime Minister Kato Takaaki. He immediately established the foundational principles of his China policy in a speech to the Diet on July 1st, 1924, stating that Japan needed to "avoid getting overly involved in China's internal affairs" and do its utmost to facilitate the two countries getting "closer together economically."[85] He also stated the need for Japan's actions to be "consistent with the spirit of the Washington Treaties" including the Nine-Power Treaty.[86] Shidehara would remain true to these words throughout his tenure, as the principles of nonintervention in China's internal affairs and the promotion of economic cooperation would be the bedposts of his "new diplomacy."[87] The cultivation of good economic relations between China and Japan was Shidehara's foremost objective as Foreign Minister, as he wanted China to serve as an export market for Japan's industrial goods.[88] While he believed in defending Japan's treaty rights and interests in the Three Eastern Provinces, his understanding of what defined a threat to those interests was not as wide as that of more hawkish figures in the IJA and the Kwantung Army.[89] He even believed that Japan should treat the Three Eastern Provinces as a part of China, and that Japanese citizens in China were not entitled to have special protection from the Imperial Japanese Army.[90] These positions would fuel tensions between Shidehara and the armed forces, which sought more intervention in China's northeast.

While Shidehara pronounced a hands-off policy towards China, Zhang Zuolin's desire to reunify the country by force put Shidehara's pledge to the test. Dating back to 1923, Zhang had been making preparations for another war with the Zhili Clique, asking the Japanese for arms through a representative of the Kwantung Province Governor and then through his advisor, Honjo Shigeru. Although Honjo rebuffed Zhang's desire for a "complete arms transfer," he noted that in the event of a disturbance to Zhang's position from elsewhere, Japan was "determined always resolutely to put a stop to it."[91] He also indicated to Zhang that his forces would be allowed to venture south of the Great Wall in the event KMT aligned forces moved north against the Zhili Clique.[92] If Zhang capitalized on such an occurrence, the Anglo American-supported Zhili forces would be squeezed on two fronts between the KMT-aligned armies to their south and Zhang's Fengtian armies to their north. Such a conundrum would gravely threaten the Zhili Clique and possibly lead to its demise.

The Kwantung Army staff concurred with Honjo's position, writing a paper in October 1923 referring to the Zhili Clique as "the greatest evil" in China at that time and criticizing Japan's policy of nonintervention in China's civil war.[93] But the views of Honjo and the Kwantung Army fell on deaf ears in Tokyo in 1923, which sought adherence to the liberal international system. While Zhang made another appeal to Shidehara in 1924, the latter was less interested in directly supporting him than previous Japanese foreign ministers. Shidehara was only willing to authorize the deployment of troops in the event of a direct threat to Japanese citizens and their property.[94] Shidehara's unwillingness to directly support Zhang with arms and supplies would continue, as he sought to avoid meddling in China's sovereign affairs and keep Japan out of wars in northeast Asia.

The Second Fengtian-Zhili War commenced in September 1924 when the Zhili Clique attacked Fengtian Clique-aligned warlord Lu Yongxiang's forces in Zhejiang province. After Lu's forces were quickly overwhelmed, the Zhili Clique was able to then focus its efforts on fighting the Fengtian Clique.[95] The initial period of the second conflict was characterized by prolonged stalemate, with both sides dug in around the area of Shanhaiguan.[96] But behind the scenes, the Zhili Clique coalition was starting to fray, as Feng Yuxiang, a Zhili Clique-aligned warlord repeatedly refused orders from Wu Peifu to deploy against Zhang's forces. Feng's insubordination was such a serious matter that Wu asked President Cao Kun for permission to withdraw forces from the front to handle Feng's disobedient forces in his rear.[97] Even though Cao refused Wu's request, the fact he had to make such a request is indicative of how much Feng's insubordination undermined the Zhili Clique's war effort.

Among the Japanese, Zhang's military advisors, Kwantung Army staff officers, and even some figures in the civilian government supported directly intervening to aid Zhang Zuolin. While Honjo and Kishi were replaced as advisors to Zhang in August and September 1924, they made sure to inform their successors that they needed to "facilitate the mutual cooperation of Japanese and Chinese armies in the event of emergency by influencing the military-related installations of Fengtian province to be modeled on those of the Japanese empire."[98] Zhang's new advisors collaborated with the Kwantung Army in promoting the cause of the Fengtian Clique, providing intelligence, arms, and advice to his forces.[99] These elements' promotion of Zhang's cause also led many in the Cabinet of Prime Minister Kato Takaaki to favor sending troops to support Zhang, with Minister of Agriculture and Commerce Takahashi Korekiyo leading the charge.[100] The influence of these elements was so great that for a brief moment it seemed Japan would send troops to aid Zhang, despite its previous commitments to not intervene in China's internal affairs.

Despite widespread Japanese support for Zhang, one man was consistent in opposing intervention in China's northeast: Foreign Minister Shidehara. Shidehara believed that Japan needed to adhere to its promises of non-intervention in China's internal affairs that it agreed to accept in the Nine-Power Treaty.[101] Consequently, he issued statements of neutrality and nonintervention on September 10th, 11th, and 22nd, and sent Consul General Kodama Hideo instructions to refrain from intervening in the Second Fengtian-Zhili conflict.[102] But Shidehara still needed to convince his contemporaries in the Kato cabinet of his position, which he attempted to do by explaining that direct intervention was unnecessary:[103]

> Even if Wu Peifu [of the Zhili clique] wins and enters Dongsan Province (the 'Three Eastern Provinces'), his forces will be fully engaged, with no reserves for deployment elsewhere… Feng Yuxiang [also of the Zhili clique] is in Zhangjiakou; however, he is not well disposed towards Wu. He is not going to just sit back and watch Wu capture more territory.

Shidehara's staunch opposition to supporting Zhang Zuolin led the Japanese government to endorse nonintervention in September 1924.[104] Even though the changing situation at the front brought this official position into question, the Japanese government still maintained its non-interventionist stance.[105] It seemed Shidehara had succeeded in preventing Japan from intervening in the Second Fengtian-Zhili conflict.

Despite Shidehara's successes in Tokyo, Japanese policy during the Fengtian-Zhili war came to be dictated by the *gekokujo* of actors proximate to the fighting. Not only had the Kwantung Army coordinated plans with Zhang's Japanese advisors, but it had also sent intelligence officers to advise Feng Yuxiang. Of the officers sent to advise Feng, Lieutenant Colonel Doihara Kenji was the most important, due to his expertise on China and connections to the Kwantung Army.[106] Doihara's connections proved instrumental in securing and funneling a "loan" worth one million yen to Feng, with an implicit message that Feng should switch his alliance from the Zhili Clique to Zhang's Fengtian Clique.[107] Feng, already dissatisfied with his positioning in the Zhili Clique, obliged, rebelling against the Zhili Clique in late 1924. The Zhili Clique, forced to respond to the threat in their rear, moved forces against Feng and his rebels. To provide better context for the effect of Feng's betrayal, the Zhili Clique's previous deployments in northern China are described below:[108]

> The Zhili Army began engaging the Fengtian Army at the watershed area near Sange Ridge along the Rehe River line.
>
> The frontlines of the Fengtian Army steadily advanced, and the main forces of both sides finally made contact, resulting in battle near Moge Ridge.
>
> However, the Feng Army's offensive gradually slowed down.
>
> The defensive positions of the Zhili Army remained ambiguous, and as the maneuvering in the mountainous areas led to mutual separation, the Headquarters of Yue retreated along the railway line and reorganized their formation at the 35-ri line.
>
> Zhili Army's Concentration Status
>
> The concentration of the Zhili Army has essentially been completed. The majority of the units have already been transported to the front lines.

2. Shanghai Gate Area
   - The 15th Division is stationed near the Shanghai Gate area.
   - The 13th Mixed Brigade is also deployed nearby.
   - The 12th Mixed Brigade is stationed in the western and eastern areas around the Shanghai Gate.
   - Artillery units and cavalry are also being deployed in the vicinity.

3. Rehe Area
   - In the Maoge Ridge direction, units of the 9th Division have advanced and are engaging in battle.
   - The 13th Division and 14th Division are pushing forward near Chifeng and Kalgan.
   - Additional forces are preparing to advance into the Rehe region via the Gubeikou area.
   - The overall disposition is one of aggressive forward movement.

4. Xuanhua—Zhangjiakou Area
   - The 20th Division is deployed in this region.
   - The 24th Division is also stationed here as a rear support and mobile reserve force.
   - These forces are expected to follow up and support the advance toward Rehe.

5. Shanhaiguan Area
   - The 22nd Division and other units are stationed and maintaining defensive lines.
   - They are reinforcing the line from the coast inland toward Rehe.

6. Cavalry Operations
   - Cavalry brigades are actively conducting reconnaissance and rapid deployments along the various flanks.
   - This includes operations in the Shanghai Gate region, Chifeng-Rehe corridor, and other sectors.

7. Naval Coordination
    - Naval forces affiliated with the Zhili Army are cooperating along the coastline and are on standby to provide support for inland operations if necessary.

The large-scale presence of Zhili troops in northeast China had made it difficult for Zhang's forces to advance in September and early October. But after Feng Yuxiang rebelled against the Zhili Clique, fewer troops were available to face Zhang Zuolin's forces in the northeast. With substantial forces diverted to the rear, Zhang Zuolin's forces were finally able to break through at Shanhaiguan and seize Beijing. Consequently, even though Feng's rebellion did not immediately produce the downfall of the Zhili Clique's forces, it divided the Zhili Clique's armies, and forced them to reckon with a war on two fronts. The primary beneficiaries of Feng's rebellion were Zhang and the Japanese, who secured victory over their opponents in the Zhili Clique and their Anglo-American backers.

Zhang's victory in the Fengtian-Zhili War helped him establish a sphere of influence in and along the eastern Chinese seaboard, where warlords friendly to Zhang rose to positions of prominence. Zhang's expansion of influence from the northeast increased his influence in China and gave him access to a series of territories that were rich in resources, people, and strategic value.[109] But many Japanese regarded his venture into North China as a potentially destabilizing enterprise, fearing that it would bring him into conflict with other warlords south of the Great Wall.[110] By and large, most Japanese preferred Zhang focus on maintaining peace and security in the Three Eastern Provinces while continuing to collaborate with Japan on economic projects.[111] Army Minister Minami Jiro explicitly stated this desire in a 1925 document entitled "Message to be Conveyed to Zhang Zuolin:"[112]

> The report indicates that the Zhang Zuolin faction has consolidated its influence over the eastern regions, focusing efforts on maintaining security and governance. Local authorities have also aligned with Zhang's control, and it has been observed that even within our areas of concern, there is growing agreement with his regime.

> As a result, in order to clarify the situation and develop effective strategies, our forces have been advised to provide limited assistance in areas where collaboration is feasible. Zhang's forces continue to advance, and if we allow further expansion, it may become increasingly difficult to contain the situation without risking our own position. We must therefore develop flexible countermeasures and prevent any disorder from spilling into regions under our jurisdiction. If circumstances allow, Japanese units in the area may be called to assist.

Minami's claim that the situation would be "difficult to contain" is a reference to Zhang Zuolin potentially becoming embroiled in a conflict with a coalition of other warlords that he could not win. If Zhang fought a losing conflict, war would come to China's northeast and threaten Japanese interests. Minami feared that Japan would be forced to commit forces to save Zhang in such a situation, which would contradict Japan's adherence to the Nine-Power Treaty. Therefore, Minami and other Japanese officials sought to prevent Zhang from expanding too much and instead focus on matters in his own backyard.

Minami was right to worry about the situation in China, as Zhang had overspent on his armed forces. Despite the income of Fengtian province only being 23-million-yuan, military expenditure amounted to 51 million yen in 1925.[113] While Zhang's government attempted to make up the difference by printing money, hyperinflation ensued, which meant that many people could not afford the basic necessities of life.[114] Naturally, this situation brought about unrest from the impoverished masses, who increasingly began to distrust Zhang's rule. The sentiments of many Chinese in the Three Eastern Provinces were summarized in a Japanese report on political conditions in the region:[115]

> Currently, local sentiment in the designated region appears unstable.
>
> Among the residents of the 14th prefectural district, distrust and dissatisfaction are growing.
>
> Newspapers and other media sources have begun reporting signs of unrest.

> This sentiment is especially prevalent around the area near the support branch.
>
> It is feared that if this situation continues, it may lead to greater opposition and further complications…
>
> According to reports, local civilians are suffering from significant hardship—many are losing their homes and families due to military operations.
>
> Additionally, the illegal transport of salt and alcohol has intensified, and there are growing cases of illness, starvation, and death among the people.
>
> The local administrative system is also beginning to fail.
>
> Immediate consideration and instruction are requested.

The dire conditions of many ordinary Chinese under Zhang's administration made the Three Eastern Provinces ripe for rebellion, especially if enemy warlords coalesced against him and threatened his armies from without. Therefore, the Japanese wanted Zhang to reduce his military spending, get his economy in order, and give up his ambitions south of the Great Wall.[116]

Zhang's expansion south along the east coast of China had already engendered great fear and animosity among the other warlords, the foremost of whom were his recent ally, Feng Yuxiang, and Sun Chuanfang, the most important remaining warlord of the Zhili Clique.[117] Control over Shanghai was a major point of contention, as it was a major port and source of revenue. Consequently, Zhang sent forces towards the Yangtze in the summer of 1925, and also appointed a Fengtian Clique-aligned warlord to govern the provinces of Jiangsu and Anhui. Zhang also took steps to reopen the Shanghai arsenal to supply the forces he had sent to the area. These moves were designed to increase the Fengtian's Clique control over Shanghai and the surrounding area.

Sun Chuanfeng reacted to Zhang's provocative moves in his backyard by attacking the Fengtian Clique forces in October 1925.[118] Sun's forces attacked on five different axes of advance, hoping to force Zhang's forces to deploy in a long, vulnerable line that could be destroyed piecemeal. But Zhang and the Fengtian Clique did not oblige, opting to withdraw to

Xuzhou after only three days. The Fengtian Clique's hasty withdrawal signaled to Feng Yuxiang, Wu Peifu, and other smaller warlords that Zhang was vulnerable; like sharks smelling blood in the water, they formed an anti-Fengtian alliance, forcing the clique to withdraw its forces further north, out of Shandong and the area around the Beijing-Wuhan railway line. By early November, 70,000 men under the command of Zhang Xueliang defended the canalized Shanhaiguan area, with 50,000 men (roughly two infantry divisions and seven cavalry brigades) advancing on Rehe, and a further 40,000 in reserve at Jinzhou.[119] The advance of 50,000 men on Rehe threatened Feng Yuxiang's base of operations in northwest China, and led him to abandon any plans to directly join the fight against Zhang and the Fengtian Clique. Instead, Feng would try to undermine the Fengtian Clique from within by fostering a fifth column.

Feng found his "mole" in the form of Guo Songling, a deputy and former tutor to Zhang Xueliang. Guo, a nationalist, was frustrated with Zhang Zuolin's capitulation to Japanese imperialism, and his ineffective governance of the Three Eastern Provinces.[120] Many civilians and soldiers in the Three Eastern Provinces felt the same way and were prepared to rise up against Zhang.[121] Because Guo temporarily ascended to command of Zhang Xueliang's 3rd Route Army on November 20th, 1925, he was in a position to give expression to these frustrations. The 3rd Route Army, numbering roughly 70,000 men in six divisions and two artillery brigades, represented the strongest force in the Fengtian Clique's order of battle.[122] Guo reached a joint agreement with Feng in which he would direct the 3rd Route Army to rebel against Zhang in return for the latter's support to take control of the Three Eastern Provinces. He also got Feng to agree with some of his broader aims, including overthrowing the warlords, establishing a new people's government, mandatory education for citizens, and worker protection.[123] Guo therefore represented the anti-imperialist, anti-feudal, nationalist trend sweeping China in the 1920s, while Zhang to a great extent represented the "old" feudal, fragmented, and subservient China that was beholden to the imperialist powers.

Guo initially sought to overthrow the Fengtian Clique quickly in a rapid coup, sending 2,000 men to Shanhaiguan from which they would proceed to Mukden and arrest Zhang Zuolin. Unfortunately for Guo, this initial plan went awry, as his forces were stopped and disarmed at Shanhaiguan.[124] The failure of Guo's coup attempt forced him and Feng to go to war with Zhang openly, with Feng declaring war on Zhang on November 27th and Guo himself moving on Mukden. While the two armies that Feng deployed against the Fengtian Clique's forces in Shandong made little progress, Guo's forces rapidly advanced up the Beijing-Mukden railway towards Mukden. The Fengtian Clique forces were in disarray, which culminated in a complete collapse of General Zuo Xiang's positions at Lianshan, two-hundred miles east of Shanhaiguan.[125] As Zuo's forces retreated to Jilin provinces, the Fengtian Clique's position in the three Northeastern provinces was in doubt, with little to stand in the way of Guo's forces advancing on Mukden.

Among the Japanese, debate once again ensued on the question of whether or not Japan should intervene to help Zhang Zuolin survive Guo Songling's rebellion. Much as in previous years, the Kwantung Army and Zhang's Japanese advisors advocated intervention on Zhang's behalf.[126] They believed that Guo's nationalist and anti-imperialist tendencies represented a threat to Japan's special rights and interests in the Three Eastern Provinces, and made sure to communicate these fears to Tokyo.[127] Shidehara, on the other hand, opposed intervention, believing that Guo in practice would be little different from Zhang as a ruler.[128] Army Minister Ugaki was caught between these two schools of thought, initially having sympathies with Guo, but later quickly shifting his support to Zhang upon recommendation from his subordinates.[129] The government, for its part, initially adopted Shidehara's non-interventionist position, opting to avoid conflict on the Asian continent and preserve its commitments to the liberal international order. But after Guo broke through at Lianshan and advanced on Mukden, the Japanese government issued a warning on December 7, 1925 through Kwantung Army commander Shirakawa to both Zhang and Guo that they must respect Japan's rights and interests in China's northeast.[130] Although this warning was issued to both commanders, it was meant in practice for Guo, as the paradigm he represented was the greater threat to Japan's special "rights and interests."

While Shidehara and the government regarded the message to Guo as sufficient to protect Japan's rights and interests, the Kwantung Army thought otherwise. Shirakawa cabled the AGS and the Army Ministry his belief that the Kwantung Army would need to be reinforced so that it could make a demonstration of force against Guo. The AGS fulfilled Shirakawa's request, telling him on December 7th that the Kwantung Army should prepare to receive more troops from Korea.[131] Soon thereafter, Shirakawa assembled forces around Mukden and moved the Kwantung Army's headquarters there.[132] These moves were made to deter Guo Songling in the event he tried to advance on the city and threaten Japan's holdings in the Liaodong Peninsula.

On December 12th, an advance guard of 2,000 men from Guo's forces broke off from the main force and moved towards the mouth of the Liao River.[133] They intended to seize the port of Yingkou, which was on the other side of the Liao.[134] But Shirakawa took action while they advanced on the city, handing Guo's forces a three-part ultimatum on December 13th mandating 1) no entry into the city of Yingkou, 2) withdrawal beyond a 30-kilometer radius of the SMR region and 3) no crossing the Liao River.[135] While Army Minister Ugaki reduced the SMR zone from a 30- to 12-kilometer radius the next day, and Shidehara indicated to Consul General Yoshida that crossing the Liao and entering Yingkou should not be prohibited, Shirakawa did not communicate his superiors' decisions to Guo's forces.[136] Instead, he sent an infantry company into Yingkou, which seemed to confirm his previous ultimatum that prevented Guo's forces from moving into the city. He also asked the Korea Army for further reinforcements and received them on December 15th, in the form of two infantry battalions and two artillery battalions.[137] These moves were designed to deter Guo's forces from entering Yingkou or crossing the SMR, and thereby deny them the most direct route to conquering Zhang's headquarters in Mukden.

Shirakawa's unilateral measures pushed Guo's forces to take a longer route—through an "immense sandy waste" where food and shelter were heard to come by—to engage Zhang's forces.[138] As Guo's forces were already cut off from their supply lines, this detour exacerbated their already existing problems.[139] The long journey of Guo's forces also bought Zhang just enough time to reinforce his forces in Fengtian, and prepare to engage Guo's rebel army. The extra time Zhang's forces got thanks to the Kwantung Army's actions proved decisive, as 12,000 cavalry brought in from Heilongjiang province delivered the coup de grace to Guo's army: an outflanking maneuver that threatened Guo's rear and supply lines.[140] The results of this attack were reported by Consul Harada on December 15th:[141]

> On the 14th, our troops launched an offensive against the opposing army.
>
> According to local reports and our own reconnaissance, enemy formations began retreating shortly after the start of the engagement.
>
> The local commander is of the opinion that the enemy withdrew in a disorganized manner.
>
> This coincides with the conclusion of our local observers that enemy morale has clearly declined.
>
> Some prisoners were taken, and a portion of the enemy baggage was seized.
>
> Our side sustained minimal losses and is now consolidating control over the area.
>
> Among the civilians in the surrounding areas, there is a growing perception that our forces have achieved victory.
>
> People are beginning to return to their homes, and some markets have resumed activity.

Harada's reference to Zhang's men as "our troops" and Guo's as the "enemy" clearly indicated where Japan's sympathies lay in the fight between the two forces. Thanks to Japan's support, Zhang was able to defeat Guo's rebellion on the battlefield, and remove the threat he presented to the Japanese Empire's interests in northeastern China. After being captured by Zhang's forces, Guo paid the ultimate price for his betrayal; his body was displayed in Mukden on Christmas Day.[142] Guo's death effectively signaled the end of the anti-Zhang Zuolin rebellion, securing Zhang's place as the chief warlord in China's northeast.

The Guo Songling episode represented the latest in a long line of unilateral actions the Kwantung Army took to secure the Japanese Empire's hegemony in China's northeast. Even though Shidehara and Ugaki had opposed the Kwantung Army's measures to prevent Guo from overthrowing Zhang, the Kwantung Army still acted on its own accord in contradiction of their desires. This pattern of Kwantung Army *gekokujo* established in the Fengtian-Zhili Wars and the Guo Songling incident greatly undermined Shidehara's "friendship diplomacy" and Japan's existing commitments to the liberal international system. More importantly, the Kwantung Army's *gekokujo* set a precedent for future unilateral interventions in the Three Eastern Provinces on the basis of protecting the Empire's rights, interests, or citizens. As the IJA became increasingly dominated by radical young officers, it was only a matter of time before these men sought to implement their own policy ideas in the Three Eastern Provinces. Thanks to the Kwantung Army's activities in the 1920s, they would have the blueprint for doing so.

Chapter 7

Reform in the IJA and the Northern Expedition

While Zhang Zuolin fought to maintain his grip on northeastern China with Kwantung Army support, important changes were occurring elsewhere that played a key role in transforming the geopolitical situation in East Asia. In Japan, one of the most important changes came in the form of the IJA's 1925 reforms, defined by a modernization program that saw the IJA upgrade its weapons systems and update its strategic philosophy to be better prepared for a prolonged "total war." But thanks to the austerity policies of the Kato Takaaki government, modernization meant that certain parts of the army had to be sacrificed to stay within the new budget constraints.[1] In turn, the IJA increasingly polarized between those who supported the modernization program and those who did not. As many younger junior and field officers greatly resented cuts to the IJA budget, they began to coalesce into an increasingly radical young officers' movement within the ranks of the IJA.

In China, the rise of the nationalist KMT occurred rapidly in the first half of the 1920s. With Soviet support, the KMT built a disciplined, united political organization based on the Leninist model, supported by a relatively strong, capable army. The KMT's newly built-up strength meant that it was able to embark on the Northern Expedition in 1926 to reunify China. The various warlord armies to the north, weakened from years of fighting, were steadily forced back by the KMT's National Revolutionary Army (NRA). The NRA's progress would eventually bring it into conflict with Zhang Zuolin and the Japanese, who feared the NRA's advance north would threaten their interests in the Three Eastern Provinces. The clash between Chinese nationalism and Japanese imperialism would only intensify in the coming years, as both sides sought to outmaneuver the other.

Ugaki Reforms the IJA

In 1925, the IJA remained a second-class force. Firepower remained its foremost problem, as it lagged far behind its potential adversaries at all levels. To illustrate the dearth in firepower, European armies that had fought in the First World War had roughly four heavy machine guns per infantry battalion, while the IJA's battalions had none.[2] European armies' squads were organized around a light machine gun, which could provide a heavy volume of fire support for riflemen in attack or defense; the IJA, by contrast, had no light machine guns in its infantry squads.[3] The IJA still depended on low-trajectory, relatively short-range field guns for artillery support, whereas European armies supplemented and/or replaced their field guns with high trajectory howitzers that could fire indirectly.[4] The IJA also had no tanks, poison gas, fighters and bomber aircraft, all of which were relatively recent inventions.[5] Such limitations meant that the IJA was in desperate need of a modernization program if it were to keep pace with its rivals and potential adversaries.

The need to modernize the IJA was especially salient in light of Army Minister Ugaki Kazushige's view that future wars would pattern after World War I as protracted, attritional struggles. To prevail in such a struggle, Ugaki believed Japan needed a technologically advanced yet relatively small standing army with large reserves that could be mobilized over a long-period of time.[6] Army traditionalists, such as Uehara, continued to oppose Ugaki's modernizing impulse, preferring to expand the armed forces at the expense of a modernization program.[7] The contrasting views between the factions led by Uehara and Ugaki came to a head at the Military Board of Councilors meeting in March 1925. At this meeting, Ugaki proposed that the size of the IJA be reduced from twenty-one to seventeen divisions, with the savings to be used for investment into new weapons systems (including aircraft, tanks, and anti-aircraft artillery) and research and development. Uehara, by contrast, continued to emphasize the importance of a large army that would overwhelm enemies in the opening phases of a conflict, which would be predicated on the principles of élan, offensive spirit, and close combat.[8] While the debate between Uehara and Ugaki was fierce, the latter's point of view won five to four when it came down to a vote.[9] Ugaki's victory meant that he would be able to go ahead with his joint program of demobilization and modernization.

On May 1st, 1925 Ugaki implemented his planned down-sizing of the IJA. He disbanded the 14th, 15th, 17th, and 18th infantry divisions—a total of 38,000 men and 6,000 horses—as well as the Taiwan Garrison Headquarters and two military preparatory schools.[10] These measures created 28.5 million yen in savings, much of which Ugaki spent on his modernization program. With the newly available funds, Ugaki created two anti-aircraft artillery regiments, the Taiwan Mountain Artillery Battalion, and a small tank unit. He also expanded the air force from two battalions to two regiments while increasing the number of radios and trucks in infantry divisions.[11] At the battalion level, Ugaki added more heavy machine guns to the order of battle, and equipped platoons with a newly developed light machine gun.[12] These measures increased the firepower and technological capabilities of the IJA while remaining within the Kato government's budgetary restrictions. Therefore, Ugaki's modernization program seemed to be a success on paper, despite the heated debate surrounding it.

Ugaki's "modernization" program reached beyond the field of armaments and into Japan's society and education system. In 1927, he reduced conscription terms from three years to two, which not only saved money but also increased Japan's pool of available reserves, as conscripts would more quickly enter the reserves due to their term of service ending a year earlier.[13] Ugaki also formed the "Young Men's Military Training Corps," which offered two-hundred hours of civics education and military training for youths between the ages of 16 to 20, and established a baseline of military preparedness among potential conscripts.[14] He even put 2,000 officers in secondary schools and colleges as military instructors, in order to "temper the minds and bodies of the students, to foster collective belief and by this means to improve the quality and character of those who are the mainstay of our people, and concurrently to advance our national defense capability."[15] While these "total war" measures were deeply controversial in Japan, they were an essential part of Ugaki's modernization program, which sought to prepare the whole of Japanese society for a protracted conflict. Ugaki's policies played a key role in radicalizing young Japanese into a militarist, bellicose worldview.

Despite the relative success of Ugaki's measures, many within the IJA still opposed Ugaki's cutbacks and reforms. At the lower levels of command, many field and junior officers did not understand why it was necessary to dissolve four divisions. All they understood was that their career path had been closed or narrowed, with precious few prospects outside the army, especially for men from depressed rural areas.[16] The lack of employment opportunities led many within the IJA to resent both the civilian politicians responsible for the cutbacks and Ugaki. Some officers also began seeking more radical solutions, experimenting with novel ideologies such as Marxism, liberalism, and national socialism.[17] Such resentment of the established order and the ideological experimentation that followed eventually gave rise to a young officers' movement later in the 1920s.[18] This officers' movement would in turn provide the catalyst for radical action, at home and abroad.

The Northern Expedition and the Continuing Ascent of Chinese Nationalism

As the 1920s progressed, the currents of anti-imperialism and reunification grew ever stronger in China. While Zhang Zuolin, Feng Yuxiang, Wu Peifu and a host of other warlords fought amongst themselves for the right to unify China, none were able to defeat the others to the degree necessary to reunify China under one banner. Consequently, these warlords exacerbated the conditions of division and fragmentation that had hindered previous Chinese modernization and unification efforts. A new paradigm outside the realm of warlordism would be necessary to unify China and martial the ever-growing support for anti-imperialism, anti-feudalism, and modernization into a coherent effort.

In Guangdong Province, the Kuomintang (KMT) represented a force that had the potential to reunify China under one banner. To a great extent, the KMT's potential stemmed from the efforts of Sun Yat-sen, who had worked tirelessly throughout the first half of the 1920s to find a sponsor for his reunification project. After failing to get the support of Japan and the United States in the early 1920s, Sun finally found a sponsor for his nationalist program in the newly formed Soviet Union.[19] The Soviets understood the KMT's nationalist orientation as a challenge to the established capitalist, imperialist powers. Accordingly, they sent funds and advisors to the KMT, seeking to transform it into an organization capable of undertaking this project and defeating the imperialist-backed warlords.[20] The Soviets also sent ample amounts of weapons to the KMT, including 157 field guns, 48 mountain guns, 128 mortars, 294 machine guns, and at least 74,000 rifles.[21] These contributions greatly strengthened the KMT, making it a viable force in the Machiavellian world of 1920s Chinese politics.

Of the Soviet advisors, Mikhail Borodin was the most influential. Under his tutelage, the KMT reformed into a Leninist-style democratic centralist organization, in which members internally debated matters of party policy but all agreed to carry out the party's decision after a vote was taken. This reform allowed the KMT to carry out unified, coordinated policy actions in a disciplined manner. The advisors also helped the KMT reform its constitution and political program, which called for the abolition of unequal treaties, debt cancellation, and "the liberation of China by the Chinese people."[22] Additionally, Borodin strengthened the instrument of the KMT's political program, the National Revolutionary Army (NRA), by founding the Whampoa Military Academy. This academy taught officers a relatively standard understanding of army doctrine, military science, and politics.[23] Borodin even managed to broker a partnership between the recently formed Communist Party of China (CPC) and the KMT, which integrated the CPC into the KMT in the name of nationalist, anti-imperialist unity.[24] These measures all greatly strengthened the KMT's cause, and transformed it from a marginally relevant entity to one that had the strength necessary to lead a revolutionary national effort to reunify China.

The rise of a Soviet-backed KMT exposed the hypocrisy of the principal imperialist powers operating in China, especially the US, UK, and Japan. Despite their pledges captured in the Nine-Power Treaty to protect China's national sovereignty and their promise not to intervene in its internal affairs, all three powers had in fact supported various warlords to keep the country divided. In turn, these warlords protected the great powers' regime of exploitation, extraterritoriality, and division in China.[25] This disparity between word and action angered many Chinese. Foreign domination combined with economic exploitation drove many to protest the exploitative and unfair conditions workers were subject to in port cities such as Guangzhou and Shanghai in 1925. Even though these protests were quelled, they signaled the rise of a nationalist and working class consciousness among ordinary people, who saw a sovereign China as necessary for their day-to-day wellbeing.[26] These protests combined with the rise of the KMT as a political vanguard spurred the imperialist powers to take collective action to secure the "liberal international order" and "Open Door" policy in China.

To conciliate the rising trend of Chinese nationalism and anti-imperialism, the great powers convened a conference in Beijing on the issue of restoring tariff autonomy to China during the autumn of 1925. At this conference, the United Kingdom decided to restore full tariff autonomy to China "in stages and in the measure in which the Chinese authorities carry out their part of the program adopted at the conference."[27] The United States concurred, and Secretary of State Frank Kellogg promptly ordered the US delegation to include proposals "looking toward ultimate tariff autonomy" for China.[28] Kellogg also told the US delegation to allow China a 2.5 percent surcharge, which was stipulated in the Washington Customs Treaty.[29] These measures would concede the issue of tariff autonomy to China's national ambitions, but still keep the broader system of extraterritorial treaties and foreign domination of the nation's domestic markets intact. Such reforms would thereby stabilize US-UK hegemony in China, strengthening their liberal-international system.

Japan, for its part, was not as willing to implement a program of tariff autonomy as the US and UK. Even though Shidehara's "friendship diplomacy," sought to foster economic cooperation between Japan and China, Shidehara still sought to uphold Japan's existing "special" rights and interests. He also sought to prevent policies that would be to Japan's economic detriment, as Japanese industry had lost significant shares in the Asian markets after World War I.[30] While Shidehara eventually did allow China a 2.5 percent tax, this concession included the condition that China not use the revenues until the conference participants agreed on the "aims and terms of collection."[31] Shidehara's stingy negotiating tactics combined with the upheaval of the Guo Songling Incident, soured the UK and US on the Beijing Conference, and it was indefinitely postponed without any consensus. The postponement signaled the beginning of the end for Wilson's "liberal international order" in China and the lofty ideals spelled out in the Nine-Power Treaty.

The imperialist powers' failure to grant China tariff autonomy was overshadowed by a much bigger event: the initiation of the Northern Expedition in July 1926. To a great extent, credit for the launching the Northern Expedition must go to Chiang Kai-shek, who leveraged his position as director of the Whampoa Military Academy to rise up the ranks of the KMT after Sun Yat-sen's death in March 1925 and promote the idea of attacking the warlord armies further north. Between March 1925 and June 1926, Chiang engaged in a political struggle with the Soviets and certain elements within the CPC that opposed such an expedition. Some probably opposed the expedition because of the combined strength of the Zhili and Fengtian Cliques, which boasted roughly 750,000 men—more than five times the size of the NRA.[32] Others likely feared that the Expedition would make Chiang too powerful.[33] Whatever their reservations, Chiang was able to overcome the doubters through deft political maneuvering and a purge of anti-expedition communists in March 1926.[34] By July 1926, the NRA was ready for action, with Chiang in overall command.

Chiang's forces were only roughly 100,000 strong, split into three route armies.[35] These three route armies were further subdivided into eight corps, ranging from 10,000-20,000 strong each.[36] In the initial phases of the Northern Expedition, these formations did not inspire confidence, as they were poorly trained and equipped. Some of Chiang's men were unarmed; others only had matchlock muskets.[37] As for their uniform and kit, a *New York Times* reporter probably summarized the state of the early NRA best when describing them as "unimpressive, straw-sandaled, mostly little men... clad in ill-fitting cotton uniforms of dirty grey or dusty yellow... Disorganization seemed complete, and the expedition appeared to be a hopeless folly."[38] These limitations meant that the KMT could not pursue large, far-reaching objectives in the initial stages of the Northern Expedition; rather, it would need to concentrate its strength against weak, isolated enemies as its best strategy to defeat the foreign-backed warlord armies.

At the recommendation of Mikhail Borodin and Red Army officer Vasily Blyukher, the KMT sought to take the offensive against Wu Peifu's Zhili Clique forces in south-central China. The KMT simultaneously sought to avoid conflict with Sun Chuanfeng's imposing forces on the east coast of China.[39] Hunan province was the first target of the KMT for two reasons: 1) many of Wu Peifu's forces were further north fighting KMT-aligned forces near the Nankou Pass and 2) the KMT had already made inroads among local warlords to rebel against Wu and the Zhili Clique.[40] These advantages allowed the KMT to take the offensive in Hunan against relatively weak forces under the command of Wu's governor of Hunan, Zhao Hengti. While Chao attempted to form a defensive line along the Lien and Lu rivers, the NRA was able to attack both flanks of the line, starting on the left on July 5th and the right on July 10th.[41] These forces achieved a breakthrough by July 11th, seizing an undefended Changsha that same day. The fall of Changsha was the first great success for the KMT's Northern Expedition, and forced Wu to divert his attention from the north to the KMT's offensive in the south.

Unfortunately for Wu, the Yangtze River flood of July 1926 impeded the redeployment of his forces.[42] Despite this setback, Wu still refused an offer from Zhang Zuolin for a joint Fengtian-Zhili alliance against the KMT and the Northern Expedition.[43] Wu's refusal, combined with the flooding of the Yangtze, left his forces in central China vulnerable to the KMT's forces, which continued advancing north from Hunan province. Even though Wu's forces attempted to dig in along the Miluo River, the KMT's Fourth and Sixth armies were still able to cross the river on August 17th, flanking the Zhili Clique defenders. These moves in turn facilitated the NRA's Seventh and Eighth armies crossing the Miluo River on the far left.[44] Two days later, on August 19th, Wu's forces were forced to abandon their trenches, with his forces diverging in two directions during the course of their withdrawal.[45]

The NRA was hot on the heels of Wu's retreating forces and continued advancing north, capturing the port of Yuezhou; the fall of Yuezhou in turn facilitated operations against Wuhan on August 22nd.[46] Due to its status as an important river port on the Yangtze River, Wuhan was an agricultural and industrial hub, and not coincidentally also home to Wu's largest arsenal.[47] These traits made Wuhan a vital strategic objective for the NRA and the Zhili Clique forces alike, with the latter digging in along the bank of the Yangtze River. But the NRA used the advantage of local guides to outflank the Zhili Clique forces defending key bridges across the river and its tributaries. The NRA also repelled a counterattack from Wu's forces aimed at the junction of the Fourth and Seventh Armies. The NRA's outflanking movements combined with Wu's failed counterattacks led to the encirclement of Wuhan.[48] By September 2nd, Wuhan fell to the NRA, with roughly 8,000 men from the Zhili Clique forces killed, wounded, or taken prisoner.[49] The fall of Wuhan represented the end of Wu Peifu as a viable warlord, as he had lost much of its vital territory to Chiang's forces. In addition, the collapse of Wu Peifu's forces meant that Chiang and the KMT could now turn their attention to Sun Chuanfeng's forces in the east.

In early September 1926, Chiang and the NRA attacked Sun Chuanfeng's forces in Jiangxi province. Sun's forces only lightly resisted the NRA's initial advance because their main body was still further north.[50] Consequently, by September 19th, the NRA seized Nanchang, the capital of Sun's fiefdom. But Sun's forces soon arrived on September 21st, routing the NRA and retaking Nanchang.[51] The resilience of Sun's army forced Chiang and the KMT to call for a halt to their offensive in September and October. During this period, Chiang and the KMT managed to stop the ongoing workers' strike in Guangzhou, which was hindering the NRA's ability to receive supplies.[52] They also came to a secret agreement with Zhejiang province governor Xia Chao to rebel against Sun.[53] These arrangements allowed the NRA to resume its offensive against Sun in October 1926.

Governor Xia Chao's rebellion against Sun commanded the attention of the latter's forces, and thereby facilitated the NRA's advance into Fujian. NRA forces advanced up the railway line that traversed the northern valley to Nanchang, breaking through the defenses of Sun's forces on October 27th.[54] By early November, Chiang's forces seized the ports of Jiujiang and Hukou on the Yangtze River before re-taking Nanchang on November 9th.[55] Simultaneously, the NRA's 1st Army crossed the rough, mountainous terrain of Fujian province. These forces continuously outflanked and outmaneuvered the forces of Zhou Yingren, the local warlord.[56] Despite Zhou's desperate attempts to resist the NRA, it was not enough, as the majority Hakka populace detested his rule and actively aided the NRA as guides and informants. By December 9th, the 1st Army captured Fuzhou, and forced Zhou's remaining 60,000 men out of Fujian province.[57]

The defeat of Sun's forces in Jiangxi and Fujian meant he looked for allies elsewhere. Thanks to the defeat of Wu Peifu and the Zhili Clique by the KMT, Sun was forced to ask Zhang Zuolin of the Fengtian Clique for support in his campaigns against the KMT. Zhang, realizing that the KMT's rising power needed to be countered, agreed to enter a coalition with Sun.[58] In December 1926, this coalition was formalized through the creation of the so-called Anguojun (National Pacification Army) over which Zhang was appointed the commander.[59] Zhang's new position leading the Anguojun once again gave him command authority south of the Great Wall, just as he desired. With roughly 500,000 troops at his command, he pledged to defend China from the "red menace" of the Soviet supported KMT.[60]

Starting on December 11th, 1926, Zhang and the Anguojun were faced with another rebellion in Zhejiang province, this time led by governor Zhou Fengqi. Once again, Sun diverted substantial forces (four divisions) to put down this rebellion, which allowed the NRA to continue its advance into Zhejiang province.[61] The 1st Army from Fujian, and the main body of the NRA from Jiangxi advanced on Zhejiang, defeating the Anguojun's forces at the towns of Lanxi and Jinhua in January 1927.[62] Then, Bai Chongxi, who oversaw the Zhejiang campaign, conducted a two-axis offensive on Hangzhou, nearly enveloping the city.[63] Bai's offensive forced the Anguojun to withdraw towards Nanjing and Shanghai where they were reinforced by Zhang Zuolin's Fengtian Clique forces. Yet the NRA forces under the command of Bai Chongxi and He Yingqin kept advancing on two axes with Bai's forces advancing towards Shanghai, and He's forces advancing towards Changzhou on the Shanghai-Nanjing railway. He's forces soon reached the Shanghai-Nanjing railway, which led to the fall of the city on March 22nd.[64] Soon thereafter, He's forces advanced on Nanjing in concert with those of Cheng Qian, finally seizing the abandoned city on March 24th.[65] The KMT's seizure of Nanjing signaled its dominance in southern China.

While the seizures of Shanghai and Nanjing were great victories for the KMT and China's resurgent nationalist movement, they also were the catalysts for a split between the KMT and their communist backers. The CPC, which at that time was most prominent in Shanghai, had played a key role in seizing the city from the warlords, fomenting strikes and rebellions that helped deliver the city to the KMT.[66] But despite these contributions, Chiang was beginning to mistrust his erstwhile Communist allies. Even though the KMT's program nominally endorsed the abolition of unequal treaties and restoration of Chinese sovereignty, Chiang and many right-wingers in the KMT did not want to antagonize the imperialist powers primarily concerned about asset seizure by the KMT.[67] This conflict between the KMT and CPC surfaced in the press in late March and early April, with both sides attacking one another.[68] But the biggest hammer blow to cooperation between the KMT and CPC came on April 1st, when the KMT government in Wuhan stripped Chiang of all his authority outside the realm of military matters at the behest of Mikhail Borodin.[69] The Wuhan government's order led Chiang to begin planning an anti-communist crackdown in Shanghai, which would commence in April 1927.

On April 12th, Chiang launched his anti-communist purge. In a violent display of murders, assassinations, and executions, the KMT destroyed communist unions, militias, and organizations in Shanghai and elsewhere. Chiang's connections with the Green Gang and its leader, Du Yuesheng, were critical to executing his anti-communist purge, as the Green Gang provided much of the local knowledge and staffing necessary for Chiang to hunt down and kill communists.[70] The Green Gang and NRA were so effective in their eliminationist objective that only 10,000 of the CPC's 60,000 members survived the orgy of violence engulfing China in the spring of 1927.[71] But these survivors soon regrouped in the countryside, where they would establish new "revolutionary base areas" amongst the peasantry in provinces such as Hunan, Jiangxi, and Anhui. The presence of these communists in the countryside would soon become Chiang's biggest threat. Once a "United Front," the forces that had undertaken the Northern Expedition were now divided, and would remain so for the foreseeable future.

Although the Northern Expedition coalition had fragmented, it still remained a threat to Japanese interests. While the KMT's pledge to end China's extraterritorial subordination was a hollow one at best, Japan nonetheless took the threat seriously, because the KMT demanded the return of the Liaodong Peninsula to China, the right to construct railroads adjacent to the SMR, and the revocation of agreements such as the Twenty-One Demands and Nishihara loans.[72] Even though Chiang allowed the foreign powers to retain their concessions in Shanghai in 1927, there was no guarantee that he would continue to be so generous once the Northern Expedition reached the Three Eastern Provinces. Therefore, many in Japan began to criticize Shidehara's "friendship diplomacy" and call for a more aggressive or "positive" China policy in 1927. This debate over Japan's foreign policy would be the most important issue in Japanese politics during the next half-decade.

Chapter 8

The Positive Policy and its Effects

The year 1927 was pivotal for Japan's foreign policy in northeast Asia. In that year, the Tanaka Giichi government adopted a "positive policy" that embraced the use of force to prevent the Chinese from threatening Japan's special "rights and interests" in the Three Eastern Provinces. Nevertheless, despite the Tanaka administration's hawkishness, the rise of the KMT continued, with the NRA advancing north at the expense of Zhang Zuolin's Anguojun. Thanks to the NRA's success, calls to end the regime of extraterritoriality and unequal treaties grew louder, especially in northeastern China. Even KMT opponent Zhang Zuolin embraced Chinese nationalism to some degree. For example, he increased his support for Chinese enterprises, sponsored nationalist political groups, and routinely contradicted the wishes of his Japanese benefactors. The ever-increasing power of Chinese nationalism, combined with the Tanaka administration's "positive policy" put Japan and China onto a collision course. Many Japanese bristled at the increasing number of Chinese who embraced nationalism, including Zhang. The contradictions between the Tanaka government's "positive policy" and Chinese nationalism would eventually boil over in 1928, during the Jinan Incident and its aftermath.

The Second Eastern Regions Conference and the Genesis of a Positive China Policy

In the spring of 1927, Japan had a financial crisis in which thirty-five banks closed.[1] This crisis forced the resignation of the Wakatsuki government, which had been unable to manage the crisis. In place of the Wakatsuki government a new one emerged led by Prime Minister Tanaka Giichi. He promised a "positive policy" in China complete with protection for Japanese nationals "on the spot" in the event of any threat to their safety.[2] As the "threats" to the safety of Japanese nationals often came in the form of anti-Japanese Chinese nationalism, many officials in the new Tanaka government embraced a more hawkish stance towards China. Yoshida Shigeru, the Tanaka government's Consul General in Mukden, was one of these officials. He outlined some of key features of the new policy in a letter to the chief of the Foreign Ministry's Asia Bureau:[3]

> 1. It is desirable that the preservation of the peace in Manchuria, whether inside the railway zone or not, should be entrusted to our forces wherever they can be deployed.
> 2. It is desirable that any strikes instigated by the Southern Army (the NRA) or by the Soviets taking advantage of the unease in economic circles and the collapse of the Fengtian p'iao should be strictly suppressed by our authorities.
> 3. It is desirable that our Manchurian administration should be implemented through the strength of the empire, in whatever circumstances and setting aside all petty stratagems…

Yoshida's recommendations that Japanese forces "preserve peace" in northeastern China and administer the region through the "strength of the empire," could be understood as a call for an effective Japanese takeover of northeastern China, due to the established historic pattern of force in determining political realities in the area.[4] Other figures, such as Mori Kaku, concurred with Yoshida's radical views, and advocated that Japan apply greater force to expand its interests northeastern China. While Tanaka was not nearly as radical as Yoshida and Mori, he still recognized northeastern China as distinct from the rest of China, a position which contradicted the KMT's view of the Three Eastern Provinces as a rightful part of China.[5] The central contradiction between the Tanaka government's "positive policy" and the KMT's nationalism would define the next two years of Sino-Japanese relations.

The Tanaka government's embrace of the "positive policy" distinguished it somewhat from Shidehara's "friendship policy" that defined the previous three years of Sino-Japanese diplomacy. Nevertheless, in many regards, Prime Minister Tanaka's understanding of statecraft was not entirely different from Shidehara's, as both men sought to avoid war and use other means to maintain Japan's interests. Tanaka's desire to avoid war is most obviously reflected in his approval of Japan being a party to the 1928 "Kellogg-Briand Pact," in which states promised not to use war and conflict as a means of resolving disputes.[6] Tanaka also expressed his desire to avoid war privately in conversations with Japanese officials. One such conversation occurred in Tokyo in April 1927, when Tanaka told Machino of his intention to secure Japanese control over northeastern China by lending Zhang Zuolin capital to build five railway lines in the region.[7] Moreover, Tanaka informed Machino of his government's desire to begin negotiations to purchase the Chinese Eastern Railway (CER). These moves would not only give Japan monopoly control over the railways of the Liaodong Peninsula, but also the central and northern part of northeastern China as well.[8] Tanaka hoped that his railway diplomacy would make a war with either China or the Soviet Union unnecessary; as he put it, "if these plans can be carried out, I can maintain order at home and we shall probably get by without a war."[9]

Although Tanaka sought to avoid war, he still appointed many hawkish figures to his government. One of these figures was Mori Kaku, who served as the Parliamentary Vice Minister of Foreign Affairs. While Mori's post had little nominal authority (as evidenced by the fact he could not access classified materials), he was still able to wield great influence over Japanese policy during his time in office.[10] Mori's influence was on display when he persuaded the Diet and the Emperor to approve the landing of roughly 2,000 troops at Qingdao on May 27th, 1927.[11] The Japanese government's ostensible reason for this deployment was to protect Japanese citizens from the troops of KMT-aligned warlords Yan Xishan and Feng Yuxiang.[12] As the Japanese Foreign Ministry put it, the deployment of troops was "but an emergency measure forced upon the Japanese government in self-defense and in order to insure the safety of Japanese residents."[13] Of course, the real reason for landing these troops was to obstruct the KMT's efforts to re-unify China.[14] To a great extent, the Japanese temporarily succeeded in achieving this objective, as the landing of 2,000 Japanese prevented NRA forces under the command of General Chen Yi-yen from advancing up the Jiaozhou-Jinan railway.[15] The KMT forces' operational pause also allowed the Japanese to send more weapons to Zhang's forces and foster a greater anti-KMT coalition.[16] In this sense, the "first Shandong Expedition" was a great success for Tanaka, Mori, and their "positive policy."

To further discuss Japan's China policy, Mori persuaded Tanaka to assemble a variety of diplomatic and military officers at the Second Eastern Regions Conference, held from June 27th to July 7th, 1927. Mori wasted no time promoting "positive" solutions to the threats posed in northeastern China. He emphasized the importance of Japan taking on the role of maintaining "peace and security" in the region. Many of the conference's attendees agreed with Mori, most notably Major Suzuki Teiichi.[17] Suzuki also added to Mori's prerogative, placing special emphasis on the threat of Russia "spreading communism" in China, which would in turn threaten Japan's hegemonic position in the Three Eastern Provinces.[18] Suzuki believed Japan needed to prepare for war with the USSR, and discussed this matter with a variety of young AGS and Army Ministry officers. They all agreed that northeastern China was vital to maintaining Japan's position in Asia, and most believed that the northeast must be separated from the rest of China and brought under Japanese control, by force if necessary. Therefore, from 1927 on forward, Suzuki and many of the others present at the Second Eastern Regions Conference sought to orient Japanese foreign, military, and domestic policy towards an eventual bid to conquer and control the Three Eastern Provinces.[19]

Suzuki was not the only figure in favor of Japan taking a stronger stance in northeastern China. Kwantung Army Chief of Staff Major General Saito Tsune also endorsed increased Japanese control and influence over the region, particularly in light of Zhang Zuolin's deviation from the empire's objectives. Tsune put forth four suggestions, to further increase Japanese control over Zhang and his region:[20]

1. A position of chief administrator of the Three Eastern Provinces shall be established, and that administrator will declare the region autonomous
2. A new agreement concerning the management of existing railways and the construction of new lines shall be concluded
3. Land development, mining, and stock raising, along with commerce and industry, shall be promoted through the principle of Sino-Japanese coexistence and coprosperity
4. Japanese advisors shall be appointed as required for the purpose of reorganizing the administration of the Three Eastern Provinces, and in particular, its finances
   Japanese advisors shall be placed in the central administration [of the Three Eastern Provinces] as well as in each province as required for the purposes of military reform

If Zhang refused these measures, Hisashi recommended finding a "suitable candidate approved by the Empire" to install as the "chief administrator" of northeastern China. He also stated that Japan must "resolve to drive out those who resist the implementation of our policies in Manchuria, and if necessary, we must be prepared to use armed force."[21] Tsune's willingness to use force in the Three Eastern Provinces reflected the sentiments of many figures within the IJA, Kwantung Army, and the foreign policy establishment which were all becoming more aggressive. As the 1920s wore on, these elements would continue gaining strength within the Japanese establishment, especially as China came closer to reunification.

On the last day of the Eastern Regions Conference, Tanaka announced an eight-point program for China that reinforced his regime's new "positive policy." Of these eight points, points 6, 7, and 8 (listed below) are the most important:[22]

6. Manchuria and Mongolia, especially the Three Eastern Provinces have vital relations with Japan from the viewpoint of our national defense and national existence. Our country not only must be especially concerned with the area but also has a duty and responsibility, as a neighboring country, to make the area a happy tranquil land for foreigners and natives alike by maintaining peace and developing the economy.
7. If an influential leader of the Three Eastern Provinces respects our special position in Manchuria and Mongolia and sincerely make efforts to stabilize the same area, our government should support him with appropriate means. (This section not to be made public)
8. If it is anticipated that the upheaval will spread to Manchuria and Mongolia, public peace and order will be disturbed, and our special position as well as our rights and interests will be violated, we must be prepared to take immediate and appropriate means to protect the area and to maintain it as a peaceful land for development by both natives and foreigners.

Tanaka also made a closing statement that summarized the conclusions from the Second Eastern Regions Conference:[23]

> As Manchuria and Mongolia, especially the Three Eastern Provinces, are of vital importance to the national defense and life of Japan, we must devote particular attention to these regions. We feel that we are responsible, because of our propinquity, for making these regions fit by the maintenance of tranquility and order and by economic development, for the settlement of Japanese and foreigners. The economic activity of Japanese and foreigners in Southern and Northern Manchuria and Mongolia, and on the basis of the principles of the open door and equal opportunities, will in our opinion stimulate the peaceful development of these regions, and we should therefore

> defend our well-earned rights and interests and solve the pending questions. I consider it best to leave political stabilization to the Chinese themselves. Any plan which represents our special position in Manchuria and Mongolia and provides the means for strengthening the political situation can be certain of the support of the Japanese government. If the disorders extend to Manchuria and Mongolia and, in consequence, undermine order and peace there in a manner which threatens our special position, rights and interests, we must be determined to defend them, no matter whence the threat comes, and to adopt the necessary measures without the loss of time to conserve these areas for peaceful residence and to be peaceably opened up by Japanese and foreigners.

Tanaka's public statements at the Second Eastern Regions Conference represented a doctrinal departure from Shidehara's friendship policy. Instead, the new "positive policy" publicly emphasized the defense of Japan's rights and interests in northeastern China. The stated willingness to deploy troops was particularly important, as many in Tanaka's government believed such deployments to be necessary to keep "peace and order" against the rising tide of anti-imperialist nationalism in China. In the coming months and years, Tanaka's government would continue to send troops to intervene in China's internal affairs, setting the stage for even more aggressive expansionist policies by the end of the 1920s.

Soon after the Second Eastern Regions Conference, more commotion broke out in Shandong, as some of the forces under Sun Chuanfeng's command defected to the KMT. Because the anti-Anguojun rebels were in Qingdao, Sun's forces were forced to leave Jinan to put down the rebellion, leaving the city exposed to the NRA. Soon thereafter, the AGS and Tanaka cabinet came to the Anguojun's aid, approving the deployment of the 8th Infantry Brigade of the 10th Infantry Division to the city.[24] Consequently, on July 7th, another 4,000 Japanese troops landed in Jinan to defend Japan's interests.[25] The landing of the Japanese forces in Jinan bought time for Sun and Zhang to reinforce their armies in Shandong, thereby allowing them to inflict a serious defeat on the NRA in August 1927. With the KMT temporarily defeated, the Japanese withdrew their forces from Shandong on September 8th, but with a warning that they would return in the event of another "threat" to Japanese citizens.[26]

Despite Japan's deployment of troops to aid the Anguojun, many Japanese officials within the Tanaka administration were starting to become dissatisfied with their client in the Three Eastern Provinces. In part, this discord was fueled by Zhang's continued embrace of economic nationalism, which threatened Japan's economic hegemony in northeastern China. His continued construction of Chinese-owned railways, including the Jilin-Hailong line, and an addendum to the Peking-Mukden Railway from Dahushan to Tongliao, also greatly concerned Japanese policymakers who sought to maintain the SMR's railway monopoly in northeastern China.[27] Similarly, even though Zhang collaborated to build the strategically important Taonan-Qiqihar and Jilin-Dunhua railways, he had not built the Changchun-Fuyu railway, which he had agreed to build alongside the two others in August 1924. Furthermore, he was not forthcoming in helping the Japanese build railways towards the northern part of his domain. Given all the help and support Japan had provided Zhang, many military men and China hawks began to view him as an ungrateful and untrustworthy figure, despite the longstanding relations of previous years.[28] Consequently, many of these figures sought to coerce Zhang into cooperating with Japan, or even remove him from the picture entirely.

At the Dalian Conferences of 1927, many participants began promoting the necessity of "coercive measures" to get Zhang to accept Japan's special rights and privileges. Many of these same figures also began to privately express support for the immediate occupation of northeastern China.[29] The more aggressive "China hawks" met staunch opposition from SMR head Yamamoto Jotaro, whom Tanaka had tapped to negotiate railway construction in northeastern China with Zhang. Yamamoto did not want the hard work of his negotiations to be destroyed by a sudden war. Therefore, he told Tanaka, Mori, and other key members of the Foreign Office that Zhang had approved the construction of two of the proposed railways.[30] This message forced Mori and the other China hawks to abandon their proposed invasion of northeastern China, buying Yamamoto time to continue negotiations with Zhang Zuolin.

In China, few were aware of the Tanaka government's desire to avoid war. To many Chinese, Tanaka's decision to land troops in Shandong represented an outrageous affront to Chinese sovereignty. Furthermore, the Tanaka administration's public embrace of the positive policy infuriated Chinese nationalists who saw it as an attempt to entrench the old order of extraterritoriality and unequal treaties. Consequently, the Tanaka administration's announcement of its "positive policy," sparked strikes, boycotts, and protests in northeastern China.[31] Many new political organizations also formed around the principles of anti-Japanese nationalism and anti-imperialism. These phenomena represented the zeitgeist of the period in northeastern China, whose residents were increasingly adopting a nationalist, anti-imperialist consciousness.

Of the anti-Japanese organizations that came into being after the announcement of the "positive policy," the "Association for Backing Diplomacy," is perhaps the most notable. Formed in the Fengtian Assembly less than a month after the announcement of Japan's new policy, it declared its intent "to resist violent oppression by foreigners," which in context clearly referred to the Japanese.[32] The association sponsored resistance in the form of militant protests on September 4th. The protests opposed Japanese imperialism and called for the "downfall of the Tanaka cabinet."[33] Even though Zhang Zuolin squelched a second round of protests upon request from the Japanese, growing anti-imperialist nationalism in northeastern China signaled a troublesome future for Japan's interests in the region.

Despite the charged atmosphere between Japan and China, Yamamoto engaged in diplomacy on behalf of Japan. In September, Yamamoto arrived in the Three Eastern Provinces, telling Chinese officials that their security and economic prosperity depended upon Japan.[34] The following month, he went to Beijing, where he intended to deliver the same message to Zhang. Yamamoto hoped that by offering "positive assistance" to Zhang, Zhang could be persuaded to allow the construction of five strategically important railways that would connect Japan's railways in the Liaodong Peninsula to the CER and its subsidiaries in the north.[35] In turn, the construction of these railways would allow the Kwantung Army and the IJA to rapidly deploy their forces north against the Soviets in the event of a conflict.[36]

On October 10th, Yamamoto began negotiating with Zhang on the construction of the five railways, spanning from Taonan to Solun, Dunhua to Tumen, Changchun to Dalai, Jilin to Wuchang, and Yanji to Hailin.[37] Zhang was somewhat disconcerted by Yamamoto's proposals, but did not reject them out of hand; indeed, he initially embraced Yamamoto's plans, working out a detailed scheme with Yang Yuting for an agreement. Zhang's willingness to embrace Yamamoto's proposal led the latter to make more proposals that would expand Japan's influence in the region. The first proposal echoed a key point in the Twenty-One Demands, demanding that China "open the resources of north and south Manchuria and eastern Inner Mongolia equally to all nations, and to facilitate their nationals' economic activities in these areas as much as possible." Another proposal essentially amounted to a defense pact that would have validated the Tanaka administration's "positive policy."[38] If Zhang accepted these proposals (Yamamoto was under the impression that he did) such an outcome would have signaled a great victory for Yamamoto's informal diplomacy and achieved the Tanaka administration's objectives without fighting. In turn, the Three Eastern Provinces would have been transformed into a Japanese satrapy, and Japan might well get to be "without a war."

Even though Yamamoto thought he got Zhang to agree to the various terms he proposed in October, the informal nature of their meeting meant that Zhang was not beholden to anything he said during the course of negotiations. He was always free to change his mind, and exercised this right on the matter of Yamamoto's proposed railways after realizing they would antagonize the USSR.[39] The informal nature of Yamamoto's negotiations also meant that Yoshizawa Kenkichi (the formal diplomatic representative Tanaka sent to negotiate with Zhang) was able to nix Yamamoto's other proposals, as he believed that the defense pact in particular represented an over-commitment to Zhang's regime. He sought instead to negotiate with Zhang on a formal basis, using the appropriate diplomatic language to secure Zhang's approval on the issues of railway construction and the acceptance of Group II of the Twenty-One Demands.[40] Unfortunately for Yoshizawa, Zhang was not prepared to give his formal approval, as the KMT government did not recognize the legitimacy of the Twenty-One Demands, and because many ordinary Chinese vehemently opposed Japan's extraterritorial regime in northeastern China.[41] Therefore, Yoshizawa failed to make progress on any of the measures Yamamoto had proposed, despite his formal diplomatic training and experience.

Yamamoto returned to Beijing in November 1927 to continue his informal negotiations with Zhang Zuolin.[42] This time, he managed to secure Zhang's formal approval for two of the five railways he had proposed in October, despite the latter's previous objections. Zhang's sudden willingness to accommodate Yamamoto's proposals stemmed from his forces' need for new weapons, which was especially pressing due to a presumed resumption of hostilities with the KMT. He also had to deal with a Soviet adversary to the north that had restored its control over the Chinese Eastern Railway and was threatening his rear. The combination of these factors led Zhang to green-light Yamamoto's railway construction scheme.[43] In February 1928, Zhang's approval was formalized in the form of a written agreement to build the two railways. This agreement was scheduled to be formally announced in July 1928.[44] The new agreement represented another step towards Japanese hegemony in the Three Eastern Provinces and pre-empted the KMT's reunification of China, thereby accomplishing the Tanaka administration's aims.

The Jinan Incident

In November 1927, Chiang Kai-shek traveled to Tokyo to discuss Sino-Japanese relations with Tanaka and his administration. Even though Chiang had resigned as commander of the NRA on August 12th, 1927, he understood that the KMT could not function in his absence and he would soon be restored to his previous position.[45] In the interim period, he tried to persuade Tanaka that Japan should not interfere with China's reunification, and thereby help correct the popular view that the scourge of warlordism was dependent on Japan. Tanaka was willing to entertain Chiang's position, but also wanted him to acknowledge Japan's special rights and interests in the Three Eastern Provinces.[46] Ultimately, the two men reached a *quid pro quo* in which the KMT would not enter northeastern China in return for Japan agreeing not to intervene in Chinese affairs south of the Great Wall.[47] Chiang believed this deal to be a greenlight from Tanaka to resume the Northern Expedition.

In January 1928, Chiang Kai-shek resumed command of the National Revolutionary Army. Chiang's leadership restored unity and coherence to the NRA, thereby facilitating a further advance against the combined forces of Zhang Zuolin and Sun Chuanfeng. The NRA's advance that began on April 7th, 1928, made swift progress against the warlord forces.[48] Forced to withdraw, Zhang's forces retreated to the city of Jinan by April 21st, 1928. Here, Sun's forces prepared to defend Jinan and its surrounding area against the coming NRA onslaught.[49]

The NRA's advance greatly worried many in the Tanaka administration.[50] In the views of these figures, the fighting in north China had the potential to disrupt "peace and security" in the Three Eastern Provinces and posed a threat to Japanese nationals despite Chiang's earlier assurances to the contrary.[51] Many of these figures advocated sending troops to flex Japan's muscles in China. Others within the administration feared the consequences of intervening in China's civil war, and sought to avoid deploying troops if possible. The ensuing debate between radical and moderate advocates of the positive policy marked Sino-Japanese relations in the next decade, as Japan sought to determine which policies to follow in China.

Initially, both Tanaka and the Army General Staff were opposed to sending forces to Shandong. Tanaka opposed deploying forces to Jinan because of his agreement with Chiang Kai-shek in November of 1927.[52] The AGS, for its part, believed that Chiang should be given a chance to make good on his guarantees to defend Japanese nationals and their property before Japan deployed any troops.[53] Osaka businessmen also opposed sending troops to Shandong, as they had significant business interests in China, and did not want to incur the profit losses that would follow a likely Chinese boycott.[54] These forces represented more moderate elements in Japan that sought to keep overt meddling in Chinese affairs to a minimum, and instead work through Zhang Zuolin to secure Japan's influence in China.

While Tanaka did not want to send troops to Shandong, others in his administration believed sending troops was necessary to ensure the protection of Japanese residents. Army Minister Shirakawa put forth this view during a Cabinet meeting on April 17th and repeated his view again the next day.[55] Navy Minister Okada Keisuke agreed with Shirakawa, and made his views known to Tanaka on April 18th. Mori Kaku also articulated his belief that Japan needed to send troops, arguing that Tanaka and his Seiyukai party had to live up to the promise made during the February 1928 presidential elections, in which they promised to protect Japanese nationals "on the spot."[56] Zhang Zuolin's retreating forces, he also argued, might harm Japanese residents, despite his nominal association with the Japanese. These arguments from figures such as Mori and Shirakawa, combined with domestic political considerations persuaded Tanaka to dispatch the 6th Infantry Division to Qingdao in April 1928.[57] Tanaka hoped this deployment would be a show of strength sufficient to circumvent a large-scale deployment of troops to Jinan.[58]

While Tanaka sought to avoid the large-scale deployment of troops to Jinan, 475 men from Qingdao already arrived by rail on April 20th.[59] Meanwhile, the 6th Infantry Division landed in Qingdao where it seemed it would remain. But on April 25th, the Japanese military attache in Jinan sent an emergency request to the commander of the 6th Infantry Division Lieutenant General Fukuda Hikosuke. After receiving this request, Fukuda ordered the forces under his command to deploy to Jinan, even though he had not received permission from a superior to do so.[60] By April 27th, 1928 a further 4,000 men from the 6th Infantry Division arrived in Jinan despite Tanaka's desire to avoid a large-scale deployment of troops there.[61] As Prime Minister Tanaka's desires had been entirely disregarded by a mere Lieutenant General, Fukuda's actions represented yet another instance of *gekokujo* on the part of a Japanese officer.

The sudden arrival of Japanese forces in Jinan led many in the KMT to suspect that the Japanese were supporting Zhang's forces. Yet the arrival of Japanese forces did not deter the KMT's planned assault on the city. In late April, the KMT's National Revolutionary Army (NRA) launched a pincer attack on Jinan and cut it off from outside reinforcement.[62] Consequently, Sun's forces began withdrawing north from Jinan, opening the way for the NRA to enter the city. By May 1st, the NRA entered Jinan with little resistance, despite the presence of the 6th Infantry Division.[63]

Even though the NRA seized Jinan with relative ease, it still had to negotiate with the Japanese forces in the city. These negotiations commenced on May 2nd, between Chiang Kai-shek and Major General Ryu Saito.[64] Chiang reiterated his promise that NRA troops would not cause any difficulties for the Japanese residents of the city or disrupt usual business. This promise led Saito to agree to withdraw Japanese troops from the city, and leave it to the KMT's custody. General Fukuda Hikosuke, Saito's superior, approved this decision, and the Japanese withdrawal from Jinan began on the night of May 2nd.[65]

Peaceful relations did not last between the IJA and NRA. Violence between the two forces erupted on the night of May 3rd, under unclear circumstances.[66] While Chiang and Fukuda agreed to yet another truce on May 4th, the Japanese alleged that their head negotiator Lieutenant General Sasaki Toichi was robbed and almost killed. Furthermore, according to the Japanese, Chiang had not completely withdrawn his forces from the Japanese district of Jinan as he had promised. As such, the Japanese felt justified in attacking Chiang's troops.[67] As Major General Tatekawa Yoshitsugu put it in the aftermath of the Sasaki incident, it was "necessary for Japan to chastise the lawless Chinese soldiers in order to maintain Japan's national and military prestige."[68] To this end, Tanaka ordered further reinforcements sent to Jinan, which began arriving on May 7th.[69]

The dispatch of reinforcements to Jinan allowed General Fukuda to issue a set of demands for the NRA and its forces in Jinan. These demands included the punishment of Chinese officers involved in the Sasaki incident, the disarming of Chinese forces still in Jinan, withdrawal from two barracks in the vicinity of Jinan, an end to all anti-Japanese propaganda, and a withdrawal of Chinese troops beyond 20 li on either side of the Qingdao-Jinan railway.[70] Critically, Fukuda only gave the Chinese twelve hours to accept his terms, knowing full well that they could not be accepted.[71] When the Chinese indeed refused to accept all of these terms in the allotted twelve hours, Fukuda commenced his assault on Jinan on May 8th, quickly overwhelming the Chinese defenders and surrounding an NRA battalion inside the city.[72] Fukuda also utilized artillery bombardments to neutralize the city's walls, which were no match for high explosive shells. This sudden assault on Jinan was successful, as the NRA was forced to retreat on May 10th and 11th.[73] In the course of the fighting, the Chinese lost roughly 1,000 dead, both military and civilian.[74]

While Japan defeated Chiang's forces in the vicinity of Jinan, Fukuda's decision to escalate the situation was unilateral. To help justify Fukuda's decision after the fact, the Kwantung Army's chief of staff sent the following telegram to the Vice Minister of War on May 12th, 1928:[75]

1. Regarding the Jinan Incident, there have been numerous reports implying that our military actions were excessive and unrelated to the core events. In reality, our army's operations are entirely for the protection of the imperial subjects and are within the scope of fulfilling that duty.
2. However, if our actions give the impression that we are conducting military operations beyond necessity, it will be problematic.
3. The initial troop deployments were meant to deter both the southern and northern armies from clashing. If this balance shifts, it would be unfortunate.
4. In particular, the Jinan Incident's origin involves the southern army's firing. The challenge lies in proving that fact.
5. As for protecting Japanese civilians near Jinan station and the bridge, we must handle it in a way that reflects the seriousness of the situation. If reports claim that our soldiers entered Jinan and the nearby railway area (within a 20-li radius) with no clear cause that is deeply concerning.

6. Our attacks on Jinan must not be interpreted as indiscriminate or unjustified. To the best of our ability, we must exercise restraint.
7. In the vicinity of Jinan, our soldiers must coordinate with local officials to maintain public order and provide protection to residents and unarmed civilians.

The claims made in this telegram are to a great extent either innuendo or false, designed to lead the Army Ministry to the wrong conclusions about the Jinan Incident. The claim that Fukuda's actions were "entirely for the protection of the imperial subjects," is disproven by the fact that he only gave the Chinese garrison twelve hours to respond to his terms before starting fighting that endangered civilians. The claim that initial troop deployments were designed to prevent Zhang's forces and the National Revolutionary Army from clashing ignores the fact that Fukuda unilaterally sent reinforcements from Qingdao to Jinan despite not having orders to do so from Tokyo. Finally, the allegation that the Jinan Incident's origin lay with the KMT firing the first shot was difficult to know, as it occurred on the night of May 3rd under unknown circumstances. While the Kwantung Army chief of staff acknowledged the difficulty of proving who fired the first shot, the boldness of his first claim makes it a blatant falsehood, as few, if any, could have known who opened fire first on May 3rd. This combination of falsehoods and innuendos suggest a broader agenda at play, in which the Kwantung Army and other Japanese officers in China sought to take unilateral action to control northern China and prevent China's reunification. Invoking the responsibility to protect Japanese civilians was merely an excuse to intervene, and would continue to be in future years.

By May 1928, the NRA continued to rapidly advance through the northern Chinese plain, rapidly pursuing the retreating Anguojun.[76] Undeterred, Yamamoto once again came to Beijing, where he negotiated with Zhang on the issue of railway construction in northeastern China. Zhang was in no position to say no to Yamamoto and the Japanese, who had supported his army with equipment, supplies, and advisors. Therefore, Zhang approved the construction of two further railways in his domain despite his initial skepticism.[77] The first was to traverse Jilin and Kainei, while the second would span from Changchun to Dalian. Yamamoto's opportunistic trip to Beijing represented Tanaka's version of the "positive policy" because it allowed Japan to gain a stronger economic and strategic foothold in the Three Eastern Provinces while avoiding outright war with China.

Even though Yamamoto managed to procure further concessions from Zhang Zuolin, the success of the Northern Expedition led many in Japan to fear that the Empire's interests in northeastern China were in jeopardy. Consequently, the Japanese Cabinet issued a declaration on May 16th, 1928, to both the KMT and Zhang's forces that Japan "may possibly be constrained to take appropriate and effective steps for the maintenance of peace and order in Manchuria."[78] Two days later, Yoshizawa told Zhang Zuolin of the Cabinet's decision and advised him to withdraw his troops back to northeastern China before the KMT defeated his forces. When Zhang refused, Yoshizawa asked Tokyo for instructions, and conveyed the ones he received to Zhang on May 20: if Zhang's forces did not withdraw his intact forces back to the Three Eastern Provinces, the Kwantung Army would block the entry of Zhang's Anguojun and the KMT at Shanhaiguan (a mountain pass along the Great Wall) and disarm the Anguojun.[79] Although this ultimatum was indeed a threat to Zhang, the Tanaka government's intent was to offer both Zhang and the KMT a compromise that both could accept: Zhang's forces would retreat unmolested to his base area in northeastern China, while the KMT would retain the lands it had conquered south of the Great Wall.[80] Japan would be the ultimate winner in this scenario, as it would not only maintain its rights in northeastern China but also divide the country into two polities north and south of the Great Wall.[81]

Despite the Tanaka government's desire to find a peaceful, effective resolution to the situation in northeastern China, the Kwantung Army favored more aggressive measures.[82] As early as April 20th, Kwantung Army chief of staff Saito Tsune advocated that the Kwantung Army disarm all Chinese forces trying to pass through Shanhaiguan. On May 3rd, a meeting of Kwantung Army intelligence officers reached a consensus that the Three Eastern Provinces had to be forcibly separated from the rest of China. Other staff officers, including Colonel Komoto Daisaku, Doihara Kenji, Itagaki Seishiro, and Okamura Yasuji, wanted the Kwantung Army to invade and conquer the Three Eastern Provinces in the name of forming "a new regime of peace, law and tranquility of the north by separating that area from the Nanking government, and also by getting rid of the warlords whose influence prevailed in Manchuria."[83] To achieve this aim, Komoto and Major General Hata Shinji advocated for the replacement of Zhang Zuolin with his son, Zhang Xueliang, as they believed that the elder Zhang was ungrateful to Japan and too independent-minded to be a reliable asset. By mid-May, Army Minister Shirakawa had come around to the views of the Kwantung Army radicals, telling Prime Minister Tanaka that Zhang Zuolin must be "persuaded" to retire.[84] The broad consensus of the Kwantung Army's officers to remove Zhang Zuolin and take action in northeastern China made it a formidable obstacle for the Tanaka government, which preferred to pursue a softer iteration of the "positive policy."

On May 18th, the AGS Chief of Staff Suzuki Soroku ordered Kwantung Army commander Muraoka Chotaro to move his headquarters to Mukden, concentrate his forces around the city, and await orders to deploy to Jinzhou to disarm any Chinese troops attempting to enter the Three Eastern Provinces.[85] Upon learning of Suzuki's decision the KMT informed the US of the Kwantung Army's redeployment. The US in turn told the Japanese government that it must alert them in advance before any attempt to disarm Zhang's forces.[86] Thanks to America's intervention, the Foreign Ministry was forced to nix the Kwantung Army's planned deployment to Jinzhou on May 19th on the basis that any foray outside the SMR zone was a violation of the Treaty of Portsmouth.[87] Although the Kwantung Army protested this decision and got Suzuki to approve preparations for deployment towards Shanhaiguan a day later, Suzuki's approval was soon undermined by Consul General Hayashi, who showed Saito a telegram from Yoshizawa mandating the Kwantung Army only move into northeastern China if the NRA pursued Zhang's forces.[88] These contradictory directives to the Kwantung Army reflected confusion in Tokyo, where Tanaka's government was divided on the question of Zhang and the Three Eastern Provinces. To try and resolve this question, Tanaka convened another "Eastern Regions Conference" between May 20th and 25th, where both sides would get a chance to present their views and lobby for their desired course of action.

The debates at the Eastern Regions Conference were confusing and longwinded, with little progress in the initial stages. As General Araki, one of the conference's key attendees put it, "Everyone had different ideas; no one agreed; we were just going in circles wasting precious time."[89] Despite the general disorganization of the conference, Mori Kaku, who was the conference's chairman, still expressed his support for the plan to disarm Zhang's troops. But he was opposed by many figures who had previously supported prompt action in northeastern China, who got cold feet after the American warning. The debate between these different elements within Tanaka's government continued throughout the five days of the conference.[90] Meanwhile, the Kwantung Army was becoming impatient, seeking approval to deploy; this impatience can be seen in Major General Saito's diary entries dated May 21st, May 22nd, and May 23rd:[91]

> May 21: On the assumption that the imperial order [to mobilize] decreed at 9 a.m. in Tokyo would become effective at 12 noon here, I advised my subordinates to that affect and at 12 summoned them. But the orders did not come through. I am beginning to think that in all likelihood the plans of the Army Command have been disrupted by some plot…
>
> Waited all day but orders have not arrived. The Kwantung Army, bursting with spirit, is piqued by the government's indecision
>
> May 22: At 10 a.m. moved Command Headquarters to the top of the Totaku building. At 3:47 p.m. the Commander in Chief [Lieutenant General Muraoka Chotaro] arrived [from Port Arthur] on schedule and immediately occupied the temporary headquarters. … I sense a growing warlike mood spreading about us.

> May 23: A Colonel Tashiro, Chief of the China Section of the General Staff Office arrived from Tokyo to brief us on the government's position. According to him, Zhang Zuolin is not to be pressed to relinquish his status... although the manifest policy of the government is that both the Northern and Southern Armies are to be disarmed alike; in practice, however, the Commanding General of the Kwantung Army is authorized to use discretion and treat the Northern Army more leniently.
>
> The Kwantung Army in Mukden, expecting mobilization orders at any moment, is in a state of animated suspense. Feeling of antagonism toward the indecisive Tokyo government is daily mounting. Even Beijing is buzzing with rumors of the Kwantung Army's impending expedition to Jinzhou and the forcible seizure of the Beijing-Mukden railway.

These diary entries indicated the Kwantung Army's desire for action and their frustration with the Tanaka government's cautiousness. Fortunately for them, Mori Kaku finally managed to get the Eastern Regions Conference to approve the Kwantung Army's deployment to Shanhaiguan on May 25th. Now all that stood between the deployment was Prime Minister Tanaka's approval.

As quickly as Mori managed to push through approval of his plans at the Eastern Regions Conference, Tanaka ended them. From his point of view deploying the Kwantung Army to Shanhaiguan was unnecessary, as Japan already established *de facto* control over the region thanks to Yamamoto's railway deals. Furthermore, on May 23rd, Zhang had agreed to peacefully withdraw from North China, with Chiang Kai-shek seemingly willing to allow such a retreat.[92] Perhaps most importantly, Tanaka did not want to risk war with the United States, which had ominously warned Japan against moving the Kwantung Army to Shanhaiguan to disarm Zhang's forces.[93] These factors ensured that Tanaka did not approve the plan to deploy the troops.

Even though Tanaka did not approve Mori's plan, many in the Kwantung Army still sought to implement the plan on their own accord. According to the Kwantung Army's plans, it would deploy to Shanhaiguan and prevent Zhang's forces from withdrawing back into northeastern China in the event of their defeat by Chiang's NRA. The Kwantung Army could then move into northeastern China and assert its control over the territory.[94] Fortunately for the Kwantung Army activists, they managed to get the higher echelons of the army to advocate this plan at a meeting between representatives of the army, navy and Foreign Office, with the army representative calling for landing an infantry division at Shanhaiguan.[95] However, Rear Admiral Yonai Mitsumasa refused to participate in the Kwantung Army's scheme and reported the proposal to his superior, Admiral Suzuki Kantaro, the Chief of Naval Operations.[96] Admiral Suzuki then reported the army's proposed plan up the chain to his superior, Admiral Okada, who was Minister of the Navy. Suzuki told Okada to prevent this proposal from gaining traction if the army brought it up at a Cabinet meeting.[97]

The next day, the army proposed the landing of an infantry division at Shanhaiguan. However, Okada pointed out that landing at Shanhaiguan would force Japan to violate preexisting international treaty commitments. Therefore, if Japan were to land at Shanhaiguan, it must be prepared to go to war with the US and UK.[98] Tanaka—not realizing the content of the treaty barred a landing at Shanhaiguan—once again used his authority to call off the whole venture on the basis of its existence.[99] But only an imperial order was sufficient to fully squelch the Kwantung Army's plans to land at Shanhaiguan, because some officers still sought to carry out their original plans.[100] While the Tanaka administration managed to temporarily thwart the Kwantung Army's attempts at unilateral aggression, the fate of Tanaka's policies ultimately depended on Kwantung Army radicals understanding Tokyo's view of the broader situation in the Japanese Empire. The central authorities' view of the situation is well-summarized by Arita Hachiro of the Foreign Ministry's Asian Affairs Bureau, who said "even if we gain one thing in Manchuria and Mongolia the loss would equal the profit if our interests were damaged along the Yangtze and in other areas of China and if our relations with the powers worsened."[101] Whether or not the provincial Kwantung Army could understand its place in the Empire's broader interests was far from certain, especially given its unilateral activities in past years. One thing was certain however: the ball was now in the Kwantung Army's court.

Chapter 9

They Behave Like Children

The KMT's advance into north China threatened to destabilize the Three Eastern Provinces. To key Japanese policymakers this scenario was completely unacceptable, as it would threaten Japan's economic and strategic interests in northeastern China. But despite general agreement on the issue of preventing the rise of Chinese nationalism, disagreement existed in the Japanese foreign policy apparatus over how to address the threat. On one side, Tanaka and his ilk sought to broker a peaceful withdrawal of Zhang Zuolin and his forces back to the Three Eastern Provinces. On the other side, the Kwantung Army and many within the IJA began to fundamentally question Zhang's usefulness as an asset due to his willingness to take actions that contradicted Japan's interests. Many of these figures favored disarming Zhang's forces at Shanhaiguan, overthrowing his government, and/or conquering the Three Eastern Provinces. While the push-pull between these elements continued throughout May and June 1928, the Kwantung Army's proximity to events on the ground gave its staff officers the necessary space to implement their desired agenda. Consequently, on June 4th, 1928, the Kwantung Army committed its greatest act of *gekokujo* to date: the assassination of Japan's erstwhile ally, Zhang Zuolin.

The Assassination of Zhang Zuolin

In Mukden, word of Tanaka's order to halt the deployment of the Kwantung Army to Shanhaiguan was a great disappointment.[1] Upon learning of the Tanaka government's orders, Colonel Komoto Daisaku went to Tokyo to ask why the Kwantung Army's plans had been nixed, and if the AGS would allow the Kwantung Army to deploy to Shanhaiguan and disarm Zhang's forces. Even though Komoto did not get his way, he remained determined to achieve his goal of removing Zhang Zuolin from power and establishing Japanese control over the Three Eastern Provinces.[2] Komoto's belief in the necessity of Zhang's removal was evident as early as April 18th, when he wrote a letter to Colonel Rensuke Isogai, one of the army's so-called China's experts, about his plans and intentions:[3]

> Given the increasing high-handedness of the Chinese side in Manchuria, there are, if we face the facts, too many things that are difficult to overlook. It can't be denied that Japanese army cliques bear some responsibility for encouraging the Chinese to become presumptuous.
>
> Resolving the Manchurian problem isn't something that anyone can do using reason. Resorting to such ploys as exchanging favors is no good. There is no road other than armed force. Of course, it is essential that anyone who uses force choose his pretext and battle standard carefully.
>
> At this point, given the slightest cause, we should deal them a great blow and force them to change their attitude toward Japan

Komoto's view that armed force was the only means of achieving Japan's objectives in northeastern China set him apart from Tanaka's government, which sought a peaceful resolution to the crisis in China. Komoto believed that the assassination of Zhang Zuolin was necessary because of the latter's "ingratitude," encapsulated in his construction of Chinese owned railways parallel to the SMR, accommodation of the rights recovery movement, and his unwillingness to accept certain elements of the Twenty-One Demands.[4] Consequently, Komoto was the Kwantung Army's strongest advocate for assassinating Zhang and conquering northeastern China, as revealed in Komoto's letter to Isogai:[5]

> Wouldn't it be advisable if Zhang Zuolin and maybe one or two others die on the road? This time, by all means, I am going to do it. Even if they try to stop me, I'll do it at all costs. To take a life in order to solve the Manmō problem, that's our greatest hope. That is the honorable thing to do. … last year and the year before, the plans to do something great were stopped in the midst. This year I'm going to score a hit by all means. My plan to cleanse Manchuria and Inner Mongolia with torrents of blood may be, I believe, a fundamental solution to the problem.

In Komoto's estimation, assassinating Zhang was absolutely necessary, even if done without approval from the Emperor.[6] Komoto was also acting independently of the Japanese government and AGS, the latter of which had ordered the Kwantung Army to stand down in May. Komoto was willing to act in contradiction to his superiors because he believed that if Zhang were assassinated, "the other generals of the Fengtian clique would scatter to the four winds," and chaos would prevail in the Three Eastern Provinces.[7] Such chaos would likely foster anti-Japanese activities and an NRA expedition to the northeast, which would force the Tanaka government to deploy troops to maintain "peace and order."[8] In light of Komoto's aims, historians such as Matsusaka Yoshihisa have understood Komoto's assassination plot to be aimed just as much at Prime Minister Tanaka as at Zhang.[9]

As early as January 1928, Komoto had informed his personal aide Captain Kawagoe Moriji and Special Service operative Nakano Yohei of his plan to assassinate Zhang Zuolin. Later in May, he informed two of his subordinate officers on the Kwantung Army staff, Colonel Ishiwara Kanji and Lieutenant Colonel Itagaki Seishiro, of his intentions.[10] But, unbeknownst to Komoto, others in the Kwantung Army—most notably his commanding officer, General Muraoka Chotaro—also sought to assassinate Zhang. Muraoka believed that Zhang needed to be removed from power for similar reasons as Komoto and sought to kill him in Beijing. To carry out the assassination, Muraoka summoned Lieutenant Colonel Takeshita Yoshiharu, a military attaché from Harbin, to his office in Mukden, where he informed him of his assignment.[11]

Just as Takeshita departed from Muraoka's office, he ran into Komoto, who was immediately suspicious of Takeshita's presence. Normally it was Komoto's responsibility to issue travel passes to military attachés; but he had not issued one to Takeshita.[12] Realizing something was afoot, he pulled Takeshita aside and was able to coax out of him that Muraoka had assigned him the secret mission of assassinating Zhang in Beijing. In Komoto's view, doing so would bring unwanted attention both in China and internationally due to the relatively heavy presence of foreigners in Beijing.[13] Furthermore, if Muraoka were associated with the assassination attempt, it would reflect poorly on his reputation. Komoto consequently told Takeshita that he would "get rid of Zhang Zuolin" so that Muraoka would not be roped into an "unfortunate situation."[14] Takeshita obliged, and went to Beijing with instructions from Komoto to identify on which train Zhang would depart. Once Takeshita identified Komoto's train, it could be blown up.

Even though Komoto thwarted Muraoka's plans to assassinate Zhang, two IJA intelligence officers, Doihara Kenji and Hata Shinji, also made plans to assassinate the "Old Marshal."[15] Their attempt to assassinate Zhang centered around his son's main bodyguard, Araki Goro, whom they had persuaded to overthrow the Zhang government to replace the elder Zhang with his son, Zhang Xueliang. Doihara and Hata needed weapons from the Kwantung Army to carry out their coup attempt, and tried to go through Komoto to obtain them. But even though Komoto broadly agreed with their agenda of removing Zhang, he distrusted their plans. He believed a plan dependent upon Zhang's primarily Chinese bodyguard was a fool's errand. As Komoto himself put it, he would not "dance by Doihara's flute," and prevented the shipment of Japanese weapons to Araki's bodyguard detail.[16] Komoto's denial foiled the plan of Doihara and Hata, allowing him to go ahead with his plot.

To ascertain the optimal place to blow up Zhang's train, Komoto assigned First Lieutenant Kanda Yasunosuke of the 20th Engineer Regiment the task of studying the terrain around the Liu River railroad bridge. But after Kanda reconnoitered the bridge across the Liu River, he informed Komoto that this area was unsuitable for an assassination because it was heavily guarded by Chinese troops.[17] In response, Komoto then suggested Kanda scout a new site near Huanggutun, where the South Manchuria Railway and Beijing-Mukden railways intersected. He knew it was highly unlikely that Chinese guards would be in this area, as the Treaty of Portsmouth forbade Chinese troops to be near the SMR.[18] As Komoto expected, there were only Japanese guards posted near but not in, the Huanggutun area; therefore, Komoto decided that he would implement his assassination plot at Huanggutun. He repeatedly sent Kanda back to the area to work out the details of his planned attack.[19]

Kanda's repeated trips to the Huanggutun area greatly agitated the commander of the local Independent Railway Guards Battalion, Major Tomiya.[20] Tomiya, like Komoto, sought to assassinate Zhang Zuolin but resented Kanda's repeated presence in his area of responsibility. Realizing Komoto was responsible for Kanda's presence, Tomiya took it upon himself to meet Komoto and explain that he was overstepping the bounds of his authority.[21] Komoto responded that on a certain day he would not have responsibility for the Huanggutun area, and Komoto would instead assign responsibility to men from the Kwantung Army and Zhang Zuolin's forces. If some sort of incident were to occur, Komoto stated that it would be his fault for assigning Chinese troops to the area. Tomiya, now realizing Komoto plotted to kill Zhang Zuolin, decided to become a co-conspirator to Komoto's plot.[22]

The day after his meeting with Komoto, Tomiya began to work with Kanda on placing explosives. They assigned some enlisted men from the engineers to haul cement bags filled with explosives to the targeted railway overpass.[23] These explosives would be triggered by an electric switch, which would bring the bridge down on Zhang's head as his train passed underneath.[24] In alignment with Komoto's plan, Tomiya would ensure that his forces would prevent anyone from intervening to prevent the plot from taking shape. He even positioned some of his men near the site of the railway overpass, with orders to kill Zhang if the initial explosion failed to do so.[25] In this manner, Tomiya and Kanda prepared to carry out the assassination of Zhang Zuolin, and thereby see through Komoto's mission.

While Tomiya and Kanda plotted the details of Zhang's assassination, Komoto attempted to create a plausible cover story that would divert attention from the Kwantung Army and transfer it to the Chinese.[26] To help craft it, he recruited ultra-nationalist coal magnate Ito Kenjiro, who was obsessed with getting rid of Zhang and establishing Japanese rule in the Three Eastern Provinces. In turn, Ito leveraged his connections with the ultra-nationalist, secret society underworld to contact a patriotic adventurer named Adachi Takamori to help him with the plot. Adachi's role would be to find potential Chinese scapegoats whose lives were in Ito's words "worthless."[27] These "worthless" men could be killed near Huanggutun and have evidence planted on their bodies that indicated ties to the KMT. Such a gambit would not only remove responsibility from the Kwantung Army but also create a *casus belli* for Japan to invade northeastern China.

Adachi rewarded Ito's faith in him by "hiring" three morphine addicts to serve as scapegoats. These men were told that they would take part in an IJA operation, but not told the details of their role in it. On the day of action, the addicts were brought to a bathhouse, bathed, clothed, and given fifty yen before being brought to Ito's lair. From there, they were to be transported to Kwantung Army Headquarters and then to the Japanese railway guards at Huanggutun. Before the last part of their fatal journey, one of them realized that Adachi and the Japanese could not be trusted and escaped; unfortunately for the others, Captain Tomiya and his men bayoneted them until they were dead. After the murders, Captain Tomiya and his men planted fake "incriminating" documents from the KMT on the bodies and hand grenades in their pockets.[28]

While Adachi and Ito managed to create the cover story, Komoto was spending his time and energy staying on top of Zhang's whereabouts and intentions. To a great extent, Komoto's information network relied on human intelligence from informants and intelligence operatives. In Beijing, Colonel Takeshita and Captain Tanaka Ryukichi were checking the scheduled departures of trains to try and discover when and if Zhang would be departing.[29] Furthermore, two Koreans were contracted to watch the Beijing train station to keep count of roughly how many troops were leaving. If a lot of troops were departing, Zhang was likely withdrawing from Beijing. On the other hand, if relatively few left it might signal Zhang's intention to defend the city.[30] Observers were also positioned at key chokepoints such as Shanhaiguan and Jinzhou to keep tabs on troop movements and any potential Zhang departure.[31] These observers allowed Komoto to stay a step ahead of Zhang and the Japanese government, and know exactly when to begin implementing his plans.

In North China, Zhang Zuolin's forces had lost control of the strategically important city of Baoding and faced pressure from certain elements of the Kwantung Army to withdraw back to northeastern China. Morale was also rapidly declining among the rank and file of the Anguojun, in no small part thanks to KMT propaganda that Zhang was a Japanese puppet.[32] Due to these factors, Zhang authorized a withdrawal from North China back to the Three Eastern Provinces on June 2nd, 1928.[33] The next day Zhang began making preparations to evacuate himself and his entourage from Beijing.[34] He also sent a telegram to the Chinese public, explaining that he could not prosecute his campaign against the "red" KMT to a successful conclusion and needed to withdraw from Beijing in order to avoid further bloodshed.[35] These acts signaled the end of Zhang's dreams to conquer China south of the Great Wall.

Rumors abounded in Beijing about a potential assassination of Zhang; consequently, his aides attempted to prepare for his departure from Beijing due to take place on June 4th. On June 3rd, Zhang's forces began taking security measures to secure the Peking-Mukden railway, where his train would travel.[36] First, they asked the Japanese to garrison fifty military police and guards on the three miles of track leading to Mukden; the Japanese granted this request. However, the Japanese denied the second request that Zhang's forces made, which sought permission to garrison Zhang's troops at the Huanggutun intersection of the Peking-Mukden Railway and SMR. Instead, the Japanese reinforced this area with their own forces in a bid to secure the Peking-Mukden-SMR railway intersection for a distance of two-hundred miles in either direction.[37] Considering that Komoto chose Huanggutun as the site of his assassination plans, its likely that he was behind the decision to prevent Zhang's forces from garrisoning the railway in this area, and reinforcing the area with Japanese. These measures would ensure Komoto's plot went ahead without interference from Zhang's forces, thereby increasing its chances of success.

On June 4th, at 1:15 a.m., Zhang's train left Beijing in the eighth car of a 20- to 40-car train.[38] Thanks to the intelligence-gathering efforts of Komoto's informant network, he knew of Zhang's arrival time in Mukden several days in advance.[39] At 3 a.m., the Japanese railway guards reported the discovery of two Chinese that sought to climb the SMR's embankment near the intersection with the Peking-Mukden railway; these men were bayoneted, and discovered to have bombs and correspondence from the KMT on their bodies. By 5:20 a.m. Zhang's train had passed Huanggutun and a report came in from Komoto's railway observers stating that Zhang's train had passed.[40] This report was the trigger to make the final preparations to assassinate Zhang.

The passengers on Zhang's train were blissfully unaware of the trap set for them ahead, as evidenced by the fact that a game of Mah-jong had started.[41] During the game, Wu Chunsheng, one of Zhang's subordinate officers, told Zhang he should "put on an overcoat" due to the cold conditions on the train. But just as Zhang did so, a powerful explosion interrupted the relaxed state of those on the train, forcing the seventh, eighth, and ninth cars into the air.[42] Zhang was struck unconscious by a steel fragment that went into his nose, and was heavily bleeding from his chest. While Zhang survived the initial concussion of the blast, he was not able to survive his severe wounds; he died a few hours later in his residence at Mukden.[43]

Although Komoto succeeded in the assassination, it was but the first step in his planned conquest of northeastern China. Komoto hoped that Zhang's assassination would lead to rioting, disorder, and chaos in the region, thereby providing the perfect justification for a Japanese intervention.[44] In anticipation, Komoto had positioned a brigade in front of the Yamato hotel in Mukden, to be deployed when the moment was ripe to invade the northeast.[45] In the interim, Komoto ordered Captain Ozaki to demand the Kwantung Army staff mobilize its forces in preparation for a full-scale invasion of northeastern China.

Zhang Zuolin's Train after his assassination (Author's Collection, Taken at the Museum of the War of Chinese People's Resistance Against Japanese Aggression)

Komoto's rationale for mobilizing the Kwantung Army was that it could prevent the KMT from invading the Three Eastern Provinces.[46] However, the Kwantung Army Chief of Staff, Saito Tsune, rejected Ozaki's demand, and refused to dispatch forces to seize control of northeastern China. Unaware that Komoto was behind Zhang's assassination, Saito did not want to instigate an unnecessary clash with Zhang's army, even though he had earlier endorsed deployment of the Kwantung Army to disarm his forces.[47] In any case, other Kwantung Army staff officers moved the brigade Komoto had left at the Yamato hotel, thereby hindering the rapid deployment of forces to the Three Eastern Provinces.[48] Both Saito's refusal to mobilize the Kwantung Army and the unauthorized movement of the brigade left in front of the Yamato hotel illustrated the lack of coordination between Komoto and other elements in the army, which was decisive in hindering the immediate implementation of Komoto's plans to conquer northeastern China.

Despite his initial failure to take advantage of the post-assassination chaos, Komoto persisted in his attempts to create a *casus belli*. These attempts were defined by repeated acts of terrorism against Japanese civilians that in turn were blamed on the Chinese. The Kwantung Army sought to capitalize on its own false flag terrorism by asking the Consulate General whether local police forces were adequate for the protection of Japanese residents.[49] These calls were designed to gain tacit approval for the mobilization of the Kwantung Army on the pretext of its responsibility to protect Japanese citizens. Once again, the Kwantung Army was rebuffed, as the Consulate General declined the Kwantung Army's implicit mobilization efforts.[50] Undeterred, General Muraoka repeated his requests to the Kwantung chief of police and the Consul General. But these figures did not give into Muraoka's persistent demands, with the Consul General in particular refusing to take part in a "conspiracy of adventurers."[51] The steadfastness of both the Kwantung police chief and the Consul General sealed the fate of Komoto's planned invasion of northeastern China and, for a time, prevented a war between Japan and China.

Komoto's lack of attention to detail and inability to coordinate with other Kwantung Army officers not only doomed his plans to failure but also ensured that others outside the circle of conspirators found out about the Kwantung Army's culpability in Zhang's assassination. One key source of information was the prospective Chinese scapegoat who had escaped Ito and Tomiya. After being taken into custody by the Chinese authorities, he disclosed the circumstances of his recruitment and mission in the plot to assassinate Zhang. Word of the Kwantung Army's involvement in Zhang Zuolin's assassination soon reached his son Zhang Xueliang, who was naturally outraged.[52] After learning that the Kwantung Army assassinated his father, it was highly unlikely that Zhang Xueliang would fall in line and collaborate with the Japanese. Therefore, far from maintaining Japan's imperialist interests in northeastern China, Komoto's decision to assassinate Zhang actively endangered these interests and increased the likelihood of political instability in future years.

Among the Japanese, elements of the Minseito Party (the opposition party to Tanaka's Seiyukai) were among the first to learn of the real origin of Zhang Zuolin's assassination. The Minseito's knowledge of the situation was in large part a product of circumstance, as six of its members had to evacuate from a train after hearing of Zhang's assassination. When one of the six Minseito passengers, Matsumura Kenzo, saw his old friend Consul General Hayashi Kijuro, the latter remarked "this is awful… the army guys have done it. What a mess!"[53] An initial investigation conducted soon thereafter validated Hayashi's suspicions: ballistics assessments confirmed the use of high-quality Japanese explosives, and the KMT "letters" on the bodies of the two dead scapegoats were written in Japanese-style Kanbun instead of traditional Chinese characters. Upon returning to Tokyo, Matsumura reported these findings to the head of the Minseito party, making it one of the first entities in Japan to learn the truth behind Zhang's assassination.[54]

After Matsumura's report, more investigations regarding the assassination of Zhang Zuolin took place. One headed by Consul General Hayashi Kyujiro was a joint Sino-Japanese venture that sought to establish the plot's origins. But even though Hayashi knew that the "army guys" were behind it, he opted not to indict them in his report to Prime Minister Tanaka. His report instead singled out Zhang's former chief of staff, Ling Yingqing, as the catalyst for assassinating Zhang, with Komoto merely giving him the Kwantung Army's blessing.[55] While Hayashi's report was a clever mix of truth and fiction, the veracity of his report was shattered by a General Kishi Yajiro, who told Prime Minister Tanaka that upon examining the scene, he concluded that the Kwantung Army was responsible for Zhang's death.[56] Upon hearing this assessment, Prime Minister Tanaka expressed his frustration with his out-of-control subordinates, saying "what fools! They [Komoto and the other Kwantung Army officers] behave like children. They have no idea what the parent has to go through."[57] He went on to tell the Navy Minister that if the army continued to take such unilateral actions, he would never be able to "develop" his plans, which revolved around the peaceful expansion of Japan's sphere of influence in the Three Eastern Provinces.[58] Therefore, it was imperative that Tanaka root out the *gekokujo* in the Kwantung Army, to prevent it from further hijacking Japan's foreign policy in northeast Asia.

To hold Komoto and the Kwantung Army accountable for its assassination of Zhang, Tanaka formed a joint committee of investigation, with elements from the IJA, the Foreign Ministry, and the Kwantung Government.[59] The chair of the military police was also sent to Mukden to conduct an on-the-spot investigation. Thanks to a letter that the military police managed to obtain, the investigators were aware of the connections between Adachi, Ito, Komoto and the three drug addict-scapegoats, and therefore knew to interrogate Adachi and the other "patriotic adventurers" in northeastern China. But despite seemingly clear evidence of Komoto's lead role, the committee did not assess him to be the ringleader, instead casting blame on Ito and Adachi. The chief of military police realized that Mori and other members of the committee were likely trying to absolve the army of responsibility for Zhang's assassination. The committee closed its investigation in December 1928 without identifying Komoto as the mastermind of Zhang's assassination.[60]

Saionji Kinmochi, one of the last *genro* still alive, advised Tanaka that he must punish the perpetrators of the Zhang's assassination severely to prevent similar *gekokujo* in the future.[61] Tanaka agreed, and recommended to Emperor Hirohito that Komoto and his associates should be court-martialed pending further investigation. Hirohito responded positively to Tanaka's suggestion, saying that such a court martial was necessary in order to "maintain strictly the military discipline of the national army."[62] With the Emperor's blessing, Tanaka summoned his Cabinet and told them of his intention to initiate another investigation.[63] He also proposed to Army Minister Shirakawa that the offenders should be court-martialed if found responsible for Zhang's assassination. Tanaka hoped these measures would establish greater discipline within the armed forces and prevent future incidents of Kwantung Army *gekokujo* in northeastern China.

Tanaka's proposed investigation and disciplinary measures for Komoto and his associates engendered great opposition from the IJA and various members of his cabinet. Vice Foreign Minister Mori Kaku and Railway Minister Ogawa Heikichi made their sympathies for Komoto clear along with their opposition to Tanaka's disciplinary measures. Army Minister Shirakawa also opposed Tanaka's proposed measures, stating that a court-martial would "expose to the public something which the army wished at the time to conceal."[64] Others in the army, including AGS Chief Suzuki, Inspector of Military Education General Muto Nobuyoshi, and Field Marshal Uehara, had similar views. These senior military figures believed that the army should handle its disciplinary measures for Komoto internally, not air its dirty laundry in front of the public and the Chinese.[65] On December 28th Army Minister Shirakawa therefore informed Emperor Hirohito that he would investigate the circumstances surrounding Zhang's death, but made no mention of a court-martial.[66] It was now up to the IJA to investigate and discipline Komoto and the Kwantung Army.

The army's "investigation" of Zhang's death was dragged out over the better part of four months, with the Emperor repeatedly inquiring about its progress.[67] By March 1929, Tanaka's cabinet agreed not to disclose its investigation's conclusions to the public in order to keep its discipline of Komoto an internal affair.[68] On March 27th, Shirakawa informed Hirohito of the army's decision; while he acknowledged that Komoto and Kwantung Army Commander Muraoka Chotaro had committed crimes, he also told Hirohito of his fear that a public trial and punishment of Muraoka and Komoto would sully the army's reputation and give the impression that the center had lost control of a field army.[69] Instead, Shirakawa recommended that the IJA issue private, administrative punishments to Komoto and his associates to avoid the potential fallout from a public show trial. Surprisingly, Hirohito was willing to go along with Shirakawa's proposed solution.[70] Thanks to Hirohito's endorsement, now all Shirakawa and the rest of the army needed to do was convince Tanaka to go along with their plan.

On May 24th, the military affairs council presented the findings of its "investigation," which suggested that the army was not involved in Zhang's assassination.[71] Upon hearing of the report's conclusion, Tanaka made his opposition clear, insisting that the army bore responsibility for what happened at Huanggutun. He also pushed for a court-martial of Komoto and the other officers involved.[72] Tanaka's stubborn adherence to the truth greatly frustrated the army, which wanted to sweep the incident under the rug. Nevertheless, Tanaka finally gave in by the end of June, realizing that if he did not assent to Shirakawa's proposed measures, the latter could resign and cause the collapse of the Tanaka government. On June 27th, Tanaka informed Hirohito of his final decision, stating his agreement with Shirakawa's position.[73] But despite Hirohito's cooperation in March, he had little patience for Tanaka's capitulation to the army, as evidenced by his pointed inquiry "is that not different from what the prime minister said last time?"[74] Hirohito further suggested that Tanaka resign, along with the rest of the government. Tanaka obliged on July 2nd, 1929, bringing his government to an abrupt end.[75]

In July 1929, the IJA finalized its punishment for Komoto: a belated retirement from active service. The army's slap-on-the-wrist punishment for Komoto despite his illegal, strategically short-sighted *gekokujo* set a bad precedent, as future conspirators would feel emboldened to implement their own plans going forward. As the IJA in the 1920s was defined by the collapse of the Choshu Clique and the rise of a new generation of young, independent-minded officers, there would be no shortage of those seeking to "go rogue" to find a solution to the "Manchurian question." Therefore, the circumstances were ripe for such incidents of *gekokujo* to multiply in future years, with the Kwantung Army as the primary incubator for new plots. The government in Tokyo would in turn be unable to shape Japan's foreign policy, with disastrous long-term and strategic consequences.

In the immediate aftermath of Zhang Zuolin's death, it was not immediately clear who would succeed him. While Zhang Xueliang was appointed governor of Fengtian province in June 1928 and became the Commander-in-Chief of Peace Preservation in the Three Eastern Provinces in July, the situation in northeastern China was still so fluid that he had to hide in Jinzhou until July 1928.[76] Even after Zhang emerged from his hiding place, questions continued to exist about the status of the other two provinces, Jilin and Heilongjiang. There were indications that the military governors of these provinces might declare independence or join the KMT government to the south.[77] Zhang also faced challenges to his nominal position as commander-in-chief from the governor of Jilin province, Zhang Zuoxiang, and his father's old chief of staff, Yang Yuting, both of whom still held significant prestige in the Fengtian Clique.[78] To secure his position, the younger Zhang needed to maneuver carefully and consolidate his power before he could take firm strategic action.

One of the obvious key questions facing Zhang Xueliang was how to deal with Japan. While Zhang Xueliang was aware of the fact that the Kwantung Army had assassinated his father, a divide existed among members of the Fengtian Clique regarding how to handle relations with Japan moving forward.[79] Some figures, such as Zhang Zuoxiang and Yang Yuting favored an accommodation with the KMT to reunify China and resist Japan. Others such as Yuan Jinkai, who later became the Vice Chairman of the Committee for Peace Preservation, favored stabilizing the northeast and working with the Japanese. This divide within Zhang's camp meant that he could not immediately take a firm position on the issue of whether to cooperate with China or Japan, despite his own nationalist sympathies. Rather, he would have to balance working with Japan and the KMT for his first six months as Commander-in-Chief of Peace Preservation in the Three Eastern Provinces.

After the assassination of the elder Zhang, many Japanese still thought his son would be a more reliable agent for the empire in northeastern China. Others realized his nationalist orientation might be an obstacle to collaborating with the empire. In the event that Zhang Xueliang needed nudging in Japan's direction, the Kwantung Army and other elements in the IJA sought to warn him against coming to an agreement with the KMT. Tatekawa Yoshitsugu, the military attaché in Beijing, put into words the views of many in the IJA regarding Zhang Xueliang "changing the flag:"[80]

> I believe it is the most necessary and urgent task for Japan to state explicitly that it will not be pleased to see the Nationalist flag raised in Manchuria and Mongolia, to direct the Manchurian government not to submit to the southern regime, by forceful and threatening measures if necessary, and as soon as possible to have [Manchurian leaders] declare themselves to be independent of the south, under the principle of "maintaining the security of the border and pacifying the people," exterminate dangerous activities internally, and devote themselves solely to the unification and maintenance of order within the Three Eastern Provinces.

Soon after Tatekawa made this suggestion, Kwantung Army commander Muraoka Chotaro took it upon himself to pressure Zhang Xueliang to work with the Japanese. Specifically, Muraoka told Zhang Xueliang that his father "stood for cooperation with Japan" and helped bring prosperity to the region, yet he was "prepared to surrender to his enemies and bring ruin to the Three Eastern Provinces."[81] Comments such as these were designed to restore Japan's relations with the new regime in northeastern China and to prepare for further Japanese penetration of the region.

Initially it seemed that Japan's attempts to pressure Zhang Xueliang might succeed, as evidenced by the fact that the younger Zhang told Consul General Hayashi that he sought to maintain the autonomy of the Three Eastern Provinces.[82] He also made clear his intent to honor the agreements his father signed earlier that year to build new railway lines from Jilin to Hoeryong and Changchun to Dalai. Zhang even expressed his willingness to consider Japanese land rights in the Three Eastern Provinces, something which his father had been hesitant to do.[83] However, despite these initial declarations, growing Chinese nationalism soon made it possible for Zhang to begin moving closer to the KMT. One of the most important conditions from Zhang's point of view was the KMT abrogation of the so-called "unequal treaties," on July 7th.[84] The announcement of the abrogation in turn fueled increased support for reunification with the rest of the country in the Three Eastern Provinces.[85] These developments encouraged Zhang Xueliang to begin talks with the KMT government to reunify the country in July 1928.[86] By July 22nd, Zhang and the KMT had agreed to a deal that on paper would raise the blue-sky, white-sun emblem of the KMT in northeastern China.[87]

While an agreement to reunify China existed on paper for five months, no moves towards raising the KMT's flag in the Three Eastern Provinces were made. To a great extent, the lack of action in this period stemmed from Zhang Xueliang's need to consolidate his own authority.[88] In the interim period, the Japanese made further warnings against reunifying with the rest of China. They also offered him incentives, including advisors and arms in exchange for him "maintaining order" in China's northeast.[89] But by December, Zhang and the KMT were both ready to complete China's reunification process: On December 29th, 1928, the KMT's emblem of the blue sky and white sun was raised over the buildings of the Three Eastern Provinces, signifying the first step towards these provinces' reintegration into the rest of China.[90]

Even after officially raising the flag in late 1928, Zhang Xueliang's government in the Three Eastern Provinces still maintained great autonomy from the KMT government in Nanjing. Nonetheless, the reintegration of the Three Eastern Provinces to the rest of China was a great boon to Chinese nationalism, especially in the northeast. Here, Zhang Xueliang doubled down on his father's sponsorship of Chinese-owned enterprises, ports, and railways, threatening the SMR's monopoly in the region.[91] To these ends, he supported the formation of the Northeast Committee on Communications in September 1929, which sought to build a Chinese-run railway network that would encircle the SMR and build a harbor on Hulu Island near Jinzhou to compete with the SMR's at Dalian.[92] Zhang Xueliang also did more to foster anti-imperialist, anti-Japanese sentiment than his father and more vehemently opposed Japan's encroachments. Zhang's actions thoroughly debunked the commonly held Kwantung Army view that assassinating Zhang Zuolin and replacing him with his son would improve Japan's position in northeastern China. In fact, the assassination made Japan's position worse, especially after Zhang reneged on his promises to recognize previous treaties and agreements. Far from alleviating Japan's "Manchuria problem," Zhang Zuolin's death only exacerbated existing contradictions within Japan's military and government, and accelerated the descent of Japan and China into war. All that was needed to ignite a regional conflict was a single spark.

Chapter 10

Ishiwara's Final War and Japan's Preparations

After Tanaka's Seiyukai government resigned in July 1929, the *genro* and Emperor Hirohito approved the leader of the opposing "Minseito" party, Hamaguchi Osachi, to become the next Prime Minister. In contrast to Tanaka, whose administration embraced the "positive policy," Hamaguchi's administration sought to uphold the liberal international order as evidenced in his reinstatement of Shidehara Kijuro as Foreign Minister. The Hamaguchi government also embraced global disarmament, as evidenced by its participation in the London Naval Conference of 1930. Economically, Hamaguchi embraced a program of fiscal conservatism, which saw further cuts in the military budget. All of these moves alienated many officers in the Imperial Japanese Army and Navy, who wanted a more hawkish China policy, a military independent of the civilian government, and a greater budget. The clash between the Hamaguchi government and the increasingly potent army radicals would define Japanese politics for the next decade, before ultimately culminating in the outbreak of World War II in the Asia-Pacific region.

The Great Depression and the Radicalization of the Army

As 1928 gave way to 1929, Japan remained mired in economic crisis. In part, this economic malaise was residual from the original crisis of 1927. But after the onset of the Great Depression in 1929, Japan's economic problems deepened. The drastic decline in rice prices, which decreased thirty-five percent between 1929 and 1931, was one of the most impactful effects of the Great Depression in Japan.[1] While the decrease in Japanese rice prices was related to a bumper rice crop in 1930, it also was the result of significantly reduced demand in the wake of the global depression.[2] The reduction in rice prices was disastrous for farmers, whose incomes drastically plummeted.[3] Many could no longer afford to pay rent, which meant landlessness, homelessness, and destitution for unfortunate Japanese farmers and their families.

Silk, like rice, was a staple of the rural farmers, who used it as a means of supplementing their income. The importance of silk to the Japanese peasantry is evidenced by the fact that two out of every five rural households were involved in silk production, and that these households depended on silk for 70 percent of their cash.[4] But the Japanese silk industry was dependent on rural export markets, particularly the United States, to survive. Consequently, when the Great Depression hit and American demand for Japanese silks greatly declined, Japanese silk exports revenues decreased from $356.1 million in 1929 to a mere $69.8 million in 1934.[5] The US Congress's passage of the Smoot-Hawley Tariff act in 1930 further contracted the market for Japanese silks, thereby making them even less competitive on the global market. Considering that silk made up 36.4% of all Japanese exports in 1929, the decline in the export of silk to the United States was a disaster for Japan, which continued to have a trade deficit into the 1930s.[6] Because barriers to trade such as the Smoot-Hawley tariff prevented the empire from remedying its economic woes through increasing exports, other solutions would be necessary to resolve the malaise of the Great Depression.

The Hamaguchi government was unable to provide a policy solution to the Great Depression, largely due to its fiscal conservatism. The Hamaguchi government's conservatism was embodied in its return to the gold standard in 1930, which had been a long-standing goal of Hamaguchi dating back to his days as Finance Minister. But this return exacerbated the depression, as it decreased the available money supply and led to further contractions in the economy.[7] Hamaguchi's austerity program—cutting government jobs, a reduction of public consumption, and perhaps most critically, a reduction in military spending—also led to further contractions of the economy, creating more unemployment and less demand for goods.[8] Additionally, Hamaguchi's approval of the London Naval Treaty in 1930 was the subject of great controversy, as many officers contended that he infringed on the armed forces "right of supreme command" elaborated in the Meiji Constitution.[9] While Hamaguchi's policies were designed to adhere to the economic and political consensus of the liberal international order, they caused greater misery for many ordinary Japanese and struck a blow to the nation's pride. Therefore, Hamaguchi's government made many enemies, especially in the armed forces. In November 1930, one of these enemies, a young ultranationalist navy officer named Tomeo Sagoya shot Hamaguchi in a bid to assassinate him. While Hamaguchi did not immediately die, he was badly wounded and hospitalized at the end of 1930. Tomeo's attempt to assassinate Hamaguchi came to symbolize increasing dissatisfaction with Japan's place in the liberal international order.

The contracting economy, shrinking export markets, disarmament, and the suffering of the peasantry were all causes of discontent within the Imperial Japanese Army's officer corps. To remedy these problems, many young officers sought radical solutions and found them in the writings of Gondo Seikyo and Tachibana Kosaburo.[10] Both writers were opponents of the capitalist system that arose in Japan after the Meiji restoration, criticizing the polarization between the rich and poor. They also criticized the alliance between the Japanese monopolies (*zaibatsu*), bureaucrats, and politicians which fostered corruption and maintained the system of exploitation that oppressed the peasantry. Both favored centering traditional agrarian ways of being, with Gondo putting forth the idea of self-rule for peasant villages. Tachibana's writings endorsed rural communalism, but with the important addition of a so-called "patriotic reform" program for the purposes of "national reconstruction."[11] In contrast to Gondo's "self-rule," Tachibana realized his "patriotic reform" movement would likely require militant action to implement, as the established authorities would not relinquish authority without a fight. Tachibana's emphasis on militant action captured the fancy of many IJA officers, who believed a revolution was necessary to overthrow the corrupt and decadent system that had destroyed the lives of so many ordinary Japanese.

The ideas of Kita Ikki were also popular among the increasingly radical officer clique. Much like Gondo and Tachibana, Kita was highly critical of Japan's post-Meiji Restoration capitalism. But where Gondo and Tachibana were primarily concerned with the peasantry and establishing an agrarian-centric society, Kita had broader domestic and geopolitical proposals and programs. The foremost measure Kita advocated for was a closer union between the people and the Emperor. To foster such a union, Kita believed Japan needed to abolish many of its pre-existing institutions such as the peerage system to "remove the wall separating the Emperor and the people, and clarify the Meiji Restoration."[12] He also believed that the Emperor should be the primary authority in Japan, and "exercise his supreme authority to suspend the Constitution for a period of three years, dissolve both houses of the diet, and impose martial law throughout the country."[13] Kita recognized that a coup would likely be necessary for the Emperor to impose his will over the established power structure, but justified it in the name of historical progress. As he put it, "during Napoleon's revolutionary era, when he dealt at arm's length with conservative forces, it was apparent to him that a coup d'etat was the only way to fulfill the revolution, even though the majority of the parliament and newspapers sought to revive the monarchy."[14]

Economically, Kita favored a socialist-inspired economic program that would place limits on private assets, land and investment, and nationalize key industries.[15] He intended to limit the amount of property a single family could hold to one million yen, with the state taking control of any excess property worth more than this amount. He also sought to limit private enterprises to a cap of 1 billion yen, with any further assets being nationalized. He even advocated basic worker rights and protections, such as an eight-hour day and prohibitions on child labor. Kita hoped that these economic policies would foster greater equality in Japanese society, which in turn would increase Japan's national strength.[16]

After a period of internal reform, Kita believed that Japan needed to expand outward into East Asia.[17] The first step of Japan's external expansion would be the acquisition of a base in northeastern China, which in turn could serve as a stepping stone to other territories. Japan could then move on the Russian Far East and Siberia, possession of which would allow it to "defend China" from outside threats. Australia would be next on Japan's target list due to its vast land endowment and potential for settlement. In Kita's conception, the conquest of these places would allow Japan to protect China from the ravenous "white" powers that had waged wars of aggression against Asian peoples for the better part of two centuries. In turn, China could become a "forward base" for Japan and part of a greater Japanese-led economic and security bloc in East Asia.

With the inspiration of Kita's newfound ideas, many young officers began forming study groups, discussion groups, and secret societies in order to determine how to put these ideas into practice. The late 1920s saw the formation of two of the most radical groups seeking internal reform: the Tenkento and the Kozakurakai.[18] The former consisted of company-grade officers and cadets who were adherents of Kita Ikki's teachings. These young men were dedicated to seeing through Kita's vision of a "national socialism," that would save Japan from corrupt oligarchs and politicians, thereby reviving the national ethos. The Kozakurakai was similar in its composition, ideology, and aims, but had a willingness to utilize terrorism. Most members of both the Tenkento and the Kozakurakai were graduates of the IJA's Military Academy (not the more prestigious staff college) and therefore were largely pigeonholed into remaining junior and field officers throughout their army careers. Nonetheless, despite the relatively junior standing of these officers, the Tenkento and Kozakurakai greatly affected the IJA at the grassroots level, as they provided a medium for radical ideas to proliferate.

Higher up the hierarchy of the IJA, more senior field officers also formed discussion groups where they could vent their frustrations with the status quo and put forth radical solutions. On November 3rd, 1928, a group of Japanese field officers formed the so-called "National Policy Association" with Major Suzuki Teiichi as chairman.[19] This newly formed group sought to implement further reforms in the IJA, resolve Japan's "population problem" and find a solution to Japan's situation in northeastern China and Mongolia.[20] Because the "National Policy Association's" agenda shared great similarities with the previously established Futabakai, both groups began meeting together in December 1928.[21] Realizing that they both shared a similar outlook, both groups saw it in their interests to merge, and formed the second iteration of the Issekikai (Evening Study) study group in May 1929. While the second Issekikai did not greatly differ from the first, its emphasis on establishing greater Japanese control over northeastern China was a distinct position that indicated the increasingly expansionist bent of the IJA's field officers.

Perhaps the most important radical group formed in the period after the assassination of Zhang Zuolin was the Sakurakai, or Cherry Society. Three lieutenant colonels—Hashimoto Kingoro of the AGS's Intelligence Division, Russia Section, Sakata Yoshiro of the Army Ministry, and Higuchi Kichiiro of the Tokyo Guards Division—founded this group at the end of September 1930. The AGS was the greatest source of membership in the new organization, with 40 out of 120 members being part of the General Staff.[22] Within the AGS, the Intelligence Division was a key feeder for the new organization as Hashimoto, Captain Obara Shigetake, Captain Tanaka Wataru, Lieutenant Amano Isamu, Nemoto Hiroshi, and Cho Isamu from the China Section all occupied key positions.[23] Other key sources of members for the Sakurakai were the Toyama Army School (ten members), the Army Ministry (nine members), and the Army Staff College (eight members). The Sakurakai offered these young, up-and-coming middle-ranking officers a repository to discuss and coordinate plans for the implementation of reform at home and expansion abroad.

The most prominent founder of the Sakurakai, Hashimoto Kingoro, began his career as an officer in 1911. By 1923, he had ascended to a position on the Army General Staff.[24] By 1928, Hashimoto became military attaché to Turkey, where he remained for two years. It was here that he began to formulate radical ideas. Seeing the rise of multiple illiberal nationalist leaders in Europe and the Middle East—Mussolini in Italy, Hitler in Germany, Reza Khan in Iran and Mustafa Kemal in Turkey—Hashimoto became convinced that Japan needed to abandon liberal democracy if it was to emerge from its economic and geopolitical crisis.[25] Therefore, when he returned to Japan, Hashimoto created several schemes to implement his plans.[26] These schemes replaced Japan's existing parliamentary democracy with a new government that would revolve around the Emperor, backed by the guns of the IJA.[27] With a new military government, Japan could achieve Hashimoto's vision of "national reconstruction," à la Hitler's Germany or Mussolini's Italy, and "grasp the reins to lead the world into a new world order."[28]

Hashimoto's influence on the Sakurakai is reflected in its radical anti-establishment sentiment. Such sentiment is clearly illustrated in its platform, which simply defined the organization's purpose as "national reorganization," by force if necessary.[29] The Sakurakai prospectus gave a more elaborate explanation for the necessity of this "national reorganization," saying "...top rulers engage in immoral conduct, political parties are corrupt, capitalists and aristocrats have no understanding of the masses... farming villages are devastated, unemployment and depression are serious. These present grave causes for our nation."[30] To resolve these problems, the Sakurakai advocated both internal reform and external expansion as part of Japan's national reorganization, with northeastern China the first and foremost target of Japan's "external expansion."[31] While the exact order of priorities, means of carrying out internal reform, and the lengths to which the reformers should go remained points of disagreement for members of the Sakurakai, almost all agreed that radical changes were necessary to rescue Japan from the Great Depression and usher in a new era of more egalitarian politics.[32]

The Sakurakai, Issekikai, and other radical groups would have vital support in their endeavors from civilian ultranationalists. Foremost among these ultranationalists was Okawa Shumei, whose emergence on the scene as a relevant political actor began when he started working for the SMR's East Asian Research Bureau in 1918. From early in his career, Okawa was a committed pan-Asianist who believed in the special destiny of the Japanese nation to lead the Asian peoples to liberation. As he put it in his 1920 paper "Otakebi," (War Cry), "We the Japanese race must be the whirling center for the liberation of mankind. Accordingly, the Japanese nation is an absolute entity that makes us fulfill our ideals of world revolution. The idealistic fulfillment and fighting organization of the Japanese nation is the work of God himself, created for this absolute aim… We must not be satisfied only with the reformation or revolution of Japan herself. As we believe in the destiny of the Japanese race as the great missionary of the war of liberation for mankind, we want to set about liberating Japan herself first."[33] He also was clear in his disdain for the European powers saying in his 1923 book "Various Problems of Revising Asia," that "every place on the vast continent of Asia had been trampled underfoot by the white man, and every race has been enslaved by him."[34] These racial views were the foundation of Okawa's ideology which sought to expel the European powers from Asia and establish Japan as the imperial leader of the East.

By 1929, Okawa became the SMR Research Section's chairman.[35] His new position did little to moderate his views, which if anything grew more bellicose. Okawa's bellicosity is evidenced in his writings from this period, in which he stated that Asia and Europe "have been considered the greatest rivals in the world today," and that due to European colonialism and Asia's awakening against it, "war between the East and West is fated indeed."[36] Nonetheless, Okawa did not simply believe in a coming "race war" between Asians and Europeans, but rather wrote that the East-West war would be fought by "the strong power representing Asia" and "the other strong power representing Europe," which he specified as Japan and the United States respectively.[37] For Japan to prepare for its coming war with the United States, Okawa believed that it needed to undergo a process of "national reconstruction" that would necessitate the end of party politics, and the implementation of a planned economy to put Japan on a war footing.[38] He also believed that Japan must become a more socialistic society, calling for the abolition of class differences and the restoration of a more direct union between the Emperor and the people.[39] These domestic reforms would strengthen Japan so that it would be in position to initiate another round of external expansion, starting in northeastern China.

To help facilitate Japan's expansionist program, Okawa leveraged his position as the SMR Research Section's head to create frameworks for the administration of northeastern China, including ideas centered around "a concord between five nationalities" (Chinese, Koreans, Japanese, Mongolians, and Manchus), and the foundation of a "Utopia based on the kingly way."[40] He also leveraged his connections with key donors, army officers, and fellow civilian ultra-nationalists to gain real-world material support for his ideas.[41] Okawa's grandiose solutions to Japan's economic-geopolitical crisis combined with his clever utilization of his position and connections made him an influential figure in Japanese politics by the early 1930s.

Kwantung Army Strategic Planning: Enter Ishiwara and Itagaki

In 1928, an Issekikai and Sakurakai member named Ishiwara Kanji was appointed to the Operations Division of the Kwantung Army staff. Born in 1889 in Tsuruoka-shi, Yamagata-ken, Ishiwara was the son of a policeman and the descendant of a samurai family.[42] In 1907, he enrolled in the IJA's Military Academy graduating in 1909; he then enrolled at the Staff College between 1915 and 1918, graduating second in his class, and being described as the "most brilliant student in the history of the College" by its assistant commandant.[43] In 1922, Ishiwara went to Germany, where he studied military history with German General Staff officers until 1925.[44] After his return to Japan in the autumn of 1925, Ishiwara taught classes on European military history at the IJA's staff college, where he gave lectures to the IJA's increasingly radical young officers.

Ishiwara's in-depth study of military history in Europe and at the staff college gave him special theoretical insight into the evolution of warfare's spatial dimension.[45] In Ishiwara's understanding, ancient Bronze and Iron Age civilizations primarily fought wars with close-or-

Ishiwara Kanji in his study
(Ishihara Kanji: Portraits of Modern Japanese Historical Figures: National Diet Library, Japan)

der "phalanxes" of melee infantry, a formation that Ishiwara conceptualized as a "point." The point only transformed into a "line" in the 1600s after the introduction of firearms to the battlefield, which loosened the existing close-order formations and lengthened the scope of the battlefield. By the twentieth century, the vast increase of men under arms, alongside further advances in industry, logistics, communications, and firepower, transformed warfare from a one-dimensional line to an "area" that was defined not only by the linear extent of deployment but also depth. The new "two-dimensional" area included multiple lines of defense to protect the vital headquarters, transportation hubs, communications assets, and depots that were necessary for armies to function coherently in the field.

Ishiwara's studies also helped him realize that a dialectic existed between two temporal types of warfare: short "decisive" wars that were largely determined on the battlefield and "protracted" wars of attrition that were determined just as much by economic, political, and social factors as they were in battle.[46] According to Ishiwara's understanding of this dialectic, a continuous cycle existed in which "decisive wars" and "protracted wars" alternated in prominence. During ancient and medieval times, European wars were "decisive conflicts" where armies of citizen soldiers fought short, sharp battles of annihilation that determined a given conflict's outcome.[47] But the paradigm shifted by the 17th and 18th centuries, as commanders prioritized exhausting and inflicting attrition upon the enemy instead of seeking battle due to the proliferation of small mercenary armies. The reintroduction of the citizen army and improvement of artillery during the French Revolution swung the pendulum back towards "decisive" battles of annihilation, which continued in the decades after Napoleon. However, by the First World War, the growth in men under arms and advances in firepower (most notably the machine-gun), meant that warfare transformed into a protracted, in-depth, and primarily defensive endeavor on the battlefield. As there were no "decisive battles" that determined the outcome of World War I, the determining factors in the Allies' victory were their mass mobilization of people and machines, and their effective coordination of military and political strategy. Therefore, World War I was a "protracted" "total" war that illustrated the necessity of states mobilizing the entirety of the population, resources, and industry at their disposal to achieve victory.

Ishiwara's studies led to another insight that informed his worldview: the rate of societal and technological advancement was exponentially increasing.[48] In his assessment, Europe's technological progress during the Middle Ages was limited, but the onset of the Renaissance brought about the advent of gunpowder, firearms, and cannon, which represented a great technological leap in the instruments of war. The

French Revolution and the onset of industrial capitalist modernity in turn brought an even greater revolution in social organization, which allowed for the modern nation-state and conscript armies to form. The Second Industrial Revolution allowed for the *en masse* production of machine guns, rifled artillery, and aircraft, which in turn led to the protracted stalemate of World War I. As each of these developments occurred in ever smaller windows of time, Ishiwara believed that the next great technological breakthrough would occur soon, likely within the next thirty years.

According to Ishiwara, airplanes would be the medium through which the next great evolution would occur.[49] Aircraft, Ishiwara thought, would radically increase their range and payload in the coming decades due to advances in refueling technology. The power of the new aircraft would usher in the development of a new third dimension in warfare, the so-called "cubic" dimension. The new bombers, equipped with colossal amounts of ordinance, could then rain destruction upon conventional armies and fleets to the point that these would be made obsolete. After the destruction of earth-bound fleets and navies, the new super-aircraft would move onto attacking cities, as well as the opponent's ability to make war. The mass bombing of cities would force every member of society to participate in conflict, thereby engulfing whole civilizations in a total war of annihilation. While this "war between peoples" would have catastrophic short-term consequences, it would usher in a new age of global peace, unity, and harmony in the long term, as all parties would come to understand the high costs associated with warfare in the new age.[50] In this sense, the "war between peoples" would have positive externalities, as the world would attain a collective Nirvana.

In a broader philosophical sense, Ishiwara believed that a clash between the East and West was inevitable in future years. The selfish, imperialist western world represented the way of hegemony, or *badao*, while Asia represented the kingly way, or *wangdao*.[51] Various western nation-states had attempted to establish ideological dominance for thousands of years, with no single power remaining hegemonic for a prolonged period. After the carnage of World War I, the United States inherited the mantle as western civilization's most powerful country.

Japan, meanwhile, inherited the mantle of the fluid, malleable eastern civilization through its process of cultural synthesis. Because Ishiwara believed that the eastern and western civilizations were diametrically opposed to one another, he foresaw a clash between Japan and the United States in the near future, which would determine the paradigm of the global order and facilitate the end of warfare as a means to settle disputes.[52]

Ishiwara's belief in the inevitability of apocalyptic war was rooted in Nichiren Buddhism, a distinctly Japanese version of Mahayana Buddhism that had its roots in 13th-century Japan. Nichiren Buddhism was in turn rooted in Buddhism's *shozomatsu* eschatology, which divided time after the Buddha's death into three periods. The last of them, the so-called "mappo" period, would be defined by great conflicts, trials, and tribulations.[53] According to Saint Nichiren, a Japanese Buddhist priest who lived between 1222 and 1282, the world already entered the "*mappo*," or Latter Law period, while he was alive. Roughly 2,500 years after the Buddha's death, Nichiren's *Senjisho* projected the world would descend into a "titanic conflict, unprecedented in human history," after which there would be an eternal period of peace and harmony where Buddhism would spread across the world.[54] Ishiwara, upon reading this passage in Nichiren's *Senjisho*, found a spiritual confirmation for his analysis of military history and science. As Ishiwara himself put it, the *Senjisho's* prophecy "gave to my military studies the most steadfast objective."[55]

Ishiwara combined his study of military history, analysis of geopolitical and technological developments, and end-times prophecies of Nichiren Buddhism to complete his theory of the "Final War." According to this schema, the "titanic conflict" Nichiren Buddhism projected would soon be at hand, not only because the 2,500 anniversary of Buddha's death was fast approaching, but also because Japan and the US were destined to clash as the representatives of the East and West. Ishiwara believed there was still some time to prepare for the conflict, because roughly 30 years still remained until the 2,500th anniversary of the Buddha's death and the worldly conditions for the "Final War" had not been met. He identified the worldly conditions to be the consolidation of US power in the West, the consolidation of Japan's power in the East, and the advent of airplanes that could circumnavigate the globe. As Ishiwara thought these conditions would be fulfilled simultaneously, the Final War of annihilation would break out soon after they came into being.

Even though the "Final War" would be a "war of annihilation," Ishiwara argued that Japan needed to prepare for a protracted conflict against the US in the near and intermediate term. To prepare, Japan needed to establish an autarkic, self-sufficient economic bloc on the Asian mainland in emulation of Napoleon's "continental strategy."[56] Ishiwara believed that Japan needed an autarkic bloc because in a war with the United States Japan would likely come under blockade from the US Navy.[57] Without the resources and strategic depth the Asian continent could provide, Japan would be unable to triumph in its "Final War" with the US. As Ishiwara himself put it, Japan's survival could only be secured when "the road to self-sufficiency in East Asia has been established and we can profitably pursue a protracted war."[58]

Northeastern China was the most important territory in Ishiwara's "Final War" schema. Not only would the iron, coal, and minerals of northeastern China facilitate the creation of a second industrial base, but the strategic position of northeastern China was such that Japan would have an ideal jumping off point to attack the rest of China or the Soviet Union.[59] Furthermore, in the event of a US bombing campaign over the Japanese home islands, Ishiwara believed Japan could move key political and economic installations to the region. These advantages made northeastern China a pivotal hub, because it was both endowed with key resources and well-positioned geographically. Therefore, Ishiwara advocated aggressive action in northeastern China, which would be the first step towards expanding Japan's empire in Asia.

As Ishiwara considered northeastern China to be of vital significance to the Japanese Empire's future, he requested to be transferred there from the staff college in 1928. That year, he joined the Operations Division of the Kwantung Army as a staff officer. In May 1929, the man who would become Ishiwara's partner in crime joined him on the Kwantung Army staff, Colonel Itagaki Seishiro. Much like Ishiwara,

Itagaki believed northeastern China's absorption into the Japanese Empire to be a fundamental prerequisite for the Empire's continued existence.[60] While Itagaki lacked the theoretical and historical acumen of Ishiwara, he did have a series of useful connections to high-ranking officers in the AGS and Army Ministry owing both to his "hearty" personality and his previous postings (like military attaché in China and service on the AGS for a time between 1926 and 1927).[61] He also had a previous history with Ishiwara: they attended the same regional military school at Sendai and spent time together as military attachés in China. Perhaps most importantly, Itagaki's personality traits complimented Ishiwara's well, as his relatively straightforward, practical nature was the perfect supplement to Ishiwara's erudite, intellectual nature.[62] Itagaki's connections, personality, familiarity with Ishiwara, and similar outlook on northeastern China all made him the perfect wingman for Ishiwara on the Kwantung Army staff. In the next three years, the Ishiwara-Itagaki pairing would leverage its connection to change the course of history in East Asia and prepare for the "Final War" with the United States.

While both Ishiwara and Itagaki believed that control over northeastern China was absolutely necessary for the continued viability of the Japanese Empire, they also believed that the Kwantung Army had to act quickly lest they lose their window for action. Two main factors contributed to their sense of urgency: the rise of Chinese nationalism after the Northern Expedition, and the burgeoning power of the Soviet Union. Both men believed that the rapid growth of China's national project and the USSR's rapidly growing economy meant that Japan had only until 1936 to implement its desired solution in northeastern China.[63] Therefore, after Itagaki's arrival in Mukden in 1929, both Ishiwara and Itagaki diligently began making plans to resolve the situation in the region, independent of their superiors.

Ishiwara and Itagaki began their planning efforts with the first step necessary for any military operation: reconnaissance. Thorough scouting throughout northeastern China was necessary because Ishiwara sought to prepare different sets of operational plans for both Zhang Xueliang's Northeastern Army and the Soviets. Therefore, starting on

July 3rd, 1929, Ishiwara, Itagaki, and a party of five or six other officers ventured from Port Arthur to Harbin, Hailar, and Manzhouli.[64] Ishiwara and his team spent ten days surveying the terrain, communication and transport network, available supplies, and time required to move forces from place to place in the event of war with either China or the Soviet Union.[65] After seeing the vast expanse of plains that defined northeastern China, the team began to understand the difficulties of moving troops in the region. Any operation to conquer northeastern China would require control over and along the main railway arteries, which were the principal means of moving troops in the region. Consequently, cities located at key points along the railroads (such as Harbin, Hailar, Changchun, and Qiqihar) would need to be seized if the Kwantung Army were to conquer northeastern China and project power elsewhere in the region.

Ishiwara also used the reconnaissance trip to lay out his worldview to the other Kwantung Army officers. On July 4th, 1929, he delivered two lectures titled the "Views of War Today" and "The Role of War in Modern History," in which he gave a survey of European military history and asserted that advances in military technology (particularly the advances of automatic weapons and rifled artillery), force participation, industrial production, and communications fundamentally transformed warfare into a protracted, primarily defensive endeavor. Ishiwara also claimed that the next war would be a "total war" of aerial annihilation that would wipe out the previous distinctions of line, zone, and front.[66] Ishiwara concluded his speech "The Role of War in Modern History" with an explanation of the "final war" that would determine the fate of global civilization. This explanation was made in terms of Nichiren Buddhism, with Ishiwara saying that "Saint Nichiren had prophesized a final great war for the unification of the world."[67]

On July 5th, Ishiwara further elaborated on his views in a discussion document titled "A Plan for the Basic Solution of the Manchurian and Mongolian Problems as Basic Policy to Revolutionize Our Country's Destiny." This document clearly stated that the United States would be Japan's ultimate opponent and that the first step to preparing for the "Final War" was to absorb northeastern China into the Japanese

Empire.[68] Ishiwara also enunciated his belief that absorbing this region into the empire would "automatically restore Japan's prosperity and solve unemployment."[69] He even believed that breaking off northeastern China from the rest of the country would allow Japan to "resolve" the "Manchuria-Mongolia issue" and "enable us to extinguish anti-Japanese agitation in China at the same time."[70] While Ishiwara's latter two statements were not supported by any tangible evidence, they nonetheless garnered the support and admiration of his fellow officers (most notably Itagaki), and established him as the leader of a Kwantung Army clique that would take action in northeastern China. In this sense, Ishiwara's lectures were "the actual inception of the Mukden Incident."[71]

During his July reconnaissance trip, Ishiwara also began making plans for the conquest and administration of a future occupied northeastern China in a document titled "Plan for the Annexation of Manchuria and Mongolia." Divided into three parts, this document established a rough outline of what the Kwantung Army needed to do to establish control over northeastern China. Part I focused on the military conquest of northeastern China and stated the necessary steps the Kwantung Army needed to take to seize control of the region: suppressing Zhang Xueliang's army, destroying his warlord clique, confiscating his property, and excluding Zhang and his clique from future political endeavors in northeastern China.[72] Part II outlined the hypothetical administration a Japanese-occupied "Manchuria" would have, and the societal structure that would be promoted. To a great extent the Chinese governmental structure would be maintained but co-opted by the Kwantung Army, with a Japanese governor general at the head of a new "Manchurian" government. As for society, the three main ethnic groups in the region—Koreans, Chinese and Japanese—would each have a different role, with Koreans concentrating on paddy agriculture, Chinese on small business and labor, and quite conveniently, the Japanese on big capital and large-scale enterprises.[73] In Part III, Ishiwara focused on the issue of defense. Interestingly, the United States—the assumed ultimate opponent in Ishiwara's "Final War" schema—was not mentioned in this section. Rather, Ishiwara only made mention of the Soviet Union in Part III, noting that four divisions would likely be necessary to deter a potential Soviet invasion.[74] This singling out reflects the fact that the Soviet Union was the most immediate outside force that could intervene to stop Ishiwara's plans, while the US was a far-off adversary. The USSR was therefore a short-term enemy and the US the long-term one.

While Ishiwara and Itagaki spearheaded plans to conquer northeastern China, the scale and scope of the project necessitated that they delegate certain tasks to other staff officers. At the end of the July 1929 reconnaissance trip, Ishiwara gave Captain Sakuma Ryozo the task of researching the details of occupying and administering a future "Manchurian" state. By September 1930, Sakuma had completed his study, which attempted to answer Ishiwara's questions.[75] Titled "Study on the Administration of Occupied Manchuria," Sakuma's study started by stating the various purposes of a Japanese administration in northeastern China:[76]

> Manchurian Governance Plan (Draft)
>
> The purpose of governing Manchuria is to bring the entire region, including parts of northern China, under our complete control. By doing so, we can fully utilize its vast territory and abundant resources under our national strength. This plan aims for long-term and stable use of the region, and through this objective, to ensure secure and permanent possession of strategic bases necessary for national defense.
>
> By stabilizing Manchuria in this way, it will also contribute to the development and prosperity of our imperial citizens. Furthermore, resolving population and food supply issues is one of the major benefits. Through the stabilization of law and order in Manchuria, we aim to guarantee the development of industry and transportation.
>
> Ultimately, the goal is to establish an economically independent region, where residents, including settlers and others, can live securely and prosper together in mutual coexistence.

From this excerpt, Sakuma seems to have created his governance plan with big-picture strategic considerations in mind, likely for the benefit of officers in the AGS and Army Ministry. Sakuma's reference to "fully utilizing" northeastern China's "vast territory and abundant resources under our national strength" echoed the desires of many officers in the Army Ministry and AGS who wanted to establish an autarkic bloc in northeast Asia. Similarly, his reference to the "permanent possession of strategic bases necessary for national defense," was a nod to the desire of many officers to secure northeastern China as an anti-communist bulwark against the USSR and a forward base of operations against China. The importance of economic factors is evident in Sakuma's reference to using northeastern China as a means to resolve "population and food supply issues," which had ailed Japan since the onset of the Great Depression. By making this statement, Sakuma was clearly advocating for Japanese settler colonialism in northeastern China, which he hoped could both alleviate Japan's landless peasant problem and establish northeastern China as a breadbasket for the Japanese metropole.

Sakuma's study went on to articulate the necessity of self-sufficiency, stating that "all expenses required for the administration of the Manchuria-Mongolia occupied territory should not only be self-supporting but supply of strategic resources and alleviation of the national economy should also be attempted through the development of resources in the occupied territory."[77] The first steps to achieving Sakuma's desired self-sufficiency were to establish military governance over unoccupied areas of northeastern China, and to "secure a stable financial base." Sakuma went on to recommend an economic development program that included the commitment of various logistics, engineering, and road and railway construction units to the area. Sakuma hoped these resources would allow Japan to build infrastructure and facilitate the autonomy of a Japanese garrison. He also recommended streamlining Zhang Xueliang's byzantine and overly complicated tax system to increase government revenues. In Sakuma's estimation, these policies would not only contribute to making northeastern China independent, but would also serve as long-term capital investments that could help Japan develop an industrial base in the region. In this way, Japan could obtain the necessary resources and hinterland to become a self-sufficient empire prepared for Ishiwara's "final war."

Sakuma's administrative recommendations clearly indicated the Kwantung Army's desire to forcibly establish a unified government over all of northeastern China. His stated principles for administration are surprisingly blunt in the need to use force to accomplish the military's goals. Sakuma's emphasis on the centrality of force is evident in his administrative principles, listed below:[78]

1. Governance and transportation agencies must be unified. Administrative rule and transportation control shall be enforced jointly by military forces and civilian officials.
2. In areas already under military occupation, such as Mukden, Changchun, Jilin, Harbin, and other key railway zones across Manchuria, administrative control shall be clearly established, and the power of governance must be firmly held.
3. Administrative authority over these regions must be formally recognized under military power. Even though this authority may appear to be civilian-led on the surface, it must be backed entirely by military force.
4. Any territories newly added to this governance zone must be acknowledged and integrated into this system, leading to the full unification of Manchuria under a single administrative structure.

Sakuma's explicit statements that "the power of governance must be firmly held" and that any new government must be "backed entirely by military force" emphasize the importance of force in establishing a new government in northeastern China. But Sakuma's statement that the Kwantung Army's administrative authority "may appear civilian-led on the surface" indicates his belief that a puppet government could help Japan economize on force in the region. Sakuma's belief in utilizing such a puppet apparatus is further evidenced in his stated desire to maintain Zhang Xueliang's administrative network at the grassroots level. In turn, the new "good government" would allow each of the respective ethnicities in northeastern China to focus on what they were good at, releasing the region's inhabitants from their "unstable life" and bad government.[79] In this sense, Japan's conquest of northeastern China would be a "benevolent" act that would better the lives of the broad masses of people, despite the violent means necessary to accomplish this goal.

Sakuma's conclusions can be compared with a 1927 Army Ministry document titled "Sending the Opinions About the Policies on Manchuria and Mongolia," which also advocated the full-scale absorption of northeastern China into the Empire in the name of national security, economic recovery, and self-sufficiency. An excerpt from this document is provided below:[80]

> The Objective of Manchurian-Mongolian Governance
>
> The objective of the unification and governance of Manchuria is to fully bring Manchuria and parts of Mongolia under our control and to absorb and utilize their abundant resources through our national power. To achieve this, it is necessary to formulate a long-term, mutually beneficial strategy, and to secure Manchuria through external operations. By stabilizing Manchuria as a strategic base, we aim to promote the development of the empire's national strength, resolve domestic issues such as population pressure and food shortages, and ensure peace and order. Furthermore, by advancing industrial production and transportation infrastructure, we will establish the economic independence of Manchuria, improve the livelihoods of local residents and their families, and ultimately realize genuine coexistence, shared prosperity, and well-being for all.

"Sending the Opinions About the Policies on Manchuria and Mongolia" emphasizes the importance of establishing an autarkic Japanese Empire and recommends pursuing a program of industrial and infrastructural development in northeastern China. In these regards, the document reflects the ideas of Ishiwara, Sakuma, and the other Kwantung Army staff officers. But it also slightly differs from those documents by putting a greater emphasis on using a lighter hand to govern northeastern China to foster regime legitimacy. Moreover, it explicitly refers to a "military occupation period," which unlike Sakuma's document, implies an eventual end to military administration. These differences foreshadowed a clash between the different approaches of the Kwantung Army and the Army Ministry to the question of northeastern China, which would come to a head in the months and years to come.

Ishiwara went well beyond Sakuma's study to plan and analyze the economic and administrative nature of a hypothetical state in northeastern China. Starting in January 1930, the Kwantung Army met every Saturday to discuss plans for the occupation and administration of the Three Eastern Provinces.[81] Soon thereafter, in March, the SMR corporation invited Ishiwara to make a presentation about the "Manchurian question" and his desired solutions. Ishiwara seized the opportunity to pontificate about his idea of the "Final War" and the importance of northeastern China to Japan's ultimate fate.[82] Ishiwara's presentation so thoroughly impressed the SMR Research Section that it made its data and personnel open to him, which allowed Ishiwara to work on a variety of topics, most notably the economic viability of a state in the region. Unfortunately for Ishiwara, the SMR's study of northeastern China's natural resources did not confirm Sakuma's conclusions about a new state being a self-supporting venture; North China's resources of coal and iron would also be needed to supplement the deposits in the northeast.[83] But this conclusion was not a deterrent to Ishiwara and the Kwantung Army staff, who continued to make plans for the eventual conquest and administration of the region.[84]

As part of the plan-making process, Ishiwara and Itagaki went on more in-person reconnaissance trips to various parts of northeastern China.[85] In October 1929, Ishiwara and a group of other Kwantung Army staff officers ventured west of Liaoyang to study the issues of coordinating attacks on Mukden and Jinzhou. They made follow-up visits in the spring of 1930 to survey Chinese defenses at Mukden, where they identified weak points in the defenses that could be breached. They also identified the location and features of key installations, such as the Mukden airfield and the houses of important Chinese officials.

Ishiwara later conducted yet more reconnaissance in eastern Jilin and Heilongjiang in the autumn of 1930, where he charted the locations of existing Chinese garrisons. Ishiwara's in-depth surveys throughout northeastern China greatly facilitated the advance-operational planning necessary for a Kwantung Army offensive in the region.

Ishiwara's reconnaissance led him to conclude that any prospective Kwantung Army invasion of northeastern China would have two key characteristics: 1) the rapid concentration of force via the South Manchuria Railway Line near the Beidaying Barracks, the center of local warlord Zhang Xueliang's strength and 2) a variety of simultaneous night attacks at key points along the SMR to take advantage of the element of surprise.[86] Ishiwara assumed that these two elements would not only make up for the Kwantung Army's severe manpower deficit (10,000 vs 250,000) but also deprive the remaining Chinese in Zhang's army of the will to fight.[87] While the Kwantung Army seized Beidaying, it would also simultaneously initiate night attacks at Mukden, Gongchangling, and Changchun, all key cities along the SMR. For Mukden, the Kwantung Army operations division created a plan called "Outline for the Seizure of Mukden Fortifications," which emphasized the importance of night attacks to compensate for the Kwantung Army's numerical inferiority, and targeted the previously identified weak points in the defenses.[88] A similar plan of attack was created for Gongchangling. In the event the Soviets tried to intervene, the Kwantung Army would seize the rail junctions of Hailar in the Northwest and Nencheng and Lungmen in the Northeast, giving the Kwantung Army the advantage of interior lines in the event of a clash.[89] In this manner, the Kwantung Army could both take and, if necessary, defend northeastern China from its two foremost adversaries on the Asian continent.

Rumblings in Tokyo

Within the upper echelons of the IJA, awareness of the Kwantung Army's plans was initially somewhat limited. But some officers were aware of these plans through the grapevine of secret societies and officer discussion groups. One of the first IJA officers to support Itagaki's and Ishiwara's plans was Komoto Daisaku, who traveled to northeastern China to discuss plans with Ishiwara and Itagaki in the latter half of 1930.[90] As a right-wing radical, Komoto enthusiastically endorsed these plans and promised to generate support for them. Soon thereafter, Komoto proceeded to Tianjin to make inroads with the ex-Chinese Emperor Puyi. Due to Puyi's presence, many Qing court officials were also in the city and served as conduits to Puyi's inner circle. Komoto realized the importance of these figures and sought to gain their support as a stepping stone to Puyi. After securing an agreement with former Qing court officials such as Luo Zhenyu and Zheng Xiaoxu, Komoto went back to Port Arthur to confer once more with Itagaki and Ishiwara before returning to Tokyo. With his sojourn complete, Komoto could now turn to securing support for the thought leaders in Japan.

Based on his own experience in the 1928 assassination of Zhang Zuolin, Komoto believed that three conditions were necessary for Ishiwara and Itagaki's plans to succeed: 1) they must ensure support from the Kwantung Army, 2) they must ensure that the Kwantung Army bureaucracy was united, and 3) they must secure the necessary funds and support for their venture.[91] Komoto believed that Ishiwara and Itagaki had already secured sufficient support and unity within the Kwantung Army for their venture but realized that they needed funding and support outside of the Kwantung Army for their plans to be successful. To secure financial support for Ishiwara and Itagaki's plans, Komoto got a China trading firm called Zhonggongshe and the head of Mainichi newspaper *Fujita Isamu* to donate to the Kwantung Army's secret funds. To secure political support for aggression in northeastern China, Komoto spoke to Okawa Shumei in November 1930 about replacing Japan's civilian elected government. Okawa, who had already toured 50 Japanese prefectures in 1929 to whip up anti-China sentiments, agreed to help Komoto with his program and continued conjuring support for action in northeastern China in 1930-31.[92]

While many in the AGS supported militant action in northeastern China, the nature of this action and how it should be prioritized relative to "internal reform" remained open questions. To try and clarify Japan's strategic outlook and priorities, Hashimoto Kingoro, head of the AGS Intelligence Division's Russia section, and Nemoto Hiroshi, the head of the Intelligence Division's China section, prepared the annual "Situation Estimate" at the end of 1930.[93] With regards to northeastern China, the estimate gave three possible alternatives for resolving the "Manchurian question:" 1) coaxing Zhang Xueliang to take a more pro-Japanese line 2) overthrowing Zhang in a coup and installing pro-Japanese proxies and 3) failing the first two solutions, the utilization of military measures.

In Hashimoto's "Situation Estimate," military measures were made a last resort, as he wanted to prioritize overthrowing the existing civilian government and replacing it with one more amenable to their agenda. Hashimoto believed that this program of "national reconstruction," would allow the Kwantung Army to enact its planned invasion of northeastern China without resistance. By contrast, if the Kwantung Army implemented its plans without the backing of the existing civilian government, there was always a chance it might be left on its own. Therefore, in Hashimoto's estimation, national reconstruction was necessary before attacking northeastern China because it would be the surest option to ensure both agendas were accomplished.[94]

In October 1930, the AGS sent Colonel Nagata Tetsuzan on a fact-finding reconnaissance mission to Korea, the Three Eastern Provinces, and China south of the Great Wall.[95] While Nagata gathered information about all three locales while on his trip, he also spoke to Ishiwara and Itagaki about how their plans fit into Hashimoto's broader schemes. Even though Nagata attempted to explain Hashimoto's logic to Ishiwara, the latter did not wish to entertain Hashimoto's views and accused Nagata of acquiescing to the Foreign Ministry's weak China policy. In response, Nagata explained that he agreed that a military solution was necessary to resolve the "Manchurian question," but also told Ishiwara that the Kwantung Army would have to wait until public support had been cultivated to implement its plans. Ishiwara remained unconvinced, telling Nagata that he intended to initiate his planned incident at Liutiaogou near Mukden in the near future. He then inquired about procuring artillery from Japan in preparation for his planned incident. Unable to convince Ishiwara to agree with Hashimoto's views, Nagata departed northeastern China and returned to Tokyo.[96]

Even though Ishiwara and Itagaki remained steadfast in their desire to invade northeastern China as soon as possible, Hashimoto, Nemoto, and other members of the Sakurakai went ahead with their own plans in Tokyo. They received their first invitation to begin making in-depth plans in the spring of 1930 through Intelligence Department chief Tatekawa Yoshitsugu's request to investigate the issue of "national reconstruction."[97] Hashimoto and Nemoto embraced this assignment and began planning a coup against the civilian government. By December 1930, Hashimoto and Nemoto were far enough along in their plans to propose overthrowing the government at a Sakurakai meeting.[98]

Word of Hashimoto and Nemoto's embrace of a coup soon got back to Tatekawa. In January 1931, Vice Minister of War Sugiyama Hajime, Military Bureau Affairs Chief Koiso Kuniaki, Vice Chief of Staff Ninomiya Harushige, and Tatekawa all began trying to nudge Army Minister Ugaki to support a coup. It seemed to these four officers that Ugaki was moving in their direction, as evidenced by the fact that he summoned a group of officers, including Koiso, Ninomiya, and Hashimoto on January 13th, 1931 to discuss "military reforms."[99] Shortly after this discussion, Ninomiya assigned Hashimoto the task of preparing a "plan for national reconstruction" to submit to Ugaki.[100]

In late January, Ugaki held another conference about the impending internal reorganization of the IJA with Sugiyama, Koiso, Ninomiya, and Tatekawa all present. At this meeting, Tatekawa mentioned Okawa's views on the issue of internal reform, specifically regarding the army taking action to assist in the "renovation of Japan." To this Tatekawa statement, Ugaki allegedly made clear feelings of agreement and asked if he had a program in mind; while Tatekawa denied having any specific program, he agreed to keep Ugaki and Ninomiya informed of any future developments.[101]

On February 7th, four of the coup plotters from the Sakurakai (Hashimoto, Nemoto, Sakata, and Hiyoshi) met at Colonel Shigeto's home in Shinagawa, Tokyo to discuss the details of the planned coup.[102] They settled on seven points that would be part of their scheme to overthrow the government. These points are listed below:[103]

1. Prior to the coup, there would be a political rally in Hibiya park
2. The day of debate over a Labor Law in the Diet would be the day of the planned coup
3. Okawa Shumei would lead roughly 10,000 people in a march on the Diet
4. The Army would encircle and block access to the Diet in the name of protecting it from Okawa's group
5. Commander of the 1st Infantry Division, Mazaki Jinsaburo would accompany Tatekawa and Koiso to the Diet, where they would force the civilian government to resign
6. A different group of officers would go to Prime Minister Hamaguchi's residence and request his resignation
7. The Army would send a delegation to Prince Saionji to request that General Ugaki be appointed as the next Prime Minister

Okawa Shumei and other civilian ultranationalists would be key to the implementation of this scheme, as they would need to rally large numbers of people to be pawns in the plotters' scheme. Indeed, Okawa had already prepared a rally at Hibiya Park to occur on March 20th, intending to use his connections to a variety of radical political entities to manufacture a suitable number of attendees.[104] The rough scheme of Okawa's plan entailed his group at Hibiya Park receiving free sake and then being incited to storm the Diet. Other rioters from the Seiyukai assembled by Mori Kaku at Shibia Park would join them.[105] These two groups of rioters combined would then present enough of a threat for the IJA to cut off the Diet and proceed with the next steps of the coup.

Throughout February, Okawa took more steps to fill in the details of the planned coup. Realizing the need for armed demonstrators to present a credible threat to the government, Okawa planned to dispatch assault teams armed with sabers to attack the Prime Minister's office and the political parties' headquarters.[106] He also went to the trouble of procuring 300 blank cartridges, presumably for firearms.[107]

Okawa even asked Hashimoto for noise bombs, receiving between eight and 300.[108] He then attempted to win the support of right-wing ultranationalist organizations, making overtures to the organizations of Shimizu Gyonosuke and Kano Satoshi, as well as coaxing 200,000 yen out of right-wing oligarch Marquis Tokugawa Yoshichika. Okawa supplemented his right-wing support with leftist support, reaching out to the so-called "Proletarian Party," Social Democratic Party, and Labor Farmer Party in late February and early March.[109] In this manner, Okawa formed a coalition of armed, dissatisfied people and organizations that would serve the ends of the coup plotters within the IJA.

For all their efforts, Okawa and company still faced significant resistance to implementing their plot. Ishiwara and Itagaki still opposed the coup on the grounds that action in northeastern China should precede any attempt at making internal reform. In the Army Ministry, Colonels Hayashi, Katsura, Nagata Tetsuzan, Imamura Hitoshi, Nakamura Kotaro, and Lieutenant Colonel Suzuki Teiichi believed that a flamboyant coup attempt would instead potentially damage Ugaki's program for modernizing the IJA.[110] Others more deeply embedded in the Sakurakai were less radical than Hashimoto and his ilk, and wanted instead to work within the system to reform it.[111] The resistance of these figures signaled an institutional disunity within the IJA that would hinder the execution of the coup plot.

Army Minister Ugaki was among the first to express his opposition to the coup plotters' plans. Even though evidence exists that Ugaki indicated his sympathies with them and disliked the state of Japanese politics, he also told Okawa that he would not provide bombs, would not prevent the IJA from crushing the planned demonstrations, and did not have any aspirations to become prime minister.[112] Other officers within the Army Ministry who had supported the coup also began to have doubts. One of these officers was Koiso Kuniaki, who upon hearing of the lack of detail in the coup plans called them "childish" and advised Okawa to nix the plot.[113] Ninomiya and Tatekawa, also came to oppose the plot, as they did not believe that it was planned well enough to succeed in March 1931.[114] This mounting opposition whittled down the plotters to the original inner circle of Sakurakai radicals, namely Hashimoto, Nemoto, and Okawa.

Despite the sudden changes of heart of many previous coup-backers, Okawa, Hashimoto, and the other Sakurakai hardliners still wanted to go ahead with the coup. The resolve of these figures to instigate a coup was evidenced by the two mock demonstrations Okawa held on March 3rd and March 10th. But these demonstrations only drew 3,000 to 4,000 people, a far cry from the 10,000 required to march on the Diet.[115] The limited turnout led Okawa to ask Koiso if the IJA could provide troops to supplement his demonstrators. While this message reached Ugaki, he denied Okawa's request.

Okawa remained persistent in his attempts to sway Ugaki to be the face of the coup attempt. On March 6th, Okawa sent a letter to Ugaki stating that "the time is just ahead for a great mission to descend upon you. Please cherish self-respect, make up your mind to be the head of a group accomplishing a great work, and do not be induced by such common people as men of the political parties to become their leader."[116] Okawa's letter, while dramatic, had the opposite effect of what was intended: Ugaki came to the realization that he needed to take decisive steps to squelch the coup attempt. To end it once and for all, he summoned Koiso and Sugiyama and gave them an order to halt the attempt. This time, there was no mistaking Ugaki's intent, as when Koiso asked him again to confirm the order on March 10th, Ugaki said "don't be a fool! Do you think you can use His Majesty's troops for such a purpose?"[117] While Okawa still intended to go ahead with the coup alone after he learned of Ugaki's order, Ugaki tasked Komoto Daisaku, who had previously supported the coup, with preventing Okawa from carrying it out. Soon thereafter, Komoto convinced his friend Marquis Tokugawa to leverage his position as a donor to Okawa to prevent the plotted coup from going forward. Tokugawa's control over the purse strings, combined with poor turnout for planned demonstrations on March 18th, finally convinced Okawa to give up the ghost of the coup.[118]

The so-called "March Plot" represented the first instance of the Imperial Japanese Army trying to influence politics by means of a coup. Notably, despite the knowledge of many prominent figures within the

Army Ministry and AGS of the coup, the plotters only received very light punishments, which served to embolden the army radicals who sought to initiate "national reconstruction." Furthermore, the root causes of army officers' radicalism—economic crisis, geopolitical troubles, and middle-ranking officer *gekokujo*—did not recede. In fact, these problems only worsened in the coming months and years. Therefore, the strongest lasting effect of the March Plot's failure was to shift the policy initiative from those that prioritized internal reforms to those that prioritized decisive action in northeastern China.

Chapter 11

Countdown to Zero Hour

While the AGS and Army Ministry were embroiled in the intricacies of the March Plot, the Kwantung Army began finalizing its efforts to prepare for an invasion of northeastern China. Plans were drawn up, more reconnaissance was undertaken, and pressure was increased on military and civilian officials in Tokyo to support an aggressive solution. All the while, local tensions continued to increase in the region, as expressions of anti-Japanese sentiment became more common. Nonetheless, many officials in Tokyo were still hesitant to endorse a full-scale invasion of northeastern China by the spring of 1931, preferring instead less radical approaches. But that summer, the Wanpaoshan and Nakamura Incidents led some in the AGS and Army Ministry to be more receptive to invasion. These incidents also radicalized many in the public, press, and politics, who sought to punish the increasingly "insolent" Chinese and assert Japan's rights and interests in the Three Eastern Provinces. The formation of this radical consensus meant that the Kwantung Army would have less resistance implementing its desired agenda when the time came. Therefore, by the end of the summer, it was no longer a question of if the Kwantung Army would invade, but when.

The Kwantung Army Promotes its Ideas and Creates Plans

In the spring of 1931 Itagaki and Ishiwara began to aggressively push for a decisive solution to Japan's "Manchurian" problem. In March, Ishiwara promoted two of his essays, "Japan's Present and Future National Defense" and "A General Outline of the History of War;" in

April, he presented a series of lectures to these staffers titled "Conclusions from Lectures on European Military History," "Outline for War with the United States" and "Outline for War Plans for Solution of the Manchurian and Mongolian Problems." Ishiwara's essays laid out his ideas surrounding the evolution of the "Final War" and the necessity of controlling northeastern China as a precondition for Japan's national survival.[1] In his lectures titled "Outline for War with the United States" and "Outline for War Plans for Solution of the Manchurian and Mongolian Problems," he made clear his belief that he preferred to isolate the United States and fight it alone.[2] He hoped to achieve "understandings" with the UK and USSR, while using "intimidation" to prevent China from engaging in anti-Japanese activities or participating in Japan's war. But Ishiwara also espoused his belief that Japan must be prepared to fight all four powers at once if need be. The fact that a war at this level would vastly overstretch Japan's meager manpower and resources was of less importance to Ishiwara than establishing control over northeastern China. In his estimation, the Japanese Empire would cease to exist without the Three Eastern Provinces, and their capture needed to be the highest national security priority.

Itagaki echoed and even expanded on Ishiwara's views in a series of lectures he gave around the same time. In the first of these lectures titled "A Military View of Manchuria and Mongolia," Itagaki plainly stated his belief that a diplomatic solution in northeastern China was unlikely and that it was an essential base for future wars with the US and USSR.[3] He elaborated on these views in the lecture "On the Solution to the Manchurian and Mongolian Problem" in May 1931, in which he echoed Ishiwara's beliefs that Japan must annex the Three Eastern Provinces to uplift itself from the throes of the Great Depression and ensure its national security.[4] But he exceeded even Ishiwara's stated aims when he also advocated for the annexation of the Soviet Maritime Province and the rest of mainland China. Itagaki justified his anti-Soviet views on the basis that the Soviet Union would expand into northeastern China, which was Japan's sphere of influence.[5] He justified his proposed invasion of China south of the Great Wall on the basis that Japan would need China's raw materials base for a future war.

Itagaki also believed that China had "no prospect of happiness and order for the Chinese people" without the help of an "outside country," due to the lawlessness, corruption, and chaos of the warlord era.[6] Itagaki's chauvinistic views were key factors that led him to support an invasion of the Three Eastern Provinces as soon as possible.

In 1931, SMR Research Section Dalian section chief Sada Kojiro addressed Kwantung Army staff officers in Port Arthur about how to execute an invasion and occupation of northeastern China. His lecture "A Scientific Look at Policies for Manchuria and Mongolia" contained four sections, entitled 1) Premises for a Manchurian and Mongolian policy 2) How can we assure ourselves economic freedom and security 3) Steps to implement our purposes and 4) Conclusions. Of these sections section three was by far the most important, because here Sada proposed the weaponization of anti-Japanese incidents to justify a seizure of the SMR and the surrounding areas. Sada believed that seizing these areas was absolutely necessary, "even if it means flouting the mother country."[7] His message to take action no matter what was in line with the ideas of Ishiwara, Itagaki, and the other Kwantung Army officers. But Sada made an important contribution to Kwantung Army plans that had previously been lacking: a justification for war. Finding a *casus belli* would allow the Kwantung Army to justify its actions to its own government and the outside world, even if it took the initiative without orders.

Prior to 1931, Ishiwara had given less thought to manufacturing a *casus belli* than he had the details of reconnaissance and operational planning for his desired invasion. As of March 1930, Ishiwara still believed that China's anti-Japanese nationalism and "intransigence" would provide Japan the necessary justification to invade the Three Eastern Provinces without any activity from the Kwantung Army. But by May 1931 Ishiwara slightly adjusted his views, realizing the need for the Kwantung Army to manufacture an incident. As he put it in a section of a memorandum entitled "Personal Opinions on the Manchurian and Mongolian Problems," he wrote: "When the military preparations are completed we do not need to go to great lengths to find the motive or occasion; all we need to do is pick the time and then proclaim to the world our absorption of Manchuria the way we proclaimed the annexation of Korea. [If necessary, the Kwantung Army could] create the occasion for this with a plot and force the nation to go along."[8] While Ishiwara still preferred a straightforward "absorption of Manchuria," his admission that the Kwantung Army could "create the occasion" for a plot indicated that he was open to other ideas. Later, in May 1931, Itagaki and a group of other Kwantung Army staff officers came up with the idea that would finally win Ishiwara over: a false flag attack on the South Manchuria Railway near the Beidaying Barracks at Liutiaogou to be blamed on Zhang Xueliang's forces. By June, Ishiwara was fully onboard with this plan, and began taking steps to make the planned "false flag" attack a reality.

In order to initiate hostilities at Liutiaogou, Ishiwara and Itagaki would need to enlist subordinates to perform key tasks.[9] Therefore, in late June 1931, Ishiwara and Itagaki told Major Hanaya Tadashi, head of the Mukden Special Service Agency and Captain Imada Shintaro, an aide to Zhang Xueliang of their plans. Soon thereafter they informed Major Mitani Kiyoshi, head of the military police in Mukden, and Captain Ono Masao, commander of the 1st Company, 2nd Independent Railway Guards Battalion of their intentions. The 2nd Independent Railway Guards Battalion's proximity to Liutiaogou, the SMR, and the Beidaying Barracks made its participation vital to the ultimate success of Ishiwara's operation; fortunately for Ishiwara, Captain Ono supported his plan. Captain Ono then took matters upon himself to gain the support of other company commanders of the 2nd Battalion, most notably Captain Kawashima Tadashi, who was in charge of the battalion's 3rd Company. After Ono asked Kawashima if he would support the Kwantung Army's plans, he and Ono went to Hanaya. To allay any doubts about the veracity of the plan, Hanaya stated that any officer who did not support the plan at Liutiaogou would be reassigned to Japan. He also told Kawashima about the plan to assault the Beidaying Barracks. Satisfied, Kawashima agreed to participate in the plot and then went about hammering out key details of the plan with Hanaya and Ono.

Hanaya, Ono, and Kawashima were all concerned with the willingness of key figures in the 2nd Independent Railway Guards Battalion to support the Liutiaogou plan.[10] The commander of the battalion,

Lieutenant Colonel Takagi Yoshito, was seen as unlikely to support the plotters, as was Captain Kawakami Seiichi, commander of the 2nd Company based at Fushun. Hanaya, Ono, and Kawashima decided to resolve these problems by entrusting Major Kojima Masanori with securing the whole battalion's participation in the plot. They then pivoted to discussing how they would conduct the explosion; eventually they settled on using a contingent from Kawakami's company under the command of Captain Komoto Suemori to set the charges and blow up the railway. After finalizing these plans, Hanaya, Ono, and Kawashima pledged to carry them out.[11] The commitment of these figures to the Mukden plot illustrated their devotion to Ishiwara's vision and their willingness to do whatever it took to carry this vision out.

Prior to the spring of 1931, the Kwantung Army's stated mission was the protection of Japanese lives and property in the SMR zone.[12] But the planned false flag on the SMR was designed to justify a *de facto* expansion of the Kwantung Army's area of operations beyond the SMR zone. Under the cover of protecting Japanese citizens and property, the Kwantung Army would invade northeastern China, and absorb it into the Japanese Empire. Therefore, Ishiwara, Itagaki, and their co-conspirators intended to turn the *de jure* limits of the Kwantung Army's mission on their head, legitimizing the assault on China's sovereignty in the name of national security and the protection of civilians. While the timing of the operation and support from Tokyo remained to be worked out, the Kwantung Army had done most of the planning necessary to invade northeastern China by the end of June 1931.

Tokyo Begins to Support the Kwantung Army's Solution

The effects of political radicalism rose to the surface in both the Japanese government and the IJA in April 1931. In that month, Army Minister Ugaki Kazushige resigned from his post and accepted a new post as Governor General of Korea. Soon afterwards, Prime Minister Hamaguchi Osachi died from an infected gunshot wound he received in the aftermath of his 1930 assassination attempt. These turns of fate signaled that the IJA radicals had taken control of Japanese politics. Realizing the ascendancy of the new generation of radicals, Ugaki believed that only a resolution to the "Manchurian question" could shield the existing Japanese establishment from their ire. As he put it in a speech to divisional commanders on April 2nd, "it is essential we work out a solution to the Manchurian and Mongolian problem... because Japan cannot survive alone as an economic unit."[13]

To replace Hamaguchi and Ugaki, Wakatsuki Reijiro assumed the mantle of Prime Minister on April 14th, and General Minami Jiro became Army Minister. Wakatsuki and his government wanted to continue the Hamaguchi government's program of austerity, arms control, and reducing the size of the IJA.[14] Minami therefore found himself between a rock and a hard place, as he tried to balance the desires of the radical officers in the Imperial Japanese Army with the liberal austerity and disarmament program of the Wakatsuki Cabinet. Minami's actions would go on to play a key role in determining whether Japan took the path of peace or war, making him one of the most important figures within the Empire.

Soon after ascending to power in April 1931, Minami was already faced with a dilemma. On one side, AGS Chief of Staff General Kanaya Hanzo rejected Wakatsuki's cutbacks, which he regarded as a slap in the face after Ugaki's cutbacks in 1925. On the other side, Wakatsuki and his government sought to implement their desired program of austerity and disarmament. Much to the chagrin of the IJA radicals, Minami capitulated to the Cabinet's reduction of the IJA, agreeing to decrease the IJA's wartime strength from 32 to 28 infantry divisions in April 1931.[15] He was also forced to agree to Japan's participation in the 1932 Geneva Arms Limitation Conference, despite the broader desire within the IJA to avoid further arms limitations.[16] Minami's endorsement of these two proposals lost him a lot of credibility with the IJA's radicals, who vehemently opposed these measures. This distrust widened the rift between the Army Ministry and the forces in the field, such that Minami remained unaware of key developments happening on his watch.

Minami also had to deal with his radical subordinates' entreaties and intrigues regarding northeastern China. After the failure of the March Plot, Hashimoto and his fellow officers in the Sakurakai pivoted to advocating the use of force in the Three Eastern Provinces, calling for "positive action" in northeastern China in a bid to win over company grade officers.[17] Meanwhile, Komoto Daisaku took it upon himself to draw up a document titled "The Advisability of a Military Solution in Manchuria and Mongolia," which he intended to deliver to Minami in person.[18] Unfortunately for Komoto, Minami denied him entry into the headquarters of the Army Ministry on the grounds he was insane.[19] But Minami's refusal to entertain Komoto did not stop others in Tokyo from making plans for a military solution in northeastern China. Going forward, Minami would have to choose between upholding the line of the Wakatsuki Cabinet or caving into the designs of his radical subordinates.

Within the AGS, the Intelligence Division led the way in advocating military action in northeastern China. Its view on the necessity of military action is evidenced in another "Situation Estimate" prepared at the behest of Intelligence Division Chief Tatekawa Yoshitsugu in the spring of 1931.[20] This document—which was the work of China Section Chief Colonel Shigeto Chiaki, Europe-America Section Chief Watari Hisao, and Hashimoto—laid out three possible options for a resolution to the "Manchurian problem:" The first of these was the overthrow of Zhang Xueliang and the creation of a pro-Japanese government, the second was the creation of a government independent of the KMT, and the third was annexation. These three options were far more aggressive than those from the 1930 "Situation Estimate," as all of them were active measures that would remove Zhang Xueliang from power. The 1931 "Situation Estimate" also did not proclaim the necessity of "national reconstruction" before taking action in northeastern China due to the failure of the March Plot. Therefore, the new "Situation Estimate" signaled that many in the AGS's Intelligence Division had become more amenable to the Kwantung Army's point of view and sought action in the region. Some figures in the Intelligence Division such as Shigeto even supported annexation, and contacted the Kwantung Army staff officers to coordinate plans for a military solution in northeastern China.[21] The increase in support for annexation represented the extent to which the situation had deteriorated and the heightened radicalism of the IJA officer corps.

Fear of the Soviet Union was a key factor motivating the Intelligence Division's desire for rapid action in northeastern China. To a great extent, this fear was rooted in the imminent completion of the Soviets' Five-Year Plan, which would put it on the road to rapid industrialization. The IJA's desire to fight the Soviets sooner rather than later is reflected in a document entitled "Views Concerning our Imperial National Defense Against the USSR":[22]

> I. Views Concerning a Russo-Japanese War and Our Policy Against the USSR
>
> 1. The Capacity of the USSR for Urging a War and Her Future Policies
>
> The present equipment and training of the Red Army have reached a fairly high standard, somewhat superior to those of the combined forces of Poland, Romania, and the other countries along the Baltic Sea; her war-operational capacity, however, in the broadest sense of the word, i.e. her economic power, her munitions industries and her national unity cannot yet be considered adequate. Even now of course the Soviet Union would find no difficulty in taking hostile action against any single one of the neighboring countries to her west, but supposing that a war against the Soviet Union is begun by all those countries in alliance and with the active backing of a Great Power—the Soviet Union is certainly not strong enough to carry on the large-scale warfare necessary for such an emergency.
>
> We still cannot even know whether or not we shall be able to see a sudden accession of power after the completion of her 5-year plan. For which at present she is making every effort (there is quite a possibility of her achieving some degree of it).
>
> 2. Japanese Policy Toward USSR
>
> As mentioned in the preceding paragraph, the USSR will carry out an active policy in proportion as her national power becomes more complete. Such being the case, if Japan is satisfied to be a small Japan,

> all well and good. But since she wants to remain a great Japan, it will be her unavoidable destiny to clash with the USSR sooner or later. However as mentioned above, considering the capacity of the USSR for national defense and the situation of the other powers, the sooner the Soviet-Japanese war comes, the better for us. We must realize that with every day that the situation develops more favorably for the USSR.
>
> In short, I hope the authorities will make up their minds for a speedy war with the Soviet Union and initiate policies accordingly. Taking into consideration the fact that the disadvantage of opening a war at present lies not on our side but on their side, irrespective of which is better, peace or war, Japan, I believe must adopt a strong policy against the USSR...

This document, which was written by a Japanese military attache in Moscow in March 1931, asserted that that Japan had a short "window" to attack the Soviet Union before its Five Year Plan allowed it to gain strength. Tatekawa echoed this point of view in an April 1931 gathering of divisional commanders, saying that "when the Five-Year Plan is completed, the increase in their national strength together with the freedom to implement policy through dictatorial government will present a grave threat to the empire together with a loss of time... When we take an overall view of the international situation an active advance into Manchuria and Mongolia will be beneficial to us in our rapid pursuance of our goals while procrastination will only add to the accumulation of disadvantages."[23] Much like the military attaché's document, Tatekawa's assertion implied that Japan only had a finite window to strike northeastern China before the Soviets were strong enough to intervene. If Japan did not attack, it would forfeit its opportunity to take action in northeastern China and possibly even be subject to a Soviet attack in the region.

While the Intelligence Division represented the most aggressive portion of the AGS on the question of northeastern China and the Soviet Union, some Operations Division officers also believed in the necessity of military action in the region. Of these figures, Major Hata Shunroku (Chief of the Operations Division) and Lieutenant General

Araki Sadao of the AGS were two of the most prominent. Both Araki and Hata were vehement anti-communists, who believed in the inevitability of a Japanese-Soviet clash. In their minds, the region's annexation was necessary for Japan to preemptively gain an upper hand in an expected Japanese-Soviet war. Hata's point of view on the matter of war with the Soviet Union was illuminated during a meeting of division and bureau chiefs with Army Minister Minami. When Minami asked Hata if the USSR was likely to make war on Japan, Hata stated that the "USSR has a five-year plan," and that "although the Russians probably won't start anything for two or three years, to say flatly that they wouldn't is a mistake."[24] Even though Hata caveated his statement to Minami, his admission that "to say flatly they wouldn't is a mistake" was designed to get Minami to approve greater planning and preparation for war with the Soviets. Minami was not convinced by Hata's line of reasoning, but Hata and the Operations Division still sought to preemptively prepare for war with the USSR by securing northeastern China. The Operations Division of the AGS would also be a key player in promoting a military solution in the region in the months ahead.

In June 1931, the Kwantung Army sent Colonel Hanaya to Tokyo to ramp up pressure on Minami, the Army Ministry, and the AGS.[25] His mission was to secretly address members of the AGS and Army Ministry and persuade them of the need for rapid military action in northeastern China. From the Kwantung Army conspirators' perspective, a sense of urgency existed because Ishiwara and Itagaki were due to be transferred, with their three-year tours of duty expiring in late 1931 and early 1932, respectively. If Hanaya failed in pitching his idea in Tokyo, Ishiwara and Itagaki would be forced to run the risk of implementing their plans without any support from the center, which might doom their efforts to failure. Therefore, Hanaya's trip to Tokyo carried great stakes for Ishiwara, Itagaki, and the future of the Japanese Empire.

When Hanaya made his appeals in Tokyo, he found a ready-made audience in Intelligence Division officers Tatekawa, Shigeto, and Hashimoto, who all agreed with the Kwantung Army's desire to initiate action in northeastern China in the fall of 1931.[26] While Head of the Organization and Mobilization Section Tojo Hideki and China Section officer Nemoto Hiroshi expressed doubts about the lack of preparation time to implement Ishiwara and Itagaki's plans, they still gave tepid endorsements. Further up the ladder of command, Hanaya informed Vice Chief of Staff Ninomiya Harushige, head of the Military Affairs Department Koiso Kuniaki, and head of the Military Affairs Section Nagata Tetsuzan of the Kwantung Army's rough plans and timetable for action in northeastern China, none of whom objected. These figures' approval of the Kwantung Army's activities meant that Ishiwara and Itagaki would have at least some support from the AGS in Tokyo once they decided to take action.

After Hanaya's successful trip, the Kwantung Army continued putting pressure on the central army authorities in Tokyo. In response to such pressure, Army Minister Minami agreed to organize a meeting among high-ranking army officers to find a resolution to the problems in northeastern China on June 11th. The attendees at this top-secret meeting were Nagata and Okamura from the Army Ministry alongside Yamawaki Mastataka, Watari Hisao, Shigeto, and Tatekawa from the General Staff. By June 19th, these men had created a document titled "General Principles Concerning the Settlement of Manchurian Problems" under Tatekawa's guidance. The document's contents are listed below:[27]

1.) The alleviation of the anti-Japanese activities of the Zhang regime shall continue to be undertaken primarily through negotiation by the Foreign Office and the army shall maintain close contact with the Kwantung Army in order to make them act with discretion
2) In spite of the above-mentioned efforts, should anti-Japanese activities be intensified, military action might be necessary
3) Internal as well as international understanding are absolutely necessary for the settlement of Manchurian problems. The Minister of War through the Cabinet, and the Military Affairs Department, and the Second Department [Intelligence Division] of the General Staff, with close contact with the Foreign Office, shall make careful preparations, such as publicizing the realities of anti-Japanese activities in Manchuria to the people

of our country and of the powers, so that in the event of military action, public opinion will support the measure and the powers will not take opposing or suppressive steps.

4) The General Staff shall make plans concerning necessary forces and the guidance of their movement in the event of military action
5) Measures with regard to cultivation of internal and international understanding shall be undertaken with a view to achieving results in approximately one year, that is, by the spring of 1932.

The "General Principles" emphasized the importance of the AGS preparations for military action in the Three Eastern Provinces and manufacturing support amongst the public. In these regards, the Kwantung Army and AGS agreed on two key ingredients of a military solution in northeastern China. But the "General Principles" outline also reveals great differences between the plotters and their furtive supporters in Tokyo. Timing was one such difference, as Ishiwara and Itagaki wanted to take action in 1931, while the document clearly states that Japan should wait until the spring of 1932. The drafters also placed a greater emphasis on international relations than Ishiwara due to their responsibility for Japan's overall national security, which forced them to consider eventualities such as war with the Soviet Union. Finally, and perhaps most importantly, the drafters put in writing their intention to "maintain close contact with the Kwantung Army to make them act with discretion." The desire of many in the Army Ministry and AGS to maintain a leash on the Kwantung Army should not be doubted, as evidenced by the fact that Minami told Kwantung Army chief of staff Miyake that his forces should maintain "patience and self-control."[28] But despite Tokyo's desires, the Kwantung Army's plans were much closer to coming to fruition than even their most ardent fellow travelers in Tokyo could have realized, due to their prodigious two-year planning efforts. As a result, the Kwantung Army would be in position to capitalize on the further deterioration of Sino-Japanese relations, which came to be embodied in a series of incidents in the summer of 1931.

Wanpaoshan and Nakamura Incidents: Further Catalysts to Action

Even though the Kwantung Army and the center in Tokyo remained at odds over the details of taking action in northeastern China, two key events at the end of June and beginning of July swung sentiment in Tokyo towards the Kwantung Army. The first of these events was the so-called "Wanpaoshan Incident," while the second incident was termed the "Nakamura Incident." Both incidents brought into question the nature of Japan's extraterritorial rights in northeastern China and the degree of sovereignty China had over its Three Eastern Provinces. In the case of the Wanpaoshan Incident, the issues in question were the right of Japanese subjects (specifically Koreans) to lease land in northeastern China and engage in commercial activities. In the case of the Nakamura Incident, the issues involved the right of Japanese citizens to travel unmolested in northeastern China, even if they were suspected of illegal activities. As both of these issues were addressed in the Twenty-One Demands, Japan believed that their citizens had the legal right to lease land, do business, and travel freely in the region. But Zhang Xueliang and the KMT continued to refuse to recognize the validity of the Twenty-One Demands, due to the coercive context of their acceptance and the onerous terms they imposed. Therefore, both of these incidents remained unresolved and illustrate the irreconcilable antagonism between Japanese imperialism and Chinese nationalism.

Wanpaoshan Incident

The roots of the Wanpaoshan Incident lay in the fact that many Koreans had migrated across the border to northeastern China in the previous five decades to escape famine and drought in Korea. After the Qing Dynasty removed the restrictions to Korean immigration in 1877, a further trickle of Koreans arrived, totaling roughly 37,000 by 1903.[29] This trickle grew after Japan established a protectorate over Korea in 1905, as many anti-Japanese Korean nationalists sought to escape the

Japanese yoke.[30] After the repression of Korea's anti-Japanese independence movement in March 1919, yet more revolutionary nationalists fled to northeastern China, turning the trickle of Korean migration into a flood.[31] Many other Koreans also emigrated in the 1910s and 1920s due to the economic destitution caused by Japanese colonialism. Thanks to this combination of factors, the Korean population in northeastern China numbered roughly 800,000 by 1931.[32]

The increase in the number of Koreans in northeastern China raised questions regarding their citizenship and governmental jurisdiction. In 1909, China and Japan attempted to resolve these questions in the Jiandao Agreement, signed in September of that year. The most relevant part of the Jiandao Agreement, Article 4, is excerpted below:[33]

> Koreans residing north of the Tumen River and engaged in agriculture shall be subject to the jurisdiction of the Chinese officials of the territory. Chinese officials shall treat the Koreans and Chinese with equality as regards payment of taxes and in the enforcement of the laws. Chinese officials shall administer Chinese law in all civil and criminal cases where Koreans are concerned. A Japanese Consular officer may at all times attend the court proceedings. In cases where capital punishment may adjudged, the Japanese Consul must be notified. If the Japanese Consul can point out any irregularities in the proceedings, he may request that another official be appointed to hold a rehearing of the case so that justice may be obtained.

The Jiandao Agreement made Koreans, who were nominally Japanese subjects at the time, subject to Chinese laws and courts. The agreement also gave Koreans the right to engage in agricultural activities and export grain from northeastern China.[34] In the official Chinese view, the Jiandao Agreement represented a *quid pro quo* that would safeguard Chinese sovereignty in the Three Eastern Provinces.[35] But Japan's assertion of the Twenty-One Demands in 1915 negated this Chinese expectation, as the terms of the new agreement made all Japanese subjects in the region accountable to Japanese law while allowing Japanese subjects to lease and purchase land in the southern part of the Three Eastern Provinces.[36] From the Japanese point of view, these new terms overrode the 1909 treaty, despite the fact that the treaty stipulated that all existing agreements between Japan and China regarding the Three Eastern Provinces should "remain in force."[37] Chinese governments upheld the validity of the 1909 agreement and not the Twenty-One Demands, especially after the rise of Chinese nationalism in the 1920s. Therefore, the situation regarding land ownership rights and legal jurisdiction for Japanese subjects in the Three Eastern Provinces was at an impasse and only grew more difficult to resolve after the assassination of Zhang Zuolin in 1928.

As Japanese subjects, Koreans would also be able to lease and purchase land in the Three Eastern Provinces. While such an arrangement was beneficial on paper for the Koreans, the status of many anti-Japanese nationalists was jeopardized, especially after the Japanese established a police presence in the region. The Japanese were also able to weaponize the presence of the Koreans against the local Chinese, who refused to recognize the Twenty-One Demands and sought to maintain some modicum of sovereignty in the region. Consequently, many local Chinese officials began to be suspicious of Koreans in their locale, and undertook discriminatory measures against them.[38]

Initially, Chinese officials in the Three Eastern Provinces sought to circumvent Japan's claim of extraterritorial rights for Koreans by forcing them to naturalize as Chinese citizens.[39] But because there were many cases of naturalized Koreans purchasing land in northeastern China and passing the rights to Japanese companies, local Chinese officials began to back off this strategy. Indeed, Zhang Zuoxiang of Jilin province went so far as to implement ordinances that prevented the naturalization of Koreans in 1929.[40] Zhang also implemented an ordinance that forced Koreans to report to local police before buying land, pledge not to transfer said land to the Japanese, and receive the approval of the provincial government for the purchase.[41] Such measures made it far more difficult for Koreans to purchase land, despite their heavy presence in Jilin province.

In 1931, Zhang Zuoxiang's ordinances were put to the test by a group of Korean farmers near the village of Wanpaoshan. Here, these farmers leased roughly 5,000 square feet of land from a man named Hao Yongde, who was the manager of the Zhangnong Agricultural company. In the first lease contract Hao gave to the Koreans, there was a provision that stipulated the transfer of land would be null and void without permission from the local Changchun governing authorities. But in a second contract the Koreans signed with Hao, there was no mention of governmental approval for the land transfer. The second contract also implied that irrigation ditches would be dug, despite the fact that no preparations were made to lease the land these ditches would traverse.[42] As the Changchun government decreed that the land near Wanpaoshan should be "carefully examined" before any transfer, Hao's second contract represented a unilateral move that contradicted the existing laws of Jilin province.[43] The stage was set for controversy to erupt due to the unclear circumstances of the contracts and the contradictory treaties that governed governmental jurisdiction.

Soon after the Koreans signed Hao's second contract, 180 Korean farmers arrived and began constructing an irrigation ditch. But the new irrigation ditch traversed the land of Chinese farmers, and was detrimental to the quality of their land. Consequently, Chinese farmers complained to the local authorities, who ordered all digging of the irrigation ditches to stop on May 24th.[44] The local authorities deployed fifty troops to the area the next day, followed by another two-hundred on May 31st. These forces ordered the Koreans to leave the Wanpaoshan area by June 1st.[45]

In response to the deployment of Chinese troops, the Japanese sent six policemen from the Changchun Consul to "protect" their subjects and allow them to keep digging the ditch.[46] After the deployment of these policemen, a series of negotiations ensued between representatives of the Chinese and Japanese governments about the general rights of Japanese citizens in the Three Eastern Provinces and the specific right of the Koreans to continue digging the irrigation ditch at Wanpaoshan. But despite over a month of talks, no progress was made, as neither side was willing to budge from their original position.[47] Consequently, on July 1st, 400 armed Chinese vigilantes arrived and drove the Koreans away from the ditch.[48] While temporarily successful, the

Japanese soon dispatched a machine-gun corps and cavalry to the site of the ditch, where they confronted at least 300 Chinese. Free of intimidation, the Koreans were able to return and continue work on the irrigation ditch, despite the presence of hostile Chinese.[49]

After the Wanpaoshan Incident, anti-Chinese propaganda took hold in both Korea and Japan. Media outlets in both countries exaggerated the incident and claimed that hundreds of Koreans died at the hands of Chinese mobs. Many Koreans took justice into their own hands as a result, targeting Chinese in Korea in a series of riots, which led to hundreds of deaths.[50] In response, the Chinese initiated boycotts against Japanese goods and also engaged in anti-Japanese activities in cities such as Qingdao, Tianjin, and Shanghai.[51] The course of events after the Wanpaoshan Incident sharpened Sino-Japanese antagonism, making negotiations surrounding the incident far more difficult. These events played just as much of a role in increasing Sino-Japanese antagonism as the incident itself, and helped pave the way towards conflict in the Three Eastern Provinces.

The Nakamura Incident

In July 1931, another incident occurred that also helped facilitate Sino-Japanese conflict. Known as the "Nakamura Incident," the episode centered around Captain Nakamura Shintaro, an infantry officer attached to the AGS who was arrested and extrajudicially executed by the Chinese in June 1931. The Japanese maintained that Nakamura's execution was an unjustified action that indicated the arrogant, insufferable attitude of the Chinese.[52] The Twenty-One Demands were an obvious reference point for this argument, as Article 3 of Group II allowed Japanese citizens to freely travel in northeastern China.[53] The Chinese alleged that Nakamura was a spy who had lied about his route prior to obtaining his traveling passport.[54] The contrasting appraisals of the Nakamura incident raise the question of who he really was and how his death impacted Sino-Japanese antagonism in the Three Eastern Provinces.

Nakamura's story begins in May 1931, when the AGS sent Nakamura to reconnoiter and survey the area west of the Greater Xing'an mountain range.[55] After stopping at the Kwantung Army's headquarters in Port Arthur and the Harbin Special Service Agency, Nakamura proceeded to Angangchi, Hailar, Manzhouli, and Taonan prior to embarking on his mission. There, Nakamura met his guide, Isugi Entaro, an ex-sergeant major who had transitioned to the role of "patriotic adventurer." He also met his two translators, one of whom was a Mongolian, and the other of whom was an anti-communist Russian. Together with these men, Nakamura set out from Iliktou station on June 6th, 1931, and proceeded south down the Chinese Eastern Railway. Their ultimate destination was Taonan, which they expected to reach roughly a month later.[56]

Nakamura's journey towards Taonan was slowed by heavy rains, which flooded many of the mountain paths they were travelling on. By the time they made it out of the mountains, it was already June 25th, with a long stretch to go until they reached Taonan.[57] But that day, they caught the attention of the Xingan Reclamation Army's Third Infantry Regiment, which could not help but notice their tall horses and copious amount of luggage. These were suspicious traits, not only because tall horses were generally military grade, but also because carrying so much luggage in a remote area was not normal. Consequently, elements of the Third Infantry Regiment decided to interrogate Nakamura.

Nakamura told the Chinese that he was an agricultural expert from the Tokyo Agricultural Society of Japan.[58] But this story was not believable, as he and his party were wearing greatcoats in the middle of summer, despite claiming to be farmers. Furthermore, no foreigners were allowed in the Taonan-Solun area due to its remoteness, which further undermined Nakamura's explanation. Consequently, the Chinese searched Nakamura's luggage and checked the two passports he had received to confirm their validity.[59] Upon review, not only did the Chinese find that Nakamura's passports were irregular, but they also found drugs, armaments, weapons, and military maps, indicating that Nakamura and his men were spies and not to be trusted.[60] Therefore, the party was taken into custody and eventually executed on June 27th.[61]

While the Japanese did not initially know that Nakamura had been executed, his failure to arrive in Taonan by early July was cause for concern.[62] Indeed, Ishiwara considered him important enough to send staff officers to inquire about his fate as early as July 8th. Soon thereafter, Ishiwara sent an investigative search party to find out what happened to Nakamura.[63] On July 22nd, Ishiwara's investigative team finally learned of their execution from the wife of an SMR employee. This information was confirmed a day later by a Chinese resident that lived near the site of Nakamura's arrest.[64] These accounts provided enough evidence for Japan to demand that China atone for the execution through apology, indemnity, punishment of responsible officers, and assurances that such an incident would never happen again.[65] While certain Chinese were prepared to accept these demands, they still needed to investigate the circumstances of his execution, which took several more months.[66] Negotiations continued through the end of the summer during this investigation and still were not completely resolved on the eve of the Mukden Incident.

At first, the Army Ministry and Foreign Ministry suppressed all mention of the Nakamura Incident in the Japanese press, as they did not want his mission to be subject to international scrutiny.[67] But by mid-August, the Kwantung Army released news of Nakamura's assassination to the Japanese press to pressure the government into pursuing a more aggressive policy in northeastern China.[68] The release of this information achieved the intended effect, as the media and general public began to demand more aggressive solutions in northeastern China.[69] Criticisms of Shidehara's "friendship diplomacy" became more common, and many Japanese began to support "using force to settle once and for all the outstanding issues relating to Manchuria."[70] Although the number of outspoken, active supporters for positive action were confined to a relatively small minority, the Nakamura Incident was a great gift to this element, as few explicitly opposed their agenda.[71] Therefore, the Nakamura Incident made Japanese public opinion "more willing to accept the forcible solution to all Manchurian problems" promoted by figures such as Ishiwara and Itagaki, and thereby helped put Japan on the path to war.[72]

Tokyo Moves in the Kwantung Army's Direction

The Wanpaoshan and Nakamura Incidents led many in Japan to support radical action in northeastern China. As a result, pressure began increasing on the Army Ministry and Minami to sanction action. The Association of Full Generals directly told the Army Ministry to take a firmer line, while the Imperial Veterans Association incited the public, demanding an immediate resolution of the Nakamura Incident.[73] In the political sphere, the Seiyukai sought to capitalize on the Incident to score political points on their adversaries in Wakatsuki's Minseito government. Throughout August, Seiyukai political candidates consistently harped on the themes of the IJA's honor being insulted, the need to protect Japan's extraterritorial rights and privileges in the Three Eastern Provinces, and the follies of Shidehara's "friendship" policy with China.[74] Mori Kaku greatly contributed to the Seiyukai's campaigns by going to Wanpaoshan, accompanied by journalists and fellow Seiyukai members. While he ostensibly sought to "investigate" the Wanpaoshan Incident, Mori really wanted to whip up support for aggressive action in northeastern China, thereby undermining Shidehara's "friendship policy." As Mori himself put it, "there really is no need for me to visit the site of the Wanpaoshan incident... I have a different idea in mind."[75]

Within the Foreign Ministry, Yoshida Shigeru also endorsed a more active posture in the region. He believed that the heightened local tensions necessitated action, and indicated this belief to his boss, Shidehara Kijuro. Yoshida's views on this subject can be seen in a report titled "Measures for Manchurian-Mongolian Problem:"[76]

> Since early June of Showa 6 (1931), the situation in Manchuria has become tense, and numerous incidents have occurred. In light of these developments, our side has been closely observing the situation, believing it to be necessary to respond carefully and with restraint. However, recently, the threat to the lives and property of our residents in Manchuria has grown more serious.
>
> Given these circumstances, it has become difficult to ignore the situation. Our consular and military authorities in the region have

> jointly deliberated and agreed that urgent action is required to ensure the safety of our people. Accordingly, they have resolved to dispatch a temporary force to maintain order and protect Japanese nationals and property in the area.
>
> We are working with local Chinese authorities and officials to avoid unnecessary confrontation. However, in the event of escalation, our troops are prepared to respond appropriately. We request understanding and cooperation in this matter.

Yoshida's report to Shidehara clearly sought to take advantage of the tense atmosphere in the Three Eastern Provinces, perhaps best exemplified in his request for "understanding and cooperation" in the event of military action. But Shidehara did not acquiesce to Yoshida's request, instead opting to continue his "friendship diplomacy" throughout the summer of 1931. Shidehara's continuation of his previous diplomatic policy illustrated his desire to avoid conflict in northeastern China—a desire that he explicitly communicated to the various consulates under his authority in the September 11th telegram below:[77]

> According to telegram No. 423 received from Mukden, the current situation on the ground remains tense. Considering the overall regional conditions, and particularly the local sentiments, it is feared that there may be unexpected civilian movements. Given that the full extent of the situation remains unclear, we believe it essential that communication and coordination between civil and military authorities be strengthened.
>
> In case military movements or other actions must be taken, we request that thorough consideration be given to avoiding unnecessary conflict. We also request that any such actions be carried out under unified command and strict coordination.
>
> In particular, we urge the utmost care in dealing with foreign nationals, and request that measures be taken to avoid inciting any further tension. We further request that, in preparation for any contingency, coordination with relevant authorities regarding supplies and troop provisions (including fuel, food, and other necessities) be completed as soon as possible. Please respond promptly with status updates.

Shidehara also continued investigating the Nakamura Incident, seeking an explanation for the Kwantung Army's actions. Shidehara's intentions can be found in a telegram he sent to Consul General Hayashi, seen below:[78]

> On the night of the 10th, Counselor Mita visited and relayed that, regarding the recent incident, it is necessary to append two points of explanation. He emphasized the need for clarification concerning the military's actions and their necessity, in order to avoid misunderstandings or misinterpretations among local residents or concerned parties. Should such misunderstandings arise, the damage and consequences could be severe. He also requested that official statements be issued accordingly.
>
> In the event that troops must be dispatched to the area for protection, the Foreign Ministry seeks clarification and coordination on such military matters. If civilian damage or conflict becomes unavoidable, the Ministry wishes to be informed of the response procedures and preparatory measures.
>
> Please consider these matters carefully and coordinate with relevant agencies, including the police, local officials, and the press. Appropriate communication methods and procedures should be determined for possible civilian evacuation and the handling of foreign residents.

Shidehara's desire to learn the full context of the Nakamura Incident put him at odds with the hawkish public, press, and army, which increasingly saw the Incident as a one-sided expression of Chinese insolence. While Shidehara continued to investigate into September, he alone could not stem the tides of war, despite his powerful position as Foreign Minister. Therefore, Shidehara needed other figures in Japan to help him prevent a conflict in northeastern China, whatever the social and political costs.

Even though the bellicose political environment in Japan increased pressure on Minami to endorse military action in northeastern China, he had something else on his mind: the Wakatsuki Cabinet's expected budget cuts to the IJA. In the spring, Minami had indicated doubts about the cuts, but still went along with them in principle. But after the Wanpaoshan and Nakamura Incidents, Minami began to voice stronger opposition to the Wakatsuki government's plans. At the Division and Army Commanders conference on August 4th, Minami attacked the Wakatsuki government on the grounds that it did not have the right to cut the army's budget.[79] Excerpts of Minami's speech are provided below:[80]

> It is a matter of sincere regret that in spite of the fact that we should be interested in Manchuria and Mongolia for existence and development of our Empire, from the standpoint of national defense, of political affairs, and of economy, the recent state of affairs there has been displaying unfavorable tendencies towards us, which in some cases have been making serious developments.
>
> The causes which brought about such a state of affairs originated in the change in the international political situation, in the selfish idea of restoring national prestige which our neighboring country has long been cultivating and propagating, and also in the development of newly-risen economic influences. We cannot help but recognize that this is not a passing phenomenon but an ever-lasting one.

Minami went onto endorse the defense of northeastern China as a "lifeline" to Japan, given the rise in tensions in the region.[81] He also ruthlessly criticized disarmament and the politics surrounding it, as he believed disarmament was undermining Japan's national interests; as he put it, "the outsiders and those who are not in responsible posts were advocating the reduction of armaments without good reason."[82] Such bold, open criticism was unusual for an Army Minister to level at his own government which he was nominally a part of. Indeed, the publication of a written version of Minami's speech was technically a violation of Article 103 of the Military Code.[83] Minami's willingness to violate the IJA's own regulations indicates his desperation to 1) prevent Wakatsuki's government from going ahead with disarmament and 2) find a solution to the crisis in northeastern China.

Minami's speech caused great stir in the liberal Japanese press about the military's influence over the government.[84] The Asahi Shimbun published several articles openly critiquing Minami for destabilizing Japanese politics and undermining his government. One of these articles is excerpted below:[85]

> The Ministerialists are greatly concerned about the speech made by the War Minister on the 4th at the conference of the Army and divisional commanders and they deem the speech improper, violent language in itself, especially the young members of parliament are extremely angry at this, thinking that the Minister openly challenged the Ministerialists...
>
> What the Ministerialists and younger members of Parliament say may be summed up as follows:
>
> "Now the national defense should not be the Army's own national defense but the people's national defense. The Army, nevertheless, being impervious to the facts, discarded the proper demands made by the representatives of the people, and besides, insulted us, regarding our words as traitorous. As we take it, the Army is trying to form a gulf between the Army itself and the people. This is indeed a matter of great regret for the future of the Empire."

Other articles went on to even more explicitly criticize Minami, asking "[w]as it not a revelation of the rash intentions of one group within the military to drag diplomacy toward Manchuria and Mongolia along in line with their thinking?" and further intimated that Minami had gone beyond his authority by making an explicitly political speech.[86] Wakatsuki's party, the Minseito, also criticized Minami's comments, which directly undermined the government of their party. But despite these criticisms, Wakatsuki himself did not censor or discipline Minami, going as far as to tell Shidehara that he hoped he wouldn't comment on Minami's speech. Wakatsuki's silence was largely the by product of his desire to prevent Minami from resigning as Army Minister, which would collapse his government.[87] Therefore, even though Minami egregiously overstepped his bounds, the army's central position in Japanese politics protected him from reprimand.

Soon after the conclusion of the Divisional and Regional Commanders Conference, rumors began to spread of the army overthrowing the cabinet, or alternatively, initiating action in northeastern China in August or September if it did not get its way on the budget.[88] In part, Minami himself was to blame due to his much-publicized August 4th address. But others at the Divisional and Regional Commanders Conference were also responsible for the proliferation of these rumors, most notably the new Commander in Chief of the Kwantung Army, Honjo Shigeru. During the conference, Honjo met with a variety of important figures in the AGS and Army Ministry, including Ninomiya, Sugiyama, Tatekawa, and Koiso to discuss future plans for northeastern China.[89] Honjo made it clear during this meeting that he favored taking military action at some point, even though he was still probably not aware of Ishiwara and Itagaki's plans for action at Liutiaogou in September.[90] He also may have asked Lieutenant General Hayashi Senjuro of the Korean Army if he would aid the Kwantung Army in the event it took action in the region. Apparently, Hayashi responded in the affirmative, assuring Honjo of his intent to support the Kwantung Army's efforts.[91] Hayashi's affirmation of Honjo's request meant that the Kwantung Army would likely have at least some support in the event it initiated an assault in northeastern China. But Honjo's request also brought another entity into play, and likely further fueled rumors of Kwantung Army activity in northeastern China.

The rumors about the Kwantung Army's plans for northeastern China also made their way to Saionji Kinmochi, who began to hear a growing crescendo of voices that seemed to confirm its intentions. Saionji had first heard these rumors in mid-June 1931 after Hanaya's trip to Tokyo; but after the incidents of July and the Regional and Divisional Commanders Conference in August, the rumors reached a new level of intensity.[92] They included implications that the Kwantung Army had come to an understanding with Yan Xishan, that northeastern China's railway stocks had depreciated in value due to whisperings of an attack, and gossip among junior officers that an attack on the

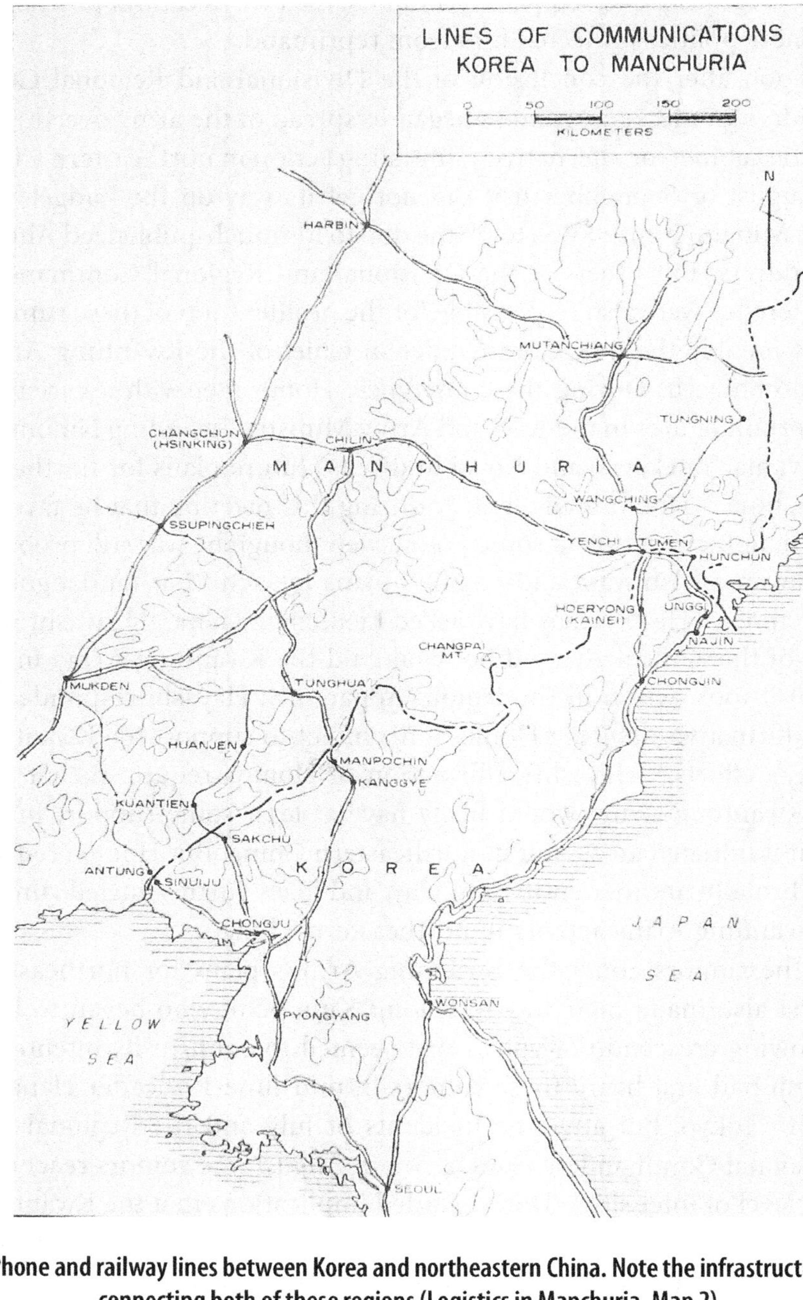

Phone and railway lines between Korea and northeastern China. Note the infrastructure connecting both of these regions (Logistics in Manchuria, Map 2).

Three Eastern Provinces was imminent.[93] The proliferation of these rumors so disturbed Saionji that he voiced his fears to his private secretary, Baron Harada Kumao, on August 18th, saying that "there is a very dangerous state of affairs in our army… I can't help but think the Imperial Army is cooking something up in Manchuria, Mongolia, and China. Admiral Okada and the Foreign Ministry are extremely worried about this too. We've just got to be more careful."[94]

Saionji's status as a *genro* gave him access to key policymakers in the Japanese government, which he used to communicate his fears. On August 19th, Harada told Wakatsuki of Saionji's fears about the Kwantung Army. On August 21st, Harada and Wakatsuki had another conversation, in which the latter told him that "I am thinking of calling the army minister and asking him to warn army circles against the tendency about which we have been concerned for some time."[95] Saionji encouraged Wakatsuki to have this conversation with Minami, and thereby confirmed Wakatsuki's belief that such a conversation was necessary. With Saionji's backing, Wakatsuki decided to arrange a meeting with Minami later on August 21st.

Once Wakatsuki met Minami, he told him that he was "very worried about the situation in our country," and asked him to help restore discipline in the Imperial Japanese Army.[96] He cited the March Plot, the radicalization of young officers, and various episodes of meddling in Chinese internal politics as evidence of the IJA's rogue tendencies. But instead of agreeing to help the civilian government restore order in the Imperial Japanese Army, Minami played coy and blamed the civilian politicians for antagonizing the army with criticism of the army's usefulness. Later, Finance Minister Inoue told Harada that Minami only pretended to comply with Wakatsuki's request and returned to making provocative statements once back at the Army Ministry.[97] Minami's evasion of the issues presented by his subordinates' radicalization was convenient for him, as he could use their plots and schemes as a form of leverage to prevent the budget cuts he opposed.[98] Nonetheless, Minami's high-wire act had swung too far in the direction of his radical subordinates, who were dangerously close to implementing their radical ideas. If not addressed, these radical activities would inevitably undermine both the Wakatsuki government and Minami's position as Army Minister.

On September 10th, Doihara Kenji of the Mukden Special Service Agency arrived in Tokyo from northeastern China. His mission was two-fold: to explain that the Kwantung Army would not tolerate any more provocations from the Chinese and ensure that the AGS and Army Ministry would support the Kwantung Army in the event of war.[99] To better promote this view in public and put pressure on his superiors, Doihara decided to give a press conference upon his arrival in Japan, where he stated that the army would have to resolve the Nakamura Incident through force if the Foreign Ministry could not reach an agreement with Zhang. More importantly, he openly stated that the Kwantung Army would have acted unilaterally in the immediate aftermath of the Nakamura Incident if not for the government's opposition.[100] That Doihara openly made these statements illustrated the extent to which the Kwantung Army's narrative on northeastern China had won over many in Tokyo. Now, it was up to him to seal the deal.

Doihara continued to push the envelope of aggression at his meetings with AGS and Army Ministry officers. At a meeting that included an audience of Minami, Sugiyama, Kanaya, Ninomiya, and Nagata, he presented a report that denied Zhang Xueliang's desire to reconcile with the Japanese on the issue of the Nakamura Incident, and urged the Kwantung Army to take the initiative.[101] Nonetheless, Minami and Kanaya both told him that the Kwantung Army should be patient, remain prudent, and avoid conflict for the time being. Doihara got a similar message from Shidehara when he met him on September 11th, as the latter sought to avoid war and continue his "friendship diplomacy."[102] But try as they might, their words were of little consequence, as the Kwantung Army was determined to follow through with its plans. Only an order from the Emperor might be able to stop the Kwantung Army, and even that may not have been sufficient.

Realizing the imminence of war in northeastern China, Hirohito summoned Minami to speak with him on September 11th. Hirohito intended to ask Minami about the state of military discipline in the IJA due to the recent rumors about the Kwantung Army's intentions in the Three Eastern Provinces. Minami anticipated Hirohito's question, and attempted to give a satisfactory answer to the Emperor's presumed inquiry. Minami stated that the young, radical officers were "criticizing the weakness of foreign policy," but that the "authorities recognize that foreign policy is national policy and do not approve their statements and actions, and they intend to adopt careful supervision."[103] Minami did concede that the young radical officers' plans and intrigues were "very undesirable from the point of view of military discipline" and that "foreign policy lay within the jurisdiction of the Foreign Ministry" and should remain unfettered from army meddling.[104] When Hirohito responded that the army's ventures into politics interfered with national policy decisions and ordered him to increase control over his subordinates, Minami reiterated that "ever since I have heard such rumors, I have been carefully controlling matters."[105] Given his address on August 4th where he openly castigated the Wakatsuki government for reducing arms spending, Minami's statement to Hirohito was misleading, as he helped fan the flames of aggression in Tokyo. But his statement was true in the sense he did not want immediate war in northeastern China, and remained unaware of Ishiwara and Itagaki's plans. Therefore, Minami's audience with the Emperor was a mix of reality and fiction reflective of his attempts to straddle the fence between the army radicals and moderate civilians.

The next day, Minami met Saionji again, and assured him of his intent to keep his subordinates under control. But Saionji was not as easily swayed as Hirohito and criticized Minami for the army's dispatch of "outlaws and fascist gangsters" to northeastern China which he believed would contribute to popular perception of the IJA as a bunch of "hoodlums."[106] He also further reminded Minami that the Three Eastern Provinces were part of China, and that Japan must not meddle so much in its affairs.[107] While Minami once again promised to "exercise my responsibility and be extremely careful," Saionji believed that Minami's lack of control over his subordinates gave the Kwantung Army the inevitable upper hand, no matter what he or Minami did.[108] Therefore, he put little stock in Minami's words, and began preparing for the fallout from military action in northeastern China.

The Kwantung Army Makes its Final Preparations

By the end of August, the Kwantung Army plotters were making their last preparations for action in northeastern China. One of their most important remaining tasks was to prime their commander-in-chief, Honjo, for the coming action. For his part, Honjo probably supported some kind of military solution in northeastern China, as evidenced by his meeting with AGS officers after the Division and Regional Commanders conference. But he was still probably unaware of Ishiwara and Itagaki's plot, which was due to take place in less than a month.[109] While Honjo's ignorance benefited the plotters, they still needed him to approve and direct offensive operations when the time came. Therefore, Ishiwara and Itagaki had the difficult task of preparing Honjo for imminent action while simultaneously not telling him of the specific nature of the events planned to unfold at Liutiaogou.

The perfect opportunity to achieve these objectives presented itself on September 1st, when Ishiwara briefed Honjo on the Kwantung Army's contingency plans. While the briefing itself was a regular meeting to get the new commander up to speed, the main action probably occurred after the briefing, when Ishiwara advocated for immediate military action to avenge Nakamura and his associates. Even though it's unlikely Honjo completely endorsed Ishiwara's proposed program, his September 1st general instructions to the Kwantung Army included lines such as "I shall act resolutely," and that the "Kwantung Army faced truly heavy responsibilities" in northeastern China.[110] Statements like these were encouraging signs to Ishiwara and Itagaki, as they realized that Honjo was primed for action. Therefore, regardless of whether or not Honjo endorsed immediate military action, his radical subordinates would capitalize on his desire to "act resolutely" and damp the fire of Chinese nationalism in the Three Eastern Provinces. All that was needed was to light the spark of war.

Even though Ishiwara had succeeded in priming his commander for the action that was to come, he still faced moments of doubt. One of these moments came on September 8th, when Consul General Hayashi asked to "discuss the situation" with Honjo during an inspection trip of the 2nd

Infantry Division.[111] Hayashi's request greatly concerned Ishiwara, as the Consul General had tried to restrain Ishiwara's plans in the past. Wary of Hayashi's watchful eye, Ishiwara told Imada, Mitani, and Hanaya that he intended to give up his previous plans to initiate a false flag attack at Liutiaogou in late September and go back to the drawing board.[112] But when Komoto arrived in northeastern China to replenish the Kwantung Army's secret funds and conveyed his disappointment to Ishiwara, the latter seemingly re-discovered his desire to go ahead with the plot. As Ishiwara put it "I wanted the rumor to spread so that it would appear that we'd given up… I haven't changed my mind for a minute."[113] Whether Ishiwara intentionally lied about his desire to nix the plot or re-discovered his desire to carry it out, the effect was the same: the Kwantung Army's plot would move forward as planned.

Ishiwara was right to worry about Hayashi, as he was privy to the rumors that were spreading about imminent Kwantung Army action in northeastern China. As early as August, Kawakami Seiichi told some Japanese civilians that the Kwantung Army was planning an incident. This information soon got back to Hayashi, who sent a telegram to Shidehara informing him of what he learned.[114] Hayashi further confirmed these rumors on September 14th, when Kawakami briefed Fushun's chief of police that he was under orders to attack Mukden airport, and that the police chief should make plans to evacuate Japanese residents. The police chief in turn informed Morishima Morita, the acting Consul General at Mukden, of his conversation with Kawakami, who then told Hayashi. Hayashi proceeded to send more telegrams to Shidehara in Tokyo trying to communicate the urgency of the situation to him.[115]

In Tokyo, Shidehara took it upon himself to tell Minami of the warnings he received from Hayashi, and added that their actions threatened to imperil Japan's national security.[116] Finally made aware of the immediate threat the Kwantung Army's plans posed, Minami called a sudden meeting at his residence including Kanaya, Sugiyama, Koiso, Ninomiya, Tatekawa, Nagata, and Imamura. Minami revealed at this meeting his belief that the Kwantung Army would not "take action without consulting army headquarters."[117] When Tatekawa concurred with Minami, the latter retorted that it seemed he had some "inside information" on the

Kwantung Army's activities. Even though Tatekawa denied this accusation, some, most notably Tanaka Ryukichi at the Tokyo trials in 1947, alleged that Tatekawa was in on the Kwantung Army's plan for action by September.[118] While it is difficult to ascertain Tatekawa's foreknowledge of Ishiwara's plans, he left the conference with the task of delivering a letter from the Army Minister and the Chief of the General Staff in-person to the Kwantung Army. This letter requested the latter remain "prudent" and warned against "rash action" in northeastern China, as government support could not be guaranteed.[119] By sending Tatekawa to the Kwantung Army, Minami was attempting to adhere to the Emperor's directives from September 11th on maintaining military discipline and keeping a tighter leash on his subordinates. Now Minami had to hope that his actions were not too little, too late.

Unfortunately for Minami, Hashimoto caught wind of his scheme and took it upon himself to send a telegram to Hanaya and Itagaki at the Mukden Special Service Agency on September 14th: "Tatekawa is expected to arrive in Mukden tomorrow, hospitable treatment will be appreciated. His mission is to prevent the incident."[120] Even though Tatekawa did not intend to arrive until September 18th, the telegram nonetheless sent the Kwantung Army plotters into a panicked frenzy, with debate raging during a meeting between the key plotters at the Mukden Special Service Agency headquarters on the night of September 14th.[121] Some, such as Hanaya, opposed action as they did not want to risk going against the AGS and being branded as traitors. Others, such as Imada, wanted to go ahead with the attack anyway. Most of those in attendance (including Ishiwara, Itagaki, Kojima, Mitani, Nagura, Kawashima, and Ono) could not decide. To break the impasse, Kawashima suggested voting. At 2 a.m. on September 15th the majority voted to hold off, but Ishiwara changed his mind after phoning Mitani later that morning and confirming the readiness of his garrison to go ahead with operations. Itagaki concurred with Ishiwara's change of heart, simply saying "Let's do it."[122] Therefore, Imada, and Kawashima were set to implement the Liutiaogou plan on September 18th, which would come to be known in history as the Mukden Incident.

Chapter 12

The Mukden Incident

By mid-September 1931, the Kwantung Army was ready to execute its false flag attack on the South Manchuria Railway. Plans had been finalized, key figures won over, and units were in position. On September 18th, Ishiwara and Itagaki gave the order to blow up the SMR and attack key cities along the railway. Two days later, the Kwantung Army ventured into Jilin province, where it received reinforcement from the Korea Army. While the AGS forced the Kwantung Army to pause its operations in October, it continued its northward push in November, attacking the Nen River Bridges and Qiqihar. In December and January, the Kwantung Army finished mopping up Zhang Xueliang's forces in the Three Eastern Provinces and seized control of the region. The Kwantung Army's control over the region was finalized with the formation of the puppet state of Manchukuo in February 1932.

During the campaign in the Three Eastern Provinces, the Kwantung Army remained a step ahead of the authorities in Tokyo. Indeed, many figures within the government and the army high command were unaware of the Kwantung Army's plans from the beginning, and likely would not have approved them if they had been aware. This reality was reflected in the immediate aftermath of the Mukden Incident, when the Japanese Cabinet passed a "non-expansion" policy on September 19th. But instead of obeying the government's directive, Ishiwara and Itagaki took it upon themselves to persuade Honjo to deploy troops to Jilin. Similar situations played out when the Kwantung Army unilaterally bombed Jinzhou and attacked Qiqihar in November. Such instances of *gekokujo* were a defining feature of the Kwantung Army's campaign in northeastern China, as Ishiwara and Itagaki were determined to absorb the whole region into the Japanese Empire. The ultimate success of Ishiwara and Itagaki's campaign not only allowed them to avoid punishment for their *gekokujo* but also set a precedent for other insubordinate officers in future years.

Division in China and Orders of Battle

Even though China had nominally been unified after the Northern Expedition in 1928, China remained quite divided. The reason China was divided was because the KMT's coalition had fractured, with one skirmish in 1929 between the KMT and the so-called "Guangxi Clique" and another between the KMT and Feng Yuxiang. While the KMT emerged victorious in these campaigns, its consolidation of power greatly alarmed both Feng Yuxiang and another former KMT-aligned warlord, Yan Xishan.[1] Consequently, fighting broke out again in 1930 between the KMT and a coalition centered around the armies of Feng and Yan. Although the combined armies of Feng and Yan greatly outnumbered Chiang's KMT forces on paper (700,000 to 295,000), the relatively poor training and equipment of these forces limited their effectiveness.[2] Furthermore, after a series of successful KMT campaigns, Zhang Xueliang joined Chiang and the KMT in fighting the Feng-Yan coalition. While Zhang's armies helped Chiang and the KMT triumph in the so-called "Central Plains War," 120,000 men were deployed away from the Three Eastern Provinces, with many of these troops remaining between Beijing and Tianjin.[3] The diversion of force away from the provinces greatly weakened Zhang's forces just as the Kwantung Army was ramping up preparations for action in northeastern China.

In addition to fighting wars with former KMT-aligned warlords, the KMT was also attempting to suppress the Communist Party of China's revolutionary "Soviets" in south-central China. Of the many Soviets that arose in south-central China, the Jiangxi Soviet was the largest and therefore became the target of repeated KMT efforts to encircle and destroy it. The KMT's first three attempts to encircle and suppress the Jiangxi Soviet came in 1930 and 1931, but failed thanks to the effective leadership of the Soviet's commander, Mao Zedong.[4] Against the converging columns of KMT divisions, Mao's forces "lured the enemy in deep," concentrated force against weaker columns, and attacked their overextended flanks and supply lines. After ambushing and encircling their unsuspecting victims, Mao's forces utilized interior lines and the cover of darkness to suddenly shift their forces and attack other KMT columns.[5] These attacks were complemented by smaller forces that performed economy of force or decoy missions. Mao's utilization of asymmetric warfare was so effective that the Jiangxi Soviet managed to survive well into the 1930s, despite considerable deficiencies in manpower, training, and equipment.

The survival of Communist resistance in south-central China, alongside the unstable relations with warlords in the far north and south, left Chiang in a difficult position. Chiang could ill-afford another conflict with Japan when he was still surrounded by his domestic enemies; consequently, in the summer of 1931, Chiang repeatedly advised Zhang to avoid provoking the Japanese at all costs.[6] After Chiang started hearing that action was imminent via Zhang's commanders in September, he issued a non-resistance order to Zhang.[7] Zhang in turn issued this order to his forces on September 6th.[8] Chiang's rationale for issuing the order was that the League of Nations would intervene to prevent Japan from seizing northeastern China if the Kwantung Army attacked.[9] Even if the League failed to prevent Japan's aggression, China would still be better off building strength in preparation for the US, UK, and USSR to enter a war against Japan as opposed to fighting alone.[10] These views led Chiang to adopt his "first pacification then resistance" policy, which prioritized the destruction of Chiang's domestic enemies before resisting Japanese aggression. While politically convenient for Chiang, many Chinese came to resent this policy, as it effectively gave free rein to the Japanese in north and northeastern China.

Despite Chiang Kai-shek's unwillingness to fight the Japanese, Zhang Xueliang's forces were still formidable on paper. While 110,000 to 120,000 of Zhang's troops were in north China to secure Hebei province from Feng's armies, Zhang still had roughly 130,000 to 140,000 soldiers in the northeast.[11] By Chinese standards, these forces had significant firepower, with 216 guns organized in nine artillery regiments.[12] They also had tanks and 4,000 machine guns.[13] Zhang's forces even had an air force with roughly 100 aircraft at their disposal.[14] The impressive number of men and machines in Zhang's order of battle gave it some combat power, despite Chiang's non-resistance order. The Kwantung Army was therefore forced to reckon with the possibility that Zhang's forces might fight.

The Kwantung Army, for its part, was a far more diminutive force. On the night of September 18th, it numbered just over 10,000 men.[15] The 2nd Infantry Division formed the core of the Kwantung Army's strength, with four regiments under the command of Lieutenant General Tamon Jiro amounting to roughly 5,400 men and 16 field guns.[16] Of these forces, the 29th Infantry Regiment was garrisoned in Mukden, while the 4th Regiment had its headquarters in Changchun, the 16th in Liaoyang, and the 30th at Port Arthur.[17] The latter three regiments' strength were dispersed throughout the SMR area, including Yingkou and Andong. The division's greatest weakness was in artillery, as the division only had 16 75mm field guns at its disposal. Other weaknesses included a lack of trucks and engineers, which greatly limited the division's logistical capabilities.[18]

The Kwantung Army also consisted of six railway guards battalions—one battalion had four companies each—under the command of Lieutenant General Mori Muraji.[19] These forces numbered roughly 5,000 men.[20] Of the railway guards battalions, the 2nd Railway Guards Battalion was the most important because it was earmarked to execute Ishiwara's planned attack on the Beidaying Barracks with three of its companies. Of these companies, the 1st and 4th would come from the area near Mukden, while the 2nd Company would come from the area around Fushun. These forces would converge on the Beidaying Barracks and seize them in a surprise night assault. The new commander of the 2nd Railway Guards Battalion, Lieutenant Colonel Shimamoto Masaichi, would oversee the assault, as he had taken over command from Lieutenant Colonel Takagi Yoshito on August 1st, 1931.[21]

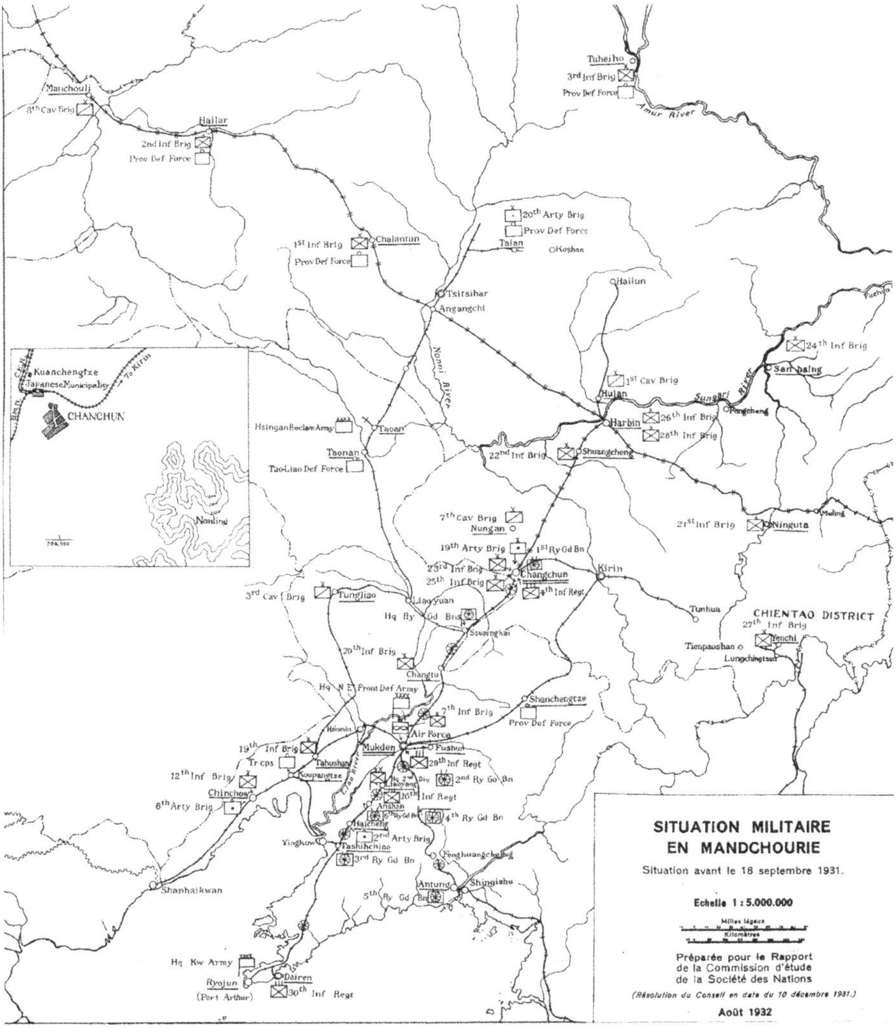

Disposition of Chinese and Japanese forces on the night of September 18th, 1931. (Library of Congress).

Last Preparations for Action

On September 16th, Ishiwara, Itagaki, and the rest of the Kwantung Army staff accompanied Honjo on a train from Mukden to Liaoyang where they would inspect the 2nd Infantry Division's exercises the next day.[22] On the surface, the Kwantung Army's inspection tour went as planned, with Honjo and his staff officers observing the 2nd Infantry Division's maneuvers. Honjo also gave his planned address to the 2nd Infantry Division.[23] The next day, September 18th, Honjo made his last comments to Tamon, stating that "with the present uneasiness in Manchuria and Mongolia growing daily more pronounced, we cannot be complacent… If violence should break out, I trust each unit will take resolute action and adhere to its purpose with unwavering devotion."[24] Little did Honjo realize how soon his words would become a reality.

While Honjo was on his inspection tour, the various Kwantung Army plotters were maneuvering behind his back to rapidly implement their plan of action. On September 17th, Imada told Hanaya of the next day's plan to commence action; then, Hanaya began making his last preparations.[25] These included contacting patriotic adventurers like Wada Tsuyoshi to conduct attacks on Japanese residents and property outside the SMR zone. Captain Amakasu Masahiko would also play a critical role in the Kwantung Army's efforts, and received 30,000 yen from Komoto for the purposes of creating disturbances in Harbin.[26] These disturbances would provide further justifications for the Kwantung Army to expand hostilities outside the immediate sphere of the SMR zone in the name of protecting Japanese civilians and their property.

In Liaoyang, Honjo learned from his chief of staff, Miyake, that Tatekawa would arrive in northeastern China the next day. This information soon made its way to Ishiwara and Itagaki who had a late-night conversation in the former's hotel room on the night of September 17th. During this conversation, they both agreed that Itagaki would meet Tatekawa and assume responsibility for events in Mukden, while Ishiwara would be in Port Arthur with Honjo.[27] This division of labor ensured that Itagaki would have the task of overseeing events near the site of action, while Ishiwara would be back at headquarters attempting to persuade his boss to support his venture.

On the morning of September 18th, Itagaki left Liaoyang by train to meet Tatekawa. Later that day, at 2 p.m. Honjo, Ishiwara, and the rest of the Kwantung Army staff officers with Honjo boarded a train back to Port Arthur. Before Itagaki left Liaoyang, Ishiwara told him, "[i]n view of the decision reached at Mukden last night, we cannot very well back out now. We will have to go through with it." To this, Itagaki replied, "[t]hat is the way I feel about it. I am glad that we both see it the same way. I'll take care of matters in Mukden, so will you look after matters in Port Arthur?"[28] Thus Ishiwara and Itagaki finalized two years of collaborative planning for an incident in northeastern China.

Tatekawa Arrives

At 11:29 a.m., Tatekawa arrived at Penchi, where he awaited Itagaki's arrival.[29] Itagaki did not arrive at Penchi until later in the afternoon, as he was busy making last-minute preparations for the explosion to occur later that night. When Itagaki finally arrived at Penchi, he was met by a beleaguered Tatekawa who briefly informed Itagaki of his superiors' concern about the young officers' plans and intentions. Itagaki assured him that he did not need to worry about such issues. Then, citing fatigue from the long journey, Tatekawa said that he needed rest before he could move to discuss important business. Itagaki therefore proposed meeting again the next day to discuss the business Tatekawa had with the Kwantung Army.[30]

Itagaki and Tatekawa boarded the train to Mukden at 5:18 p.m. and arrived at the Mukden station at 7:05 p.m., where they were greeted by Major Hanaya.[31] From the station, Itagaki, Hanaya and Tatekawa, went to the Kikubun inn, where Tatekawa could have an evening of rest and relaxation. Once they reached the inn, Tatekawa made a last reference to his business in northeastern China when he asked Itagaki if the young Kwantung Army officers were "stirred up tremendously by the Nakamura Incident." Itagaki once again denied this charge, to which Tatekawa said "I'm certainly glad to hear that. I'm worn out now, so let's go on tomorrow." This conversation, which was a nod towards Tatekawa's real business, was likely for Hanaya's benefit, to signal the purpose of Tatekawa's visit while also reiterating his desire to have an evening of rest and recreation.[32]

Tatekawa's fatigue played into the hands of the Kwantung Army plotters, who intended to distract him while their plans were going into effect. As part of their plan, Itagaki and Hanaya had dinner with Tatekawa at the inn, after which Itagaki took leave to arrive at the Mukden Special Service Agency Headquarters at 9 p.m.[33] Itagaki's departure left Hanaya with Tatekawa, who entertained themselves with alcohol and geisha girls. According to Hanaya's account after the war, it was at this time that he tried to ascertain whether or not Tatekawa would intervene to prevent the Kwantung Army from taking action in northeastern China. While Tatekawa did not give him a straight answer, Hanaya came away with the belief that he did not intend to intervene.[34] Regardless of Tatekawa's intentions, he was heavily drunk and asleep by 9 p.m., content with food and sake.[35] Tatekawa's drunken state was a great success for Hanaya, who succeeded in distracting Tatekawa from giving his letter to Honjo and preventing the attack.

While Tatekawa's exact motives and intentions on September 18th remain unclear, there is evidence to suggest that he deliberately turned a blind eye to the Kwantung Army plotters' activities. According to the International Military Tribunal for the Far East (IMFTE) judgement, Tatekawa later told a friend that he allowed himself to be diverted to the inn because he did not want to put a stop to Ishiwara and Itagaki's schemes.[36] Tanaka Ryukichi's affidavit at the IMFTE goes even further than the IMFTE judgement, as Tanaka said that Tatekawa knew Ishiwara and Itagaki's plans were about to go into motion and fully supported them. He even went as far as to say that Ishiwara and Itagaki suggested to Tatekawa that he spend the night at a "teahouse" (presumably the Kikubumi inn) so that he did not give his letter to Honjo, which Tatekawa apparently agreed to do.[37]

Even though Tanaka's affidavit takes liberties with some facts (e.g. Ishiwara and Tatekawa being together on September 18th), his allegation that

Tatekawa supported some form of military action in northeastern China was true. As a member of the Sakurakai and AGS, Tatekawa had used his position to advocate for military action in the Three Eastern Provinces, drawing up documents and studies that included such a measure among its recommendations.[38] He had also agreed with Hanaya's proposal for the Kwantung Army to initiate military action in June. Therefore, some historians believe Tatekawa abetted the execution of the Kwantung Army's plot by not promptly delivering Minami's letter to Honjo.

Regardless of Tatekawa's intentions, Hanaya and Itagaki had succeeded in preventing him from delivering Minami's letter to Honjo in Port Arthur. But Tatekawa's presence in northeastern China became a subject of controversy when he was discovered by a Japanese civilian while on the train from Penchi to Mukden. Even though the civilian did not succeed in getting Tatekawa to acknowledge his identity, the Andong consular police soon heard of the incident and sent a dispatch to Morishima in Mukden.[39] Similar incidents followed, as Tatekawa met the director of the South Manchuria Railway corporation in a hotel, and a Taiwanese recognized him on a train during his return trip.[40] Consul General Morishima and SMR Director Kimura Eiichi both feared that Tatekawa's presence in the northeast foreshadowed a Kwantung Army offensive, especially given the various rumors they had been hearing in the previous days and weeks. They urged Consul General Hayashi to act and prevent the Kwantung Army from going through with its plans.[41] While Hayashi decided to write Honjo a request to remain vigilant in the face of his scheming subordinates, his message fell into the hands of Katakura Tadashi, who supported the plotters. Consequently, Honjo did not receive Hayashi's message until well after the evening of September 18th.[42]

A Single Spark: Action at Liutiaogou

Near the site of action at Liutiaogou, Imada finally told his subordinate lieutenants—Komoto, Noda Sukuo, and Tamura Tadashi—of the planned explosion they were to trigger along the SMR later that evening.[43] Of these men, only Komoto knew of the plan in advance, as he was the one charged with laying the explosives. Soon after Imada informed his lieutenants of the plan, they inspected the equipment of Kawashima's 3rd Company, which then took up positions facing Wenkuantun, roughly three kilometers away from Liutiaogou. Kawashima took 105 men and set up a headquarters in the town of Wenkuantun itself.[44] The 3rd Company's role was to deploy to the scene of the action at Liutiaogou, and help assault the Beidaying barracks from the north. The company's rapid response would be critical to the execution of Ishiwara's plan, as its success was predicated on rapidly overwhelming the garrison at the Beidaying Barracks.

By 10 p.m. on September 18th, Komoto and his contingent of seven or eight men were at Liutiaogou with their explosive powder. The next train to Mukden was set to pass Liutiaogou around 10:30, where it would arrive around 10:40.[45] Ishiwara, Itagaki and the other plotters intended for this train to be uprooted by the blast so that the Kwantung Army would have a *casus belli* to start its attack. But Komoto and the others proximate to the action did not want a large blast, because the fallout from the destruction of the train would inevitably lead to civilian casualties and divert the 2nd Independent Railway Guards Battalion away from its assault on the Beidaying Barracks. Therefore, Komoto hoped that the explosion he would trigger did not overturn the Japanese train, whatever the desires of his superiors.[46]

At roughly 10:20 p.m. on September 18th, 1931, Komoto lit the fuse that triggered an explosion on the South Manchuria Railway. Much to Komoto's relief, his explosion did not even damage the tracks enough to derail the southbound train due to arrive at Mukden at 10:40.[47] After the explosion, Komoto sent a runner, Private First Class Konno, to inform Kawashima of the explosion.[48] Some of Komoto's men also called the Special Service Agency and the headquarters of the 2nd Battalion from field telephones to tell them that Chinese troops from the Beidaying Barracks had blown up the SMR.[49] Simultaneously, a firefight broke out between Komoto and some Chinese who had come out of their barracks due to the commotion of the explosion. This firefight and the explosion that preceded it initiated a conflict between China and Japan that was to last fourteen years.

Action at Beidaying

The precise nature of the events surrounding the assault on the Beidaying Barracks remain hazy, as different accounts give different explanations as to what happened. Nevertheless, according to all accounts, events after the explosion moved quickly. After learning of the explosion via telephone, Colonel Shimamoto ordered three companies from south of the Barracks at Mukden (the 1st and 4th) and Fushun (the 2nd) as well as one from the north of the Beidaying Barracks (the 3rd) to move along the SMR to and seize the Barracks, as per Ishiwara's plan.[50] The 3rd Company, already moving south after hearing the explosion, arrived around 10:50 p.m. The 1st and 4th Companies from Mukden arrived just after midnight. These companies numbered 500 men in total. They were facing a total of roughly 7,000 Chinese in the Beidaying Barracks and its immediate vicinity.[51]

Even though the Chinese outnumbered the Japanese forces by a significant margin, the Japanese were able to seize the offensive initiative due to the unpreparedness of Zhang's forces. In line with Chiang Kai-shek's "first pacification, then resistance" policy, Zhang Xueliang ordered his forces not to put up resistance in the event of an attack on the Beidaying Barracks; instead, the 7th Brigade—the unit quartered within the Barracks—was to withdraw. The fact that most weapons were locked in storage, to the point where even on-duty sentries were only "armed" with dummy rifles illustrates the extent to which the Chinese were not prepared for an attack on the night of September 18th, 1931.[52] The pattern of Chinese unpreparedness would continue in other parts of northeastern China in the coming days and weeks.

A few minutes before 11 p.m., the 3rd Infantry Company began its assault on the left (north) flank of the Beidaying Barracks.[53] At first, the assault was impeded by water patches that occupied much of the 250 yards between the SMR and Beidaying Barracks. But inside the barracks, the lights were on, which illuminated the silhouettes of the defenders and exposed them to fire from their Japanese adversaries, who by comparison had good cover in the darkness.[54] The silhouetted interior of the Barracks helped guide the 3rd Infantry Company's firing, and helped them establish fire control over the barracks. Thanks to the company's withering fire, the Chinese in the Barracks withdrew in a panic. Consequently, the 3rd Infantry Company was able to occupy the left wing of the Beidaying Barracks.

After arriving at the Beidaying Barracks around midnight, the 1st and 4th Companies joined the 3rd Infantry Company's assault, with the 1st Company attacking on the right and the 4th Company in the center.[55] Like their 3rd Company compatriots to the north, the 1st and 4th companies were able to fire into the illuminated barracks complex and establish effective fire control over their assault sectors. At 1 a.m., the Japanese used a cannon the Chinese left in an outhouse to directly fire two rounds into the south gate of the barracks. These two shells blew it open, and the 1st Infantry Company stormed through soon thereafter. Now inside the Barracks Complex, the Japanese began securing the exits. While this process was mostly without resistance, Japanese forces in the northeast corner of the barracks blocked the exits nearest the Chinese 620th Infantry Regiment, which was still attempting to withdraw. Consequently, the 620th Infantry Regiment was forced to fight its way out from the Barracks. Even though the regiment was able to fight its way through, the Japanese secured the Barracks, establishing complete control by 6 a.m. on September 19th.[56]

Assaults on Key Cities

As the 2nd Independent Railway Guards battalion attacked the Beidaying Barracks, events proceeded elsewhere along the SMR. At 10:40 p.m., Colonel Hirata was notified about the explosion on the SMR from Lieutenant Colonel Shimamoto via telephone.[57] Shimamoto then told Hirata of his intention to begin an attack on the Beidaying Barracks. After receiving Shimamoto's phone call, Hirata ordered his command, the 29th Infantry Regiment, to mobilize and prepare for action. Soon thereafter, he saw Shimamoto riding towards him on horseback, and had an exchange with him in which the former shouted "I am going to attack the enemy at Beidaying," to which Hirata retorted "I am going to attack the enemy at the Mukden installations."[58] This call-and-response illustrated the coordination between different units within the Kwantung Army, a key reason why they were able to attack multiple objectives almost simultaneously.

Mukden represented an important objective for the Kwantung Army, as it lay along the SMR and was the gateway to the rest of the towns and cities along the railway to the north. For this reason, the Kwantung Army's "Basic Plan of Operations for Manchuria" stated that "the army will press towards Mukden with all its strength, deliver a united attack against the center of the Mukden army, settle that garrison's fate, and resolve the matter with the utmost expedition."[59] But Hirata and Shimamoto's joint decision to attack Mukden from within represented a slight deviation from this plan. Soon after they mobilized Japanese forces in Mukden (part of the 29th Infantry Regiment), they received a summons to the Mukden Special Service Agency Headquarters where Itagaki awaited them. Much to the surprise of Shimamoto and Hirata, Itagaki issued an "army order" that affirmed Shimamoto and Hirata's decision. He also arranged for the dispatch of the 3rd and 5th Independent Railway Guards Battalions, which would fall under Shimamoto's command. Itagaki's order provided Shimamoto and Hirata with the necessary manpower to overwhelm the unsuspecting Chinese garrison and seize the city, from within and without.

The Chinese garrison in the vicinity of Mukden was roughly 15,000 strong with 40 guns, significantly outnumbering Hirata's 29th Infantry Regiment. But despite their numerical inferiority, Hirata's regiment took advantage of the element of surprise to attack the Chinese garrison. The Chinese forces within the city, shocked by the 29th Infantry Regiment's sudden offensive, did not offer much organized resistance.[60] The Japanese had scaled Mukden's walls by 2:15 a.m., seized the city center by 3:40 a.m., captured the Eastern wall by 6:00 a.m., and finally captured the arsenal and aerodrome by 7:30 a.m. The fall of these critical installations in such a short time illustrates the effectiveness of Ishiwara's prior reconnaissance in identifying key points in Mukden and the weaknesses in the Chinese defenses. More importantly, the Kwantung Army had secured the seat of Zhang's power in the region, and

The Kwantung Army's entry into Mukden. Note the presence of armored cars. (Author's Collection, taken at the Museum of War of Chinese People's Resistance Against Japanese Aggression)

the critical node that would allow it to continue advancing in northeastern China.

While the 29th Infantry Regiment conducted its attack in Mukden, the rest of the 2nd Infantry Division began moving its forces towards Mukden in line with the Kwantung Army's "Basic Plan of Operations."[61] The 2nd Infantry Division's Headquarters, the 16th Infantry Regiment (Liaoyang) and the 30th Infantry Regiment (Port Arthur) all departed for Mukden in the wee hours of September 19th and arrived in time to conduct an assault on the East Barracks, five miles northeast of Mukden.[62] With the help of an artillery battery, they forced the Chinese to withdraw from barracks east of the city at 2:30 a.m. on September 19. But soon after the artillery of the 2nd Infantry Division forced the Chinese to withdraw from the "East Barracks," Lieutenant General Honjo (who had established his headquarters in Mukden to better control the fighting) ordered the Japanese units outside the city to come into the city and defend Japan's railway installations. Honjo's order was carried out on September 19th, 1931.[63]

After the fall of Mukden, many Chinese greatly feared the new status quo as evidenced by the fact that 100,000 civilians fled the city.[64] Furthermore, many small businesses closed down and many key facilities stopped operating.[65] In some cases, scattered acts of resistance broke out, as some in Mukden sought to resist the Japanese on their own accord.[66] These acts of resistance directly led to the deaths of six government officials and eight civilians in the immediate aftermath of Japan's occupation.[67] More importantly, the continuing hostilities ensured that many of the local residents, both Chinese and foreigners did not resume business as usual. If this state of affairs persisted, Japan would have a difficult time administering Mukden, thereby putting its control over the city in jeopardy. Therefore, it was of paramount importance for the Kwantung Army to establish control over the city and establish a municipal government to oversee everyday functions.

Fortunately for the Japanese, Chinese resistance in the city after September 19th was scattered and ineffectual, allowing Kwantung Army's forces to avoid protracted fighting.[68] Instead, these forces conducted routine patrols, and evacuated Japanese civilians.[69] The Kwantung Army's relatively peaceful restoration of order helped them win over some elite influential Chinese, who began to willingly collaborate with their new overlords.[70] In turn, life slowly began to return to normal in the city, especially after Doihara Kenji was appointed municipal governor on September 21st.[71] The combination of Doihara's military administration and the co-optation of local elites allowed the Kwantung Army to make a relatively rapid transition to governing their newly seized territories, despite the feelings of many local people. Thanks to this method's success, it would be repeated in other cities and provinces later on.

Further north at Changchun, the Kwantung Army also took the offensive on the night of September 18th. Here, the Kwantung Army had a garrison of roughly 900 men, including the 1st Railway Guards Battalion and two battalions of the 4th Infantry Regiment, 2nd Infantry Division under the command of Major General Hasebe. Even though they were far outnumbered by the Chinese, who numbered roughly 5,000 men and 36 guns, Hasebe ordered his units to attack after receiving notice of mobilization.[72] But because Changchun and its surrounding area had multiple important objectives, Hasebe had to split his force in multiple directions.[73] He therefore assigned two companies of the 2nd Battalion, 4th Infantry Regiment (roughly 200 men in total) to attack the so-called Nanling Barracks (4km south of Changchun) while he assigned the rest of the 4th Infantry Regiment to attack the Kuanchengtze Barracks. The forces attacking the Nanling Barracks would be opposed by a Chinese garrison of roughly one 2,300-strong infantry regiment, which had a supporting artillery regiment of 30 guns; the forces attacking the Kuanchengtze would be opposed by roughly 600 infantrymen. Once again, even though the Japanese were outnumbered, they would rely on the advantages of stealth, surprise, and night combat to overwhelm their enemies, who were not prepared to fight.

The initial assault on Nanling went well, as the Japanese were able to surprise the Chinese in the northwest part of the Barracks complex and destroy 16 guns in a night attack.[74] But interlocking fire limited their progress towards adjacent barracks complexes, and the two Japanese infantry companies were forced to retreat to a nearby village. In the early morning hours of September 19th, the 2nd Battalion, 4th Infantry Regiment received reinforcements in the form of two companies from the 1st Independent Railway Guards Battalion.

After receiving reinforcements, the Japanese renewed their assault on the Nanling Barracks, with the 2nd Battalion, 4th Infantry Regiment's two companies attacking on the left and the 1st Independent Railway Guards Battalion's two companies attacking on the right.[75] Once again, the two companies from 2nd Battalion, 4th Infantry Regiment ran into heavy rifle, trench mortar, and gun fires. While these two companies were able to slowly close the distance between themselves and the defenders of the Nanling Barracks, they were not able to get close enough to storm the complex. By contrast, their compatriots on the right were able to advance within 100 yards of the southeast corner of the barracks relatively quickly, in large part due to the cover and concealment offered by tall stalks of sorghum. Even though the 1st Independent Railway Guards Battalion's two companies could not completely close the distance between themselves and the defenders of the Nanling Barracks on the right (owing in large part due to a 3 meter high, 40cm thick wall) their advance put significant pressure on the defenders of the barracks, forcing the Chinese to withdraw at 4 p.m. on September 20th.[76] Soon thereafter, the 2nd Battalion, 4th Infantry Regiment seized the Nanling Barracks, thereby accomplishing its objective.

The rest of the 4th Infantry Regiment also had a difficult time securing its objective at Kuanchengtze.[77] To a great extent, the 4th Infantry Regiment's difficulties were rooted in the cover brick buildings afforded the Chinese defenders. The thick bricks protected the Chinese from Japanese small arms fire, and thereby made it possible for them to rain fire upon the Japanese advancing over open ground. For the Japanese, the combination of these factors meant that the 4th Infantry Regiment took heavy losses in the course of their assault.[78] While the 4th Infantry Regiment finally took a corner of the barracks by 6 p.m. on September 19th, it stopped its advance at roughly 10 a.m. on the 20th. During this pause, it brought up mountain guns to within 500 meters of the barracks, allowing the infantry to advance under the cover of artillery. With the help of direct artillery fire, the 4th Infantry Regiment finally seized the Kuanchengtze Barracks at 11:10 a.m.[79]

Command Controversy

Despite the persistent rumors surrounding the Kwantung Army, most officials in Tokyo were caught off guard by the sudden outburst of fighting on September 18th. As the premise of Ishiwara and Itagaki's attack was that the Chinese had attacked the SMR, the high command, Foreign Ministry and the Cabinet were forced to parse through a litany of information to try and get to the bottom of what actually happened. Simultaneously, they were forced to respond to the rapid pace of events on the ground, which Ishiwara, Itagaki, and much of the rest of the Kwantung Army staff ceaselessly tried to accelerate. While key figures at all levels of the army, government, and foreign ministry tried and, in some cases, succeeded in halting the Kwantung Army's attempts to expand the incident, these successes were by and large fleeting. From September 18th until February 1932, the Kwantung Army had the initiative, and would press it until it brought the whole of northeastern China under the yoke of the Japanese Empire.

At roughly 10:40 p.m. on September 18th, Consul General Morishima received a phone call from the Army Special Service Agency in Mukden that told him an explosion occurred on the SMR and summoned him to the Army Special Service Agency headquarters in Mukden.[80] Once Morishima arrived at the Army Special Service Agency headquarters, he found Itagaki, Hanaya, and the rest of the office staff busy at work. Itagaki told Morishima that the Kwantung Army had been mobilized because the Chinese had attacked the SMR, and that the Japanese needed to take military action to counter their aggression.[81] When Morishima then asked on whose authority the Kwantung Army had been mobilized, Itagaki responded that "since it was an emergency situation and the commanding officer was in Port Arthur, I issued the order on his behalf."[82] Morishima countered this reply by reminding Itagaki of the need to pursue a peaceful diplomatic settlement and avoid conflict. But Itagaki was not willing to accept Morishima's point, shouting "does the Consulate General wish to interfere with the prerogative of the Imperial Command after it has been invoked?"[83] At this point, Hanaya, who had been watching the exchange between Morishima and Itagaki, drew his sword and said that he would not tolerate any interference with the imperial command, threatening to kill anyone who dared intervene.[84] Morishima, realizing that Itagaki and Hanaya were prepared to see through an offensive in northeastern China, gave up trying to reason with them, and went back to the Consulate General, where he told Hayashi of his experience.[85]

Hayashi, who had been slow to realize the danger of the Kwantung Army's plans, did not immediately inform Shidehara of Morishima's experience. He instead put forth the Kwantung Army's narrative about the attack on the SMR and subsequent events. In part, Hayashi's initial willingness to put forth the Kwantung Army's narrative in his communications with Shidehara can be attributed to the suddenness of the Mukden Incident and the fighting at Beidaying. Consider the following telegrams below:[86]

> Showa 6 [1931], September 19—Secret
> To: Minister of Foreign Affairs, Shidehara
> Telegram No. 619 (Urgent)
> From: Mukden Consul General, Hayashi
> Re: Telegram No. 616 (Top Priority)
>
> At 10:00 p.m. on the 18th, in response to the instructions from the Kanto Army Headquarters, the Special Services Agency in Mukden dispatched Lieutenant Colonel Itagaki and others.
>
> At 10:30 p.m., 400 soldiers from the Beidaying garrison advanced under the direction of the commanding officer toward the southwestern railway line and blew up the tracks near Liutiaogou.
>
> Shortly thereafter, our patrolling troops discovered the enemy troops and engaged in battle near the Tiger Stone area. Our forces (600 troops from the Beidaying garrison) engaged with 500 enemy soldiers from the northwest direction and seized part of the Tiger Stone sector.
>
> Currently, all available garrison troops are being reinforced and dispatched to the front line.
>
> Preparations for dispatching to both Nanjing and Beiping (Peking) are underway.
>
> Message:
>
> At 11:15 a.m. on the 18th, Japanese consular police in Beidaying, Mukden, were surrounded and attacked. According to a report from the Japanese Consul, in response to the attack by Chinese troops, the Japanese side engaged in defensive combat and expelled the enemy.
>
> Japanese forces were dispatched from the Mukden garrison, and it appears they are now en route to both the northern and southern branches (Beiping and Nanjing).

The urgency of the crisis and need to communicate with Shidehara likely led Hayashi to merely regurgitate the Kwantung Army's narrative of what happened, because it was the only information available. But as the night went on and Hayashi received more information, he probably began to question the Kwantung Army's narrative. The most important information Hayashi received came in the form of a series of calls from Zhang Xueliang's advisors indicating that the Chinese were not fighting and pleading with him to get the Kwantung Army to stop its onslaught. Hayashi was less content to regurgitate the Kwantung Army's narrative after receiving these calls, and sent the following telegrams to Shidehara:[87]

> Showa 6 [1931], September 19—Secret
> To: Minister of Foreign Affairs, Shidehara
> From: Mukden Consul General, Hayashi
> Telegram No.: 620 (Top Priority)
> Re: Telegram No.: 619 (Second Report)
>
> Following the previous report, the Chinese military and the Fengtian side (Zhang Xueliang's forces) began to withdraw to the west side of Mukden.
>
> In response, our army and the Fengtian side jointly began to occupy the eastern side of the city.
>
> We are now in a situation where there is no active resistance.
>
> I plan to monitor whether the Chinese side intends to expand its response and, if necessary, proceed accordingly.
>
> I will report further developments.
>
> Showa 6 [1931], September 19—Secret
> To: Minister of Foreign Affairs, Shidehara
> From: Mukden Consul General, Hayashi
> Telegram No. 623
> Re: Telegram No. 618, Second Report
>
> At around 5 a.m. on the 19th, Japanese forces surrounded Beidaying.

> According to the Japanese military attaché, Chinese forces adopted a non-resistance policy and abandoned Beidaying, retreating outside the North Gate.
>
> Consequently, Japanese forces occupied Beidaying without resistance.
>
> However, there are still a number of Chinese soldiers and civilians inside the city, and there is a risk that nighttime disorder and looting could occur.
>
> We must absolutely prevent harm to foreign nationals and their property.
>
> Accordingly, I have requested that Japanese forces and the Fengtian side jointly ensure the protection of such persons and property.
>
> To avoid incidents, I also advised that troops remain stationed on the Japanese and Fengtian side near their respective consulates and business districts.

Both telegrams indicated Zhang Xueliang forces' non-resistance policy, in effect at both Beidaying and Mukden. Hayashi's inclusion of these nuggets subtly undermined the Kwantung Army's narrative that the Chinese were aggressors, because if the Chinese had initiated hostilities they likely would have been inclined to fight. Instead, the Chinese retreated in haste from Beidaying, and put up little resistance in Mukden. In light of Hayashi's discoveries, the Kwantung Army's whole narrative surrounding Zhang Xueliang's supposed aggression came into question.

Hayashi attempted to call Itagaki and communicate Zhang Xueliang's desire to "settle the incident amicably" and the need to not "aggrandize the incident" due to China and Japan not officially being at war.[88] But Itagaki simply responded that because the situation concerned the "prestige of the State" that it was the "army's intention to see it through thoroughly."[89] This statement from Itagaki essentially ended all talk of peacefully settling the Incident. No matter what Hayashi or anyone associated with the Foreign Ministry said, the Kwantung Army would continue prosecuting its campaign until northeastern China was completely conquered.

In light of the Kwantung Army's determination to continue fighting, Hayashi decided to send a series of telegrams to Shidehara describing his exchanges with Itagaki and his assessment of the situation.[90] One of these telegrams, sent on September 20th, can be seen below:[91]

> When I met with Commander Itagaki and Chief of Staff Ishimoto, I conveyed the Foreign Ministry's position on the matter.
>
> They both listened attentively, but insisted that considering future developments, certain measures would inevitably be required.
>
> Commander Itagaki emphasized that from now on, in the event of any disturbances involving Chinese forces or civilians, he would not hesitate to take military action.
>
> He also stated that once Haicheng (海城) was occupied, they would need to coordinate with the civil authorities regarding protection of residents.
>
> If resistance occurred at Fengtian Station, retaliation would be unavoidable.
>
> They are currently mobilizing troops accordingly.
>
> In the meantime, we have observed that preparations are being made to lay railway track up to Haicheng.
>
> I will continue to monitor and report the situation in detail.

In subsequent telegrams, Hayashi also told Shidehara of Itagaki's unwillingness to stop the incident despite repeated requests, saying that the latter failed to show even the "slightest sign of complying with my request."[92] Hayashi's understanding of this reality (largely due to his conversation with Machino) led him to later communicate to Shidehara that the Kwantung Army "seems to follow the policy of immediately embarking upon positive action along the Manchurian railway zone" and that the army's policy was to "fight it out."[93] Hayashi also noted how the Kwantung Army forbade the SMR from sending repair men to the site of the explosion, concluding that "it is considered that the recent incident was wholly an action planned by the army."[94]

In Port Arthur, Lieutenant General Honjo Shigeru, commander of the Kwantung Army, only learned of the explosion on the SMR from newspapermen around 11 p.m.[95] Miyake, Honjo's Chief of Staff, learned about the Mukden explosion even later than Honjo did, at 11:46 p.m. via telephone (he did confirm the occurrence of the explosions to Honjo).[96] After receiving Itagaki's first report about action at Beidaying, Miyake gathered a variety of Kwantung Army staff officers, including Arai Masao, Takeda Hisashi, Nakano Ryoji, Katakura Tadashi, and Takeshita Yoshiharu. Katakura and Arai, who were aware of Ishiwara's plot, told their compatriots that they should not "let Itagaki and Ishiwara down the way we did Komoto Daisaku," to which the others agreed.[97]

After their brief gathering, the various Kwantung Army staff officers joined Ishiwara in trying to convince Honjo to expand the incident into a full-scale invasion of the Three Eastern Provinces. Honjo was initially not prepared to commit to such a drastic step and refused to go along with Ishiwara's proposed program. As Katakura put in his diary, "the commander in chief had not yet decided upon an offensive, and planned only to mobilize the Army in Mukden and to undertake disarmament."[98] While Honjo's staff insisted on the necessity of a more sweeping offensive, extending to Yingkou, Fenghuangcheng, and Changchun, Honjo merely ordered the concentration of Japanese units in the Mukden area as rapidly as possible.[99] Honjo's staff feared that his inability to seize the moment would hamstring Ishiwara and Itagaki's plans, as they would be forced to obey orders from the AGS and government in Tokyo the following day.[100]

At first, it seemed that the deployment to Mukden to disarm the Chinese forces would be the limit of what Honjo would approve. However, at roughly 12:20 a.m. on September 19th, Honjo received a call from the Mukden Special Service Agency that the Chinese were moving up reinforcements and that the fighting had spread to other areas. Realizing that this was their chance, Ishiwara and the rest of the Kwantung Army staff bombarded Honjo with calls to action, telling him he should "mobilize the whole of the Japanese military might to seal the fate of the enemy in the shortest possible time."[101] Ishiwara himself told Honjo that the Kwantung Army needed to "resolutely make ourselves the master of the enemy garrison."[102] After thinking about his officers' refrains for a brief time, Honjo uttered the fateful words: "yes, let it be done on my responsibility."[103] With these words, Honjo gave his approval to a full-scale invasion of northeastern China that would put Japan on the path of war.

After he gave his approval to escalate, Honjo made a request to the commander of the Korea Army, Lieutenant General Hayashi Senjuro, to reinforce the Kwantung Army.[104] Honjo, Ishiwara, and much of the rest of the Kwantung Army staff then left Port Arthur at roughly 3:30 a.m. for Mukden, where they would set up their headquarters.[105] They arrived around noon, and once they did Consul General Hayashi Kyujiro castigated Honjo for allowing the Kwantung Army to take action despite the warning he had sent him previously.[106] Honjo, who had not received this warning because Katakura had intercepted it, sent Shimamoto to Port Arthur to find out why he hadn't received this telegram. But despite Honjo's tinge of skepticism, he still approved sending messages to the Army Ministry and AGS stating that the Kwantung Army needed to "assume responsibility of maintaining public order over the entirety of Manchuria" and further requesting three additional divisions to achieve this objective in the afternoon on September 19th.[107] These messages put the onus on the Army Ministry and AGS to react to the situation in northeastern China and formulate a policy.

In Tokyo, the Army General Staff first heard of the explosion on the SMR line at 1 a.m. on September 19th. Subsequent telegrams blamed the Chinese for the attack and requested reinforcements.[108] At 5 a.m., General Umezu Yoshijiro called an emergency meeting of the AGS in which acting Operations Division Chief Imamura took the proverbial bull by the horns and proposed what eventually emerged as the consensus of everyone in attendance: that the Kwantung Army should implement its plans to seize control over the southern part of northeastern China.[109] Simultaneously, Vice War-Minister Sugiyama convened a meeting of important Army Ministry officials, who reached a different conclusion: the situation should be clarified and until then, the Kwantung Army should not take action.[110] At 7 a.m., the most important AGS and Army Ministry officers had a combined meeting, including Sugiyama, Koiso, Ninomiya, General Affairs Division Chief Umezu Yoshijiro, Intelligence Division Chief Hashimoto Toranosuke, and acting

Operations Division Chief Imamura. Koiso stated that "[t]he Kwantung Army's action is entirely just" and suggested sending reinforcements, to which all in attendance agreed. The willingness of army moderates to go along with Koiso's more hawkish solution illustrates the influence AGS radicals had on key policy decisions.[111]

While the AGS and Army Ministry attempted to formulate a coherent policy regarding reinforcements for the Kwantung Army, Korea Army commander Lieutenant General Hayashi Senjuro had already taken steps to aid his comrades to the north. Upon receiving a call from Honjo requesting reinforcements between 1:20 and 2 a.m., Hayashi dispatched an aerial detachment to Mukden and various elements of the 20th Infantry Division to the Sino-Korean border, where 4,000 men of the 39th Infantry Brigade were already positioned.[112] Hayashi informed Tokyo of his unilateral decision a little bit before 7 a.m., sending a telegram stating "[i]n view of the situation in the Mukden area, I have dispatched two companies from the 6th Air Regiment. Preparations are also being made for sending one mixed brigade from the 20th Infantry Division [the 39th Infantry Brigade]. I have in addition ordered the 19th Division to make plans for sending as many troops as possible."[113] But because sending troops from the Korea Army into northeastern China represented an act of war outside the realm of immediate self-defense, the Army Ministry determined that the Emperor's authority as commander in chief would be necessary to implement such a move. Minami therefore cabled Hayashi that "arbitrary crossing of the border is inadmissible. Standby until imperial sanction is secured."[114]

Prime Minister Wakatsuki first heard of the Mukden Incident between 6 and 7 a.m. on September 19th, and then called a Cabinet meeting for 10 a.m.[115] At this meeting, Wakatsuki asked Army Minister Minami whether the Kwantung Army's actions were defensive, following up with a question about what Japan should do if it turned out that the Mukden Incident was a Kwantung Army conspiracy.[116] He further stated his desire to avoid enlarging the scale and scope of the conflict, if possible. Minami responded by perpetuating the Kwantung Army's narrative that the Chinese had fired first and that the Japanese had merely engaged in an act of self-defense.[117] But Foreign Minister Shidehara completely undermined Minami's claims by reading some of the telegrams Consul General Hayashi had sent, which included details about the 2nd Company, 2nd Independent Railway Guard's Battalions attempts to secure a train carriage and the Kwantung Army headquarters' preparations for action before the incident.[118] These telegrams to some extent exposed the Kwantung Army's duplicitousness and provided support for Wakatsuki's position of avoiding expansion of the incident.

Shidehara's exposure of the Kwantung Army's skullduggery left Minami feeling "dispirited," depriving him of the desire to advocate for the dispatch of the Korea Army to reinforce the Kwantung Army.[119] Instead, Minami merely informed the Cabinet that certain elements within the army believed reinforcements for the 10,000-strong Kwantung Army would be necessary to protect Japanese civilians and property against 130,000 Chinese. But this appeal was not convincing to the rest of the Cabinet, and Minami was forced to concede that he would not allow the situation in northeastern China to escalate.[120] Thereafter, the Cabinet decided on a general policy of non-expansion of the Mukden Incident, which Wakatsuki announced to the press at 12:20 p.m. on September 19th.[121] The AGS reiterated this policy in a cable to Hayashi at 12:30 p.m., with AGS Chief Kanaya ordering the latter to await the Emperor's sanction before deploying troops to the Three Eastern Provinces.[122]

Vice-Minister of War Sugiyama, Vice Chief of Staff Ninomiya, and Inspector of Military Education Araki Sadao had a separate meeting on the morning of September 19th.[123] These figures, representing the "big three" of the IJA's administrative apparatus, all agreed that Japan should use the Mukden Incident to find a solution to the "Manchurian question." But unlike the Kwantung Army radicals, who sought a wholesale occupation of the region, these men merely sought to maintain Japan's existing treaty rights and replace Zhang Xueliang with a more amenable Japanese puppet. This view was more in line with the first option of Tatekawa's April 1931 "Situation Estimate" document and the consensus that the many AGS and Army Ministry personnel had arrived at in their early morning meetings on September 19th. Going forward, the staff officers in the AGS and Army Ministry would try to bridge the divide between the desires of the Kwantung Army radicals and the civilian government's policy.

On the afternoon of September 19th, Chief of the General Staff Kanaya expressed his desire to adhere to the Cabinet's position of non-expansion and an eventual return to the status quo at a meeting of AGS section chiefs. But Kanaya's subordinates were not on board with this policy, as evidenced when Imamura Hitoshi responded "[t]he die is cast. To halt it now and restore the old state of affairs will not only have an unfortunate effect on the morale of the Kwantung Army but will also be a serious blow to the nation's army as a whole. It is necessary now to overcome all obstacles, uphold the prestige of the nation and the army and endeavor to achieve our great goal."[124] Imamura subsequently drafted a telegram to the Kwantung Army with the following contents:[125]

> 1. Your determination in handling the matter after the night of September 18 is most appropriate and has enhanced the prestige of the imperial army.
> 2. In view of the Chinese attitude since then, and in deference to the cabinet decision not to complicate the matter unnecessarily, please keep this principle in mind when planning further action by the army.

Imamura's praise of the Kwantung Army's actions on September 18th indicated his approval for implementing a military solution in northeastern China. But the second point was more in-line with the AGS and Army Ministry's views, which did not endorse the Kwantung Army's conquering the entire region. Instead, Imamura and other Operation's Division officers sought to implement plan one of Tatekawa's April 1931 situation estimates. They put their intentions into writing in a document titled "Settlement of the Current Crisis in Manchuria," which called for the overthrow of Zhang Xueliang and the construction of multiple strategically valuable railways. By September 20th, the "big three" of the IJA (Minami, Kanaya, and Muto Nobuyoshi) accepted these ideas and agreed to try to force the government to put them into practice. As they put it, “if the government does not accept this draft plan, the army will not bestir itself even if the cabinet falls as a result.”[126]

In the Imperial Palace, the Emperor learned of the Kwantung Army’s actions from newspaper reports that put forth the Kwantung Army’s narrative.[127] Hirohito was nonetheless predisposed towards avoiding an escalation of the conflict. His position is evidenced in his September 19th meeting with Chief of Staff Kanaya, in which Hirohito expressed his displeasure with Hayashi’s unilateral decision to move elements of the Korea Army towards the Sino-Korean border. Given the Emperor’s position as the nation’s sovereign, Kanaya was forced to promise that he would look into the matter and expressed regret over Hayashi’s actions. This meeting with the Emperor would shape the AGS’s future interactions both with the Kwantung Army and the Korea Army on the subject of expanding the incident in the days ahead.

Elsewhere in Tokyo, Wakatsuki sought the counsel of Harada to make sense of the situation and how he should move forward.[128] Wakatsuki told Harada that he hadn’t received any reports from the Army Ministry or Foreign Ministry about the incident. The lack of readily available information meant that Wakatsuki could not ascertain what exactly happened, and thereby left him at the mercy of the Kwantung Army and its fellow travelers in Tokyo. Nonetheless, Wakatsuki was still determined to prevent the Kwantung Army from expanding the Mukden Incident, whatever the cost.

Wakatsuki’s fears were not welcome among the Emperor’s court advisors, most notably Secretary to Lord Keeper of the Privy Seal Kido Koichi. When Harada told Kido of Wakatsuki’s fears, Kido, as he put it in his diary, “was annoyed that the Prime Minister was relying on outside assistance, so to speak, in solving this crisis. [Kido] insisted that [Kido] believed the cabinet had to hold conference after conference, and thereby attempt to unify public opinion and reveal the unswerving determination of the cabinet itself.”[129] Harada, in his conversation with Kido, denied that Wakatsuki was trying to get help from the Emperor or his advisors, but Kido remained adamant that the Cabinet needed to be the force to resolve the crisis in northeastern China. Harada brought this assessment back to Wakatsuki, and told him of Kido’s view that "no other way seemed to exist than to control the army through cabinet meetings repeatedly held."[130]

Despite Kido's frustrations with Wakatsuki, he agreed to meet him at Harada's residence on the evening of September 19th along with Saionji, Harada, and a variety of Foreign Ministry personnel.[131] All agreed that the Kwantung Army was not following orders from the AGS and Army Ministry, and that the Emperor shared their desire to limit the hostilities in northeastern China. But Kido believed that the Kwantung Army would not obey orders as long as Saionji and the other advisors to the Emperor influenced his thoughts and statements. Therefore, Kido recommended that Saionji, a voice for peace and reason, "not come to Tokyo unless the situation changes," and for the Emperor not to speak up about the situation in the Three Eastern Provinces.[132] In this manner, Kido abdicated the responsibility of the throne for events in northeastern China and left it up to the Cabinet to resolve the issue on its own.

Kido's abdication of responsibility severely hindered the likelihood that the Kwantung Army's actions could be stopped. While the Cabinet had influence over policy and the budget, command of the army and navy were exclusively the emperor's domain according to the Japanese Constitution. The emperor also had the sole right to declare war. Therefore, in the event that the Kwantung Army or Korea Army initiated action outside the SMR zone of their own accord, the Imperial Throne was technically the only entity that could put a stop to such improprieties. In this sense, Kido's suggestion that the Emperor not speak on the situation in northeastern China gave the Kwantung Army and Korea Army a blank cheque to seize the initiative.

On the afternoon of September 19th, Lieutenant General Hayashi was still trying to convince the AGS to allow the Korea Army to cross the Sino-Korean frontier. He sent a telegram to the AGS at 3:50 p.m. that read "First, the bulk of the expeditionary force has already departed and will cross the border by tonight. Second, due to the situation in which the Kwantung Army is placed, I feel it is imperative to assist it immediately."[133] Kanaya responded with a telegram at 6 p.m. reiterating the non-expansion policy to Hayashi. At 8 p.m., telegrams were dispatched to General Muro, the commander of the 20th Infantry Division, and General Yoshimura, commander of the 39th Brigade, to "stop mobilizing" and report their present positions.[134] But Hayashi and the Korea Army remained persistent, sending a telegram to the AGS that stated Hayashi's belief that it was "necessary to dispatch a mixed brigade in view of the Kwantung Army's request for reinforcements and other general conditions."[135] By this point, many Operations Divisions officers considered Hayashi's constant requests as blatantly insubordinate.[136] Nevertheless, slightly after midnight on September 20th, the AGS sent Hayashi a telegram under Kanaya's name that they hoped would force Hayashi to await imperial sanction before committing his forces, once and for all. The contents of this telegram were as follows:[137]

> Your unilateral action this morning is not necessarily objectionable, in view of then-existing conditions. However, as the situation around Mukden has somewhat improved, I have given my consent to the cabinet decision not to extend hostilities for the time being, unless unexpected conditions should develop. Your army's movement out of Korea must wait upon imperial approval. It is necessary to gather and hold the mobilized troops at Shingishu (Sinujiu) until further notice. Please telegraph immediately the present whereabouts of the mixed brigade.

While the opening line indicated the sympathies of some officers on the AGS Operations Division, the rest of the message clearly ordered Hayashi not to deploy his forces. The 39th Mixed Brigade thus had to detrain at Sinujiu and await further instructions.[138] Therefore, Hayashi's attempt to unilaterally reinforce the Kwantung Army was temporarily thwarted.

Plotting the Next Moves

Starting at 6 p.m. on September 19th, Minami and Kanaya sent Honjo a series of messages informing him of the Cabinet's non-expansion policy.[139] In response, the Kwantung Army sent a telegram at 7 p.m. stating "This is the best moment to solve the Manchurian question. Should our army fail to grasp it and adopt a conservative attitude, it will forever become impossible to solve the question."[140] This second message reflected the attitudes of Ishiwara, Itagaki, and the rest of the Kwantung Army plotters, who wanted to continue expanding their operation and were in the process of making plans to do so. They had already identified Jilin as another logical target because it threatened the flank of the SMR and had a significant Korean population that the Kwantung Army could justifiably intervene to "protect." But Jilin was by no means the limit of the army's plans, which also extended to seizing Harbin. Because of its strategic position astride both the SMR and the Soviet-owned Chinese Eastern Railway (CER), Ishiwara hoped the Kwantung Army could use Harbin as a launching pad to the areas north of the Chinese Eastern Railway; then the army would not only have brought most of northeastern China under their control, but also expelled Soviet influence from the region. In this sense, Harbin represented the region's crown jewel, as it would allow the Kwantung Army to achieve multiple strategic objectives.

To plot their next moves, Ishiwara, Itagaki, and many other Kwantung Army staff officers decided to hold a conference at the Shinyokan hotel, where Ishiwara took responsibility for the actions that transpired in the previous 24 hours and claimed he was "guilty."[141] Katakura responded that this admission of guilt was not productive and suggested that Tatekawa be brought to the hotel, which Itagaki supported. While Katakura went out to find Tatekawa, a long discussion commenced at the hotel about the state of affairs. By 2 a.m., the officers in attendance reached the consensus that Japan needed to settle the "Manchurian question" with force, come what may.

When Tatekawa arrived, Itagaki and Ishiwara conveyed their consensus to him, with the added stipulation that it would be necessary to seize control of all of northeastern China to eradicate Soviet influence from the region. They also stated that there was no need to fear intervention from the United States, United Kingdom, Soviet Union, or League of Nations, as they were all preoccupied with their own problems. But Tatekawa was not convinced by this line of reasoning, as he believed that the Soviet Union might intervene if the Kwantung Army advanced north of the CER. Instead of going through with a full-scale invasion of northeastern China, Tatekawa proposed that the first plan of his April 1931 "Situation Estimate"—the forcible overthrow of Zhang Xueliang and installation of a pro-Japanese government—be implemented.[142] The Kwantung Army staff officers refuted Tatekawa's proposal, reiterating that it was unnecessary to fear war with the USSR and that a full-scale invasion and occupation of northeastern China was necessary to resolve the existing issues there. This impasse persisted throughout the night; Tatekawa and the Kwantung Army staff officers could not come to an agreement. Consequently, they resolved to resume their discussion the following day.[143]

Once morning arrived on September 20th, Tatekawa went to see Honjo to explain his views in lieu of talking with the Kwantung Army officers.[144] In his conversation with Honjo, Tatekawa said that a new government in northeastern China was necessary and that the Kwantung Army should seize control of Jilin and Taonan. But he remained adamant that no military action should take place north of the CER, lest the Soviets be antagonized. Honjo agreed with Tatekawa's assessment, sending a cable to Tokyo with the contents of their consensus. Then, Honjo departed to tour the sites of action at Liutiaogou and Mukden, where he spent much of the rest of the day.[145]

Even though Tatekawa succeeded in getting to Honjo before Ishiwara and Itagaki could, the latter two had an ace up their sleeve in the form of Lieutenant Colonel Osaka Michisada. An advisor to the Jilin Province government, Osaka played a key role in fomenting anti-Japanese disturbances in Jilin, including riots, threats, and arbitrary acts of violence.[146] These anti-Japanese incidents led to the Kwantung Army receiving a stream of messages asking for protection from Japanese civilians.[147] One of these messages, from the head of the Jilin resident's association can be seen below:[148]

> The situation in Jilin has become tense. Some fired at the houses of the Japanese residents. A few women and children were evacuated, but the greater part of the residents are unable to leave the city and the danger is increasing every minute. We request the commander,

> by decision of the association, to take a measure promptly for perfect protection of the residents."

Conveniently for Ishiwara and Itagaki, they already had troops in position to deploy to Jilin because they convinced Honjo to send the 2nd Infantry Division to Changchun.[149] Therefore, all that remained for Ishiwara and Itagaki was to persuade Honjo of the need to deploy it to Jilin to "protect" Japanese civilians.

Ishiwara and Itagaki began trying to convince Honjo on the evening of September 20th. Honjo was initially not amenable to their pleas, stubborn in his adherence to the Army Ministry's instructions. Honjo's stubbornness was so great that a frustrated Ishiwara told Captain Katakura that he couldn't "do anything more to budge the commander and so I'm giving up responsibilities for the direction of operations. Katakura you take over."[150] While this statement seemed to indicate Ishiwara was giving up his leadership role, it was in fact a deft manipulation to galvanize the various officers on the Kwantung Army staff behind him.[151] Ishiwara knew that if he, the tireless and persistent thought-leader seemingly "gave up," the Kwantung Army staff officers with whom he planned the Mukden Incident would step in and do their utmost to help him convince Honjo of the need to attack Jilin.

Ishiwara's statement had the intended effect as various Kwantung Army staff officers, including Chief of Staff Miyake, Arai, Itagaki, Ishiwara, and Katakura, all visited Honjo in his quarters at the Shinyoku hotel. Arai explained the situation in Jilin, Ishiwara laid out the tactical and strategic necessity of the operation, and Itagaki hammered home the need to resolutely pursue the objective.[152] Honjo was initially angered by Itagaki's emotional appeal for action, which implied that he was wilting in the face of having to make a difficult command decision. Consequently, Miyake decided to kick out other staff officers besides himself and Itagaki, both of whom proceeded to have a long conversation (between two and three hours) with Honjo on the subject of attacking Jilin.[153]

In their private conversation with Honjo, Miyake and Itagaki reiterated the need to cover the SMR's flank. They also explained the fact that defending Northern Korea from Korean nationalists in Jiandao would be more difficult if the Kwantung Army did not attack Jilin. Perhaps most importantly, an assault on Jilin would provide a reason for the Korea Army to send reinforcements.[154] These explanations, combined with the united consensus of various Kwantung Army staff officers led Honjo to finally approve an assault on Jilin at 3 a.m. on September 21st.[155] Soon after he approved the offensive, Honjo sent orders to the 2nd Infantry Division for its execution, while delaying his notification of the AGS in Tokyo to 6 a.m.[156]

After Honjo sent his orders to the 2nd Infantry Division at Mukden and Changchun, Ishiwara saw to it that the division rapidly embarked on an advance towards Jilin by train.[157] While the 2nd Infantry Division's departure left Mukden virtually undefended, the Chinese garrison of Jilin had largely retreated from the province, making the 2nd Infantry Division's flight from Mukden a non-issue.[158] In a rapid advance, the Japanese seized Jilin city on September 22nd; soon thereafter, on September 26th, a government official in Jilin province named Xi Qia declared "independence."[159] According to a Japanese official announcement, their presence was "not with a view to military occupation but only for the purpose of removing a menace to the SMR on its flank."[160]

After Xi Qia declared Jilin's independence, the Kwantung Army continued its advance in the region. It had finished conquering the province by mid-October through a series of rapid assaults. The nature of the Jilin campaign can be seen in reports from the Consul General of Mukden to the Foreign Ministry in Tokyo, one of which is excerpted below:[161]

> According to reports from the local area and internal sources, on October 7, the 8th Battalion of the Independent Garrison Force under the command of Staff Officer Ikenaga crossed the Songhua River and began advancing. Their movement was aimed at securing the designated area. They proceeded with the occupation of the government offices in Jilin Province.
>
> At the same time, the 4th Battalion of the Garrison Forces under Staff Officer Kumagai crossed from the eastern side and entered the

> city. Internal forces were combined with those already stationed inside Jilin.
>
> While this operation was underway, the local police and other officials offered no resistance and have shown cooperation.
>
> It was decided that full control over Jilin would be established, and as of now, the military has taken over all administrative functions.
>
> Local reaction has been largely passive, and resistance is not expected at this point. However, care is being taken to monitor for any possible unrest.
>
> Civil administration will be handled jointly with the garrison until further notice.

The lack of Chinese resistance was no surprise given Chiang Kai-shek's non-resistance order and Xi Qia's defection to the Japanese. Nonetheless, the Kwantung Army's attack on Jilin represented a great transgression of previous Japanese policies. Not only had Ishiwara and Itagaki violated the Cabinet's previous non-expansion policy, but they also violated the Wakatsuki Cabinet's resolutions from September 21st, which decided against outside reinforcements for the forces in northeastern China and declaring war on China. Furthermore, they also contradicted the Emperor's wishes, which were to maintain "strict adherence" to the Cabinet's non-expansion policy and thereby avoid war.[162] As Ishiwara and Itagaki's invasion of Jilin was designed to facilitate receiving reinforcements from the Korea Army and increased the likelihood of a broader Sino-Japanese conflict, they clearly contradicted government policy on two counts. Therefore, the Jilin episode, even more than the Mukden Incident, illustrates the fact that Ishiwara and Itagaki were the primary drivers of events in northeastern China, and Tokyo was merely along for the ride.

Just as Ishiwara and Itagaki intended, the Kwantung Army's assault on Jilin facilitated the Korea Army's dispatch of forces across the Sino-Korean border. Hayashi, upon requests for aid from the Kwantung Army, decided to seize the opportunity, moving his 39th Infantry Brigade across the Sino-Korean border in the early afternoon of September

The 2nd Infantry Division's entry into Jilin. Note the Japanese flag on one of the buildings. (Museum of War of Chinese People's Resistance Against Japanese Aggression)

21st. At 2:40 p.m. the AGS received a telegram in arrears from Hayashi referencing his decision, saying "the 39th Infantry Brigade planned to leave Sinujiu at 1 p.m. today." Before the AGS could respond, it received another telegram from Hayashi that explained his rationale for dispatching the 39th Infantry Brigade. This telegram is excerpted below:[163]

> As the Kwantung Army began movements in the region of Jilin, attention was called to the insufficiency of the force and strong appeals for half of the Korean Army were made. Not being restrained because it is a righteous cause, I the commander have allowed a mixed battalion to cross the border at Sinujiu. Please note.

Hayashi went on to apologize for violating the supreme command's orders, and that he was exclusively responsible for dispatching the 39th Brigade to northeastern China.[164] But his apology and reference to the "righteous cause" of the Kwantung Army were mere veneers over the fact that he cared little for the opinions of his superiors or even the Emperor despite his words. Indeed, Hayashi and the Korea Army had coordinated to some extent with the Kwantung Army on offensive plans even prior to the Mukden Incident, knowing full well that the Emperor may not approve the Korea Army's deployment. Therefore, the fact that Hayashi deployed the 39th Mixed Brigade despite knowing that the "sacred and inviolable" Emperor did not give his approval indicates the extent to which *gekokujo* had overtaken the IJA's chain of command. With such a compromised command structure, it would be difficult for the government, AGS, and Army Ministry to reassert control over policy.

After the 39th Infantry Brigade crossed the border, it rapidly advanced into Fengtian Province. This advance was detailed in a Japanese operational history of the brigade's activities:[165]

> On September 19, after the 39th Mixed Brigade completed its mission near Fengtian, it began moving its main force toward Changchun in the afternoon. However, due to ongoing unrest in the Fengtian area, the Army Staff deemed that the situation was not yet sufficiently stabilized. Lieutenant General Hayashi, commander of the army, determined that it was necessary to suppress the local unrest before proceeding. As a result, it was decided that the brigade would remain stationed in the Fengtian region for the time being. At around 2:30 p.m., the unit began its operations to cross the Hun River. That evening, the brigade entered Xiaozhen Village and launched military operations. The unit pacified the area, prepared defensive positions in coordination with nearby units, and worked to restore calm to the wider surroundings. Consequently, order in the Fengtian region was gradually reestablished, and operations within the combat zone resumed promptly.
>
> Afterward, the enemy forces fled in disarray, scattering to various locations throughout Liaoning Province. Our troops pursued them swiftly, engaging in a vigorous campaign to eliminate any resistance. During this pursuit, many civilian homes were destroyed, and numerous casualties and losses occurred among the local population. In response, to protect civilians and restore order, the Eastern Gar-

> rison entered the affected areas and established control. Given that many civilians had lost their homes and property, they were temporarily sheltered in nearby army warehouses. Meanwhile, the remaining elements of the 39th Mixed Brigade advanced to Qiansuo Village and continued their rapid push toward Maocun by the evening. Under the command of General Hayashi, the unit took control of the southern area of the battlefield. At the same time, the Eastern Garrison took proactive steps to restore order, secure key routes, and establish a firm presence in the region. With the brigade regrouped in Qiansuo, they prepared for a decisive push toward the enemy's southern stronghold and launched a rapid pursuit in the direction of Huinan.

The 39th Infantry Brigade's great success hastened the Kwantung Army's conquest of Fengtian and Jilin, which in turn provided the coveted flank protection the Kwantung Army sought for its corridor along the SMR. Therefore, Hayashi's deployment of the 39th Infantry Brigade was a great boon to Ishiwara and Itagaki's plans, allowing them to focus their efforts towards Harbin in the north.

In Tokyo, there were some young staff officers on the AGS who sympathized with Hayashi's decision. In their view, Hayashi's decision was necessary because the Kwantung Army was running low on manpower after dispatching forces to Jilin. Consequently, soon after they heard of the 39th Infantry Brigade's dispatch, these officers set about drawing up a document to secure ex post facto approval from the Emperor. They hoped that General Kanaya, the chief of the General Staff, would deliver it directly to Hirohito and make the case for Hayashi's actions.

Army Minister Minami also provided ex post facto justifications for the actions of the Kwantung Army and the Korea Army. With regards to the Kwantung Army's expansion of hostilities into Jilin, Minami told the Cabinet that the Kwantung Army's pursuit of Zhang Xueliang's forces was necessary for "strategic and technical reasons" on September 21st.[166] On the issue of the Korea Army's deployment of the 39th Mixed Brigade into northeastern China, Minami made an ex post facto justification for its actions at an October 7th meeting of the Privy Council:[167]

> The Korean Army which was waiting for orders on the frontier on the 19th, advanced into Manchuria on the 21st without receiving any order to that effect. It was because the main force of the Mukden Army had advanced to Changchun and Jilin, leaving Mukden in the hand of scanty troops. And as there was a danger of Chinese troops making a counter-attack, the Commander-in-Chief recognized the tenseness, admitting no further delay and dispatched his troops on his own responsibility. It is true that there was an omission of procedure, but it is not so serious as an infringement on the prerogative. If there was no previous sanction of the operation plan, and if it were in the peace-time, it would have been outrageous. However, when the situation at that time is taken into consideration, there is some foundation for the move. The Chief of the General Staff reported to the Throne, viewing the matter in the same light.

Minami's justification for Hayashi's actions was rooted in the Imperial Japanese Army's principle of "field initiative," *dokudan senko*, which gave commanders in the field the ability to make on the spot decisions without interference from above. While this command principle was well-suited to being applied tactical or operational level settings, the question of moving troops from Japanese territory into a different country was a strategic decision that was qualitatively different than most other command decisions. Therefore, following procedure was not merely a bureaucratic formality but a necessary step to ensure that the army served the interests of national policy as determined by the Emperor. Minami's failure to acknowledge this reality belies his inability to control subordinates, and the general lack of discipline in the IJA.

Despite Minami's ex post facto justifications, many others in the Army Ministry opposed the Korea Army's deployment of the 39th Brigade on the grounds that it was inappropriate to put the Emperor in a difficult position. As Nagata Tetsuzan put it, "[i]t is improper to resort to the expediency of appealing to the emperor directly, without going through the cabinet, since troop increases involve expenditures and thus require cabinet approval. Should the cabinet fail to sanction the dispatch of troops, the ultimate decision would have to be carried to the emperor. This is not the way to conduct ourselves as his subjects."[168] From this quote, it seems Nagata did not want to put the onus of legitimizing Hayashi's decision on Emperor Hirohito. He instead wanted the decision to be routed through the Cabinet, which was unlikely to support Hayashi's decision. Fortunately for Nagata, Operations Division chief Imamura decided against using Kanaya's audience with Hirohito to gain ex post facto approval for Hayashi's decision. Therefore, Kanaya did not present the document the AGS had drawn up, and merely informed Hirohito of Hayashi's actions during his visit.[169]

On September 22nd, Wakatsuki, Shidehara, and the rest of the Cabinet members met and deliberated on whether or not to approve Hayashi's actions.[170] Even though they ultimately decided not to approve his decision to move troops into northeastern China, they did vote to fund the 39th Mixed Brigade's expenses.[171] Wakatsuki's logic for funding the 39th Brigade's expenses was simple: if he did not provide funds, the brigade would be "wiped out" by the numerically superior Chinese.[172] Furthermore, Wakatsuki and many others in the Cabinet believed that denying funds would not prevent the 39th Infantry Brigade from participating in hostilities, as it had already crossed the Sino-Korean frontier.[173] The Emperor shared this view, as when Kanaya came to meet him to try and inform him of Hayashi's decision, he apparently said that the matter "could not be helped."[174] While the emperor and the cabinet did not give explicit approval for the 39th Brigade's deployment their lack of staunch opposition effectively greenlit Hayashi's *gekokujo*. The lack of action taken against Hayashi and the Korea Army set a precedent for the future that would allow commanders in the field to abuse their initiative and make policy decisions far beyond their pay-grade.

The invasion of Jilin marked a great victory for the Kwantung Army, as it not only absorbed a new territory but also got away with an egregious act of *gekokujo*. Naturally, the Kwantung Army wanted to continue pressing offensive momentum, and found the perfect pretext for doing so in a series of anti-Japanese terrorist acts in Harbin. The most notable incidents were instigated by Amakasu Masahiko, who managed to get Chinese hooligans to throw bombs into the Japanese consulate, the office of the Bank of Korea and the headquarters of the

Harbin Daily.[175] As the 4,000 Japanese civilians in Harbin considered these actions a threat to their livelihood and safety, Consul General Ohashi Chuichi and OSS chief Colonel Komatsubara Michitaro asked the Kwantung Army headquarters to send forces to Harbin.[176]

Upon hearing Ohashi and Komatsubara's requests, Kwantung Army headquarters sent multiple telegrams to the AGS and Army Ministry asking if it could deploy troops to Harbin. But this time the AGS and Army Ministry both explicitly refused permission, with Kanaya mandating that any advance on Harbin must first be approved in Tokyo and Minami saying government policy prevented the Kwantung Army from advancing north of Kuanchengtze.[177] The AGS and Army Ministry's firmness stemmed from the desire of key figures in Japan (most notably the Emperor, Saionji, and the cabinet) to avoid expanding the incident, and perhaps more importantly, fear of war with the USSR.[178] Ishiwara and Itagaki, for their part, were still not worried about war because they believed that the Soviets could not intervene due to "internal conditions." They also believed that an offensive towards Harbin would help them avoid war with the USSR, which would not be able to respond to a rapid strike. As the Kwantung Army put it, war with the USSR could be avoided through "careful planning and instantaneous execution of our plans, giving the Russians no opportunity to respond except with "passive tactics."[179]

Despite the AGS's and Army Ministry's previous replies, the Kwantung Army continued to press them to approve an advance on Harbin. On September 22nd, Itagaki telegrammed Nagata that the Kwantung Army needed to "take advantage of this great opportunity to solve the Manchurian question."[180] Honjo followed up Itagaki's statement with another request to advance on Harbin. But the AGS and Army Ministry stood firm, not giving the Kwantung Army an inch on the question of deploying forces. On September 24th, Kanaya cabled the Kwantung Army that "troops were not to be sent to Harbin even if the situation suddenly became critical."[181] Similarly, Minami informed the Kwantung Army that the Prime Minister told the Emperor that "no on-the-spot protection will be provided for the Japanese in Harbin, and that if necessary they will be evacuated without sending troops."[182] These repeated refusals finally got the message across to the Kwantung Army that the AGS and Army Ministry would not allow them to advance further. Therefore, the Kwantung Army was temporarily forced to halt its attacks and reorient for the days and weeks ahead.

Chapter 13

Intermission in Northeastern China: Waves Beneath the Surface

During an operational pause in late September and October 1931, the Kwantung Army took advantage of the lull to contact various separatist figures in northeastern China and win their support. One of these figures, Zhang Haipeng, began moving his forces into Heilongjiang province in a bid to assert control over the province on behalf of the Kwantung Army. Meanwhile, Ishiwara greenlit the bombing of Jinzhou on October 8th in an attempt to re-start operations in the region. Even though the bombing did not lead to the resumption of operations, it did get the attention of the League of Nations, which attempted to negotiate a settlement between China and Japan. But neither party was ready to compromise on their respective positions, which made the resumption of hostilities an inevitability in the near future.

Meanwhile, the political scene in Tokyo was unraveling due to the Kwantung Army's invasion of northeastern China. With the Wakatsuki government on the back foot, Hashimoto Kingoro of the Cherry Society seized the moment and started making plans to overthrow the government again. In what became known as the "October Incident," Hashimoto and his small group of conspirators got close to assassinating multiple key government figures and implementing their *coup d'etat*. Hashimoto hoped that this gambit would allow Japan to undertake "national reconstruction" and untie the Kwantung Army's hands. While Hashimoto's plans ultimately fizzled, his near coup indicated the IJA's increasingly powerful position in Japanese politics, which had created an atmosphere of political instability. This atmosphere would facilitate further Kwantung Army offensives, and the ultimate conquest of the Three Eastern Provinces.

Operational Pause and Administering Northeastern China

After the Kwantung Army halted its advance at Harbin, many in the government, Cabinet, Foreign Ministry, AGS, and Army Ministries assumed that hostilities were over. These figures began trying to negotiate a peaceful settlement to the Mukden Incident. This supposed desire for peace was made clear in the "Herald of Asia," a "semi official" Japanese newspaper that stated military operations were completed.[1] The Japanese ambassador to the US also communicated the government's desire for peace to US Secretary of State Henry Stimson saying "it may be superfluous to repeat that the Japanese government harbors no territorial designs in Manchuria."[2] On September 26th, the Japanese representative to the League of Nations corroborated this sentiment when he noted that Japanese troops had been withdrawn to the railway zone, Japan had no territorial designs on northeastern China, and merely wished to protect its rights and interests.[3] On October 5th, Baron Shidehara stated that the Japanese were gradually withdrawing their troops into the railway zone and intended to negotiate with the Chinese (despite the latter's hesitancy).[4] These statements assumed that the Kwantung Army was done making attacks into northeastern China and would adhere to government directives going forward.

Despite the good intentions of Shidehara and others in the government, the Kwantung Army was not keen to go along with their plans. A variety of Kwantung Army staff officers had already met at a policy conference on September 22nd to figure out how to administer and occupy northeastern China.[5] They reached a consensus that emphasized the importance of establishing a pro-Japanese government in the region. Colonel Doihara Kenji of the Special Service Agency provided the framework for this new government, advocating the formation of a multiracial, "autonomous," "Manchurian" state that would give Japan control over defense, foreign policy, transport, and communications, hence limiting its actual autonomy.[6] Additionally, the new state would also give the Kwantung Army a veneer of legitimacy over its new domain in northeastern China, and perhaps allay criticism both at home and abroad. The new puppet state would therefore serve multiple purposes and help the Kwantung Army transform northeastern China into a hub in Japan's autarkic empire.

To help create the new "Manchurian" puppet state, the Kwantung Army officers identified various figures who could fracture Zhang Xueliang's domain: Xi Qia of Jilin, Zhang Haipeng of Taonan, Tan Yulin of Rehe, Yu Zhishan of Tunglientao, and Jiang Jinghui of Harbin were the most prominent. Soon thereafter, the Kwantung Army began contacting these figures, with Itagaki visiting Zhang Jinghui's home on September 22nd, Oya Shinkai visiting Yu Zhishan on September 24th, and Imada contacting Zhang Haipeng on September 25th.[7] The Kwantung Army hoped that these figures would declare their provinces "independent" and thereby splinter Zhang Xueliang's Northeastern Governmental Affairs Council.

While the various separatist leaders in northeastern China would help the Kwantung Army divide Zhang's government, the former Qing Emperor Puyi represented a centripetal force that could bring the Three Eastern Provinces together. Doihara was the foremost advocate for impressing Puyi to lead a Japanese puppet state, due to his status as a Manchu and former Emperor. Luckily for Doihara, Puyi longed to be Emperor again and saw Japan as the most likely entity to help him achieve his goal.[8] Puyi's desire is illustrated in the period after the Mukden Incident, when he sent his adviser Liu Xiangye to meet multiple important Japanese in northeastern China. He also sent another adviser, Shang Yenying, to meet various members of the former Qing Dynasty that were still in the region.[9] Puyi hoped that these moves would assure the Kwantung Army of his willingness to serve their interests, and thereby help him become Emperor once more.

On September 30th, Puyi's wishes came true when the Kwantung Army sent Kawasumi Tosiichi to talk to him about becoming Emperor again.[10] Kawasumi came bearing a letter from Xi Qia, who was a distant relative of Puyi; in this letter, Xi told Puyi to come back to the "land where our ancestors arose," and become Emperor once more. A Chinese advisor to Puyi, Luo Zhenyu, corroborated Xi's sentiments, saying that "restoration" would occur in the coming days and that Puyi's "subjects" were "longing" for his return. He also praised the "unselfish assistance" of the Kwantung Army for paving the way for Puyi's restoration. Nevertheless, despite Luo's words and Puyi's desire to become Emperor again, he did not commit to going to northeastern China immediately, saying that he would think the matter over. Puyi wanted to be certain of the Kwantung Army's control over northeastern China before he committed to reclaiming the title.

Puyi was right to not immediately commit to being "Emperor," as not everyone in Japan was on board with him being a Kwantung Army figurehead. Indeed, when AGS Chief of Staff Kanaya learned of the Kwantung Army's attempts to contact Puyi, he told the Emperor he would not approve the former Qing Emperor's "restoration."[11] Similarly, Prime Minister Wakatsuki told Army Minister Minami that the army should not be involved with creating a new state in northeastern China, as such a move would constitute a contradiction of the Nine-Power Treaty.[12] Army Minister Minami concurred, sending a cable to the Kwantung Army that prohibited "participation in the movement to establish a new regime in Manchuria," and dispatched secret police to prevent the Kwantung Army from taking action on the ground.[13] The AGS and Army Ministry's disapproval of Puyi's "restoration" meant that the Kwantung Army's plans to co-opt him were temporarily postponed. Nevertheless, the army would remain in contact with Puyi in the coming weeks and months, waiting for the moment to bring him to the Three Eastern Provinces.

While Minami had squelched the co-optation of Puyi, the various separatist leaders in the provinces began seceding from Zhang Xueliang's government.[14] Xi Qia declared Jilin's independence on September 26th; that same day, Yuan Jinkai became the chairman of the Liaoning Committee for the Maintenance of Peace and Order.[15] In the strategically vital Chinese Eastern Railway district, Zhang Jinghui formed the Emergency Committee of the Special District on September 27th and also declared independence.[16] On October 1st, the commander of the Taonan garrison, Zhang Haipeng also broke his ties to Zhang Xueliang, declaring himself the "commander of the peace preservation at the frontier."[17] These figures' declarations of independence represented the fracturing of Zhang Xueliang's coalition, and signaled the beginning of the end of his rule in the Three Eastern Provinces.

After the various provincial separatists declared independence, they began taking steps to form new organs of administration with the help of the Kwantung Army. The new "independent" provincial governments took the place of Zhang Xueliang's Northeastern Governmental Affairs Council. While a broader regional government would eventually be necessary (hence Doihara's courting of Puyi), the Kwantung Army did not intend to immediately merge these provincial governments into a larger body. Instead, they would wait until they got greater popular support for their occupation of northeastern China.[18] Only then would the Kwantung Army begin making moves to create an "independent" government for all three provinces.

While the Kwantung Army consolidated its position in the Three Eastern Provinces, Zhang Haipeng sought to expand his influence. He had attempted to become the governor of Heilongjiang in previous years; he still coveted this position in the fall of 1931. To fulfill his ambitions, Zhang Haipeng began moving his forces up the Taonan-Angangchi railway to seize the governorship of Heilongjiang by force.[19] But soon after Zhang Haipeng deployed his forces, Heilongjiang governor Wan Fulin ordered his forces to assemble at Qiqihar on October 7th. Wan made Ma Zhanshan, who had overseen the Heilongjiang Garrison Army, the commander of the forces set to face Zhang's advance.[20] The redeployment of Ma and his forces was duly noted by the Japanese, evidenced in the telegram below:[21]

> According to information received from the Harbin and Dalian consular districts, it has been confirmed that General Ma Zhanshan is currently absent from his headquarters in Hailun, and command has been delegated to his deputy. It is believed that on the 8th, General Ma

> departed his headquarters and moved to Baiquan, located approximately 80 kilometers southeast. Since the Manzhouli incident, unrest has been observed among both the Chinese military and local civilians in the region.
>
> In the central area, the Harbin military district commander has issued orders to reinforce positions along the Songhua River. Troops are expected to be concentrated in Hailar, Qiqihar, and Harbin. At present, no large-scale troop movements have been confirmed in the southern Manchurian Railway zone. However, depending on future developments, reinforcements may be considered.
>
> Additionally, local Mongol tribes and their leaders appear to be cooperating with Japanese forces, and some have already begun reconnaissance operations.

Although the Kwantung Army had paid lip service to the idea of not expanding the Mukden Incident, the telegram contains references that indicate the Kwantung Army's preparations to continue its attacks. Cooperation with Mongol tribes facilitated the Kwantung Army's reconnaissance and intelligence activities, and helped prepare it for future operations. Furthermore, the Kwantung Army had armed Zhang Haipeng's forces, and sought to use him as a vessel to advance their interests further north. Therefore, despite the fact the Kwantung Army did not immediately send forces north to Heilongjiang, these activities indicate their intention to attack the province.

While the Kwantung Army began preparing to move towards Heilongjiang, Ishiwara and the other Kwantung Army staff officers began eyeing other places to resume military operations. They concluded that the city of Jinzhou was the best target due to its strategic location along the Beijing-Mukden railway and its status as the site of Zhang Xueliang's new headquarters.[22] 200,000 of Zhang's men were also in Jinzhou and its surrounding area to defend what remained of Zhang's domain.[23] Consequently, Jinzhou quickly became the site of renewed Sino-Japanese conflict, as Zhang and his forces attempted to resist the Kwantung Army's encroachments. As one pro-Kwantung Army source put it, "pro-Japanese Chinese and local organs were never at peace, unstable characters appeared everywhere, there were constant rifle-shots in the vicinity of Mukden, shops that had resumed business closed again, people began evacuating once more, and there was tremendous insecurity."[24] According to Ishiwara and the rest of the Kwantung Army staff officers, the unstable situation meant that Japan had to destroy what remained of Zhang's regime in Jinzhou. In their words, rooting out Zhang was necessary for "securing the peace of the occupied areas, establishing a firm basis for solution of the Manchurian question, and eliminating the base of machinations and plots."[25]

On October 4th, the Kwantung Army released a statement aimed at Zhang Xueliang that accused him of disturbing the peace and attempting to incite violence against the Japanese. An excerpt of this statement is provided below:[26]

> Recently movements have been growing everywhere to establish a regime, and though the people all praise the dignity of the Japanese Army, they do not in the least attempt to uphold their former leader. This is none other than the result of indignation felt against many years of oppression by war lords. The [Kwantung] Army stands aloof from politics and diplomacy and concentrates on maintaining public peace by keeping its forces prepared and by holding quiet vigilance. ... However, it sincerely wishes to realize rapidly the paradise of coexistence and co-prosperity for the thirty million residing in Manchuria and Mongolia, and believes that from a moral point of view to promote the unification [of various independent movements] is the urgently needed means of relief that proves the neighborly friendship of our country.

In this statement, the Kwantung Army accuses Zhang of bad governance and a lack of popularity among his subjects. While there may have been a kernel of truth in these claims, the statement was designed to justify further aggression against Zhang in Jinzhou. The army hoped that its statement would justify its actions to the Japanese government at home and the western powers abroad, and thereby avoid any potential negative consequences that would emerge from disobeying orders.

Now all that remained was for the Kwantung Army to put the next phase of its plans into action.

Soon after releasing the anti-Zhang statement, Ishiwara unilaterally decided to bomb the city of Jinzhou.[27] The targets of the bombing were to be the Fengtian Province government buildings in the city and the barracks of Zhang Xueliang's 28th Infantry Division. The objective was to destroy Jinzhou as a base of "provincial authority" and, if possible, reconnoiter Chinese defenses along the nearby Daling river. Ishiwara also hoped that bombing Jinzhou would cause a crisis for the Japanese government and intimidate Chiang Kai-shek's KMT from intervening to help save Zhang Xueliang. With their internal and external adversaries in disarray, the Kwantung Army would have a free-hand to launch an attack on Jinzhou, and perhaps the rest of northeastern China in the autumn of 1931.

On October 8th, Ishiwara, clad in an odd combination of flying togs and slippers, boarded one of five-bombers due to take part in the raid on Jinzhou to take off at noon.[28] These bombers dropped seventy-five twenty-five kilo bombs on the city, causing only minimal damage.[29] The Kwantung Army sent a cable to AGS Chief of Staff Kanaya while the bombing of the city was in progress, justifying their actions after the fact. The telegram, sent around noon, said that Jinzhou was the "center of disturbances in Manchuria," and that therefore the Kwantung Army "contemplates taking some sort of action against it. Please be so advised."[30] Later on the evening of October 8th, the Kwantung Army followed up its earlier telegram with a message that night which simply read, "Jinzhou was bombed. We beg your thoughtful disposal of the matter."[31]

Jinzhou's bombardment left the AGS and Army Ministry in a difficult position. On the one hand, the Kwantung Army had clearly disobeyed orders, as it had gone ahead with bombing a target two-hundred miles north of the SMR zone. Conversely, a public statement holding the Kwantung Army accountable for its *gekokujo* would undermine the army's credibility, and make it seem as though it did not have control over its subordinate formations. Consequently, the AGS and Army Ministry opted to publicly endorse the Kwantung Army's actions, with the Vice-Chief of the AGS approving the bombing on October 14th and the

Vice-Minister of War doing so the next day.[32] But privately, the higher-ups in the AGS and Army Ministry feared the bombing would bring them unwanted international attention, and sent Hashimoto Kingoro to talk some sense into Ishiwara.[33] Despite Hashimoto's best attempts, the Kwantung Army held fast to its position, and questioned Hashimoto's interference with its command initiative.[34] A second messenger from AGS, Major Endo Saburo, also found little success in convincing Ishiwara of the AGS's point of view.[35] These episodes illustrate how little power the AGS had over Ishiwara's actions, despite its nominal authority over him.

Much as the AGS and Army Ministry feared, the bombing of Jinzhou caused a great international stir, as European and American statesmen condemned the Kwantung Army's actions. More importantly, on October 13th the League of Nations Council met to deal with the fallout. The Chinese representative seized the opportunity of an international audience to demand a restoration of the previous status quo; Yoshizawa responded by saying that Japan's actions were justified due to the endangerment of Japanese lives and property prior to September 18th.[36] Given the Kwantung Army's impropriety in bombing Jinzhou, the League found China's arguments more convincing and voted to mandate Japan's withdrawal to the SMR zone on October 24th.[37] But Yoshizawa countered the League's demands by stating that any withdrawal to the SMR zone was dependent upon "satisfactory assurance regarding the protection of the South Manchuria Railway."[38] While Aristide Briand, who oversaw the League of Nations Council, attempted to avoid the issue by convincing the Chinese and Japanese governments to enter direct negotiations, the Chinese representative at the council said that "evacuation is a precondition of direct negotiations."[39] Negotiations at the League for a diplomatic settlement subsequently bogged down, as neither Japan or China was willing to yield. Therefore, the Kwantung Army had a free hand to continue taking action in northeastern China, as it did not have to face immediate consequences for the bombing of Jinzhou.

In Tokyo, rumors spread that the Kwantung Army would declare independence if the government continued to obstruct its progress. Major Cho Isamu was likely the source of these rumors, as he told other staff officers in Tokyo that the Kwantung Army was "aroused by the irresponsible words and actions of the court and government circles as well as of the senior statesmen," and that the Kwantung Army was upset because "while they are fighting a sacred war in order to expand national power and prestige, the Emperor does not approve and the government has been obstructing every move."[40] Cho went on to state that the Kwantung Army would declare independence if "the first signal for internal reform is fired in Japan," even going so far as to abandon Japanese citizenship if need be.[41] While Cho later admitted that the rumors he spread were false, there was a kernel of truth present in them, as Ishiwara had spoken of severing relations with Japan if the Incident failed, and may have sent a telegram to Tokyo to this effect.[42] But regardless of Ishiwara's intentions, even rumors of declaring independence were a step too far for the AGS and Army Ministry, who were determined to bring Ishiwara back into line. Ishiwara's *gekokujo* had finally gone too far, and the AGS was going to ensure he did not separate himself from Tokyo.

On October 17th, the AGS sent a message to the Kwantung Army directing it to "desist from any plan detaching itself from the imperial army" and to continue trusting the AGS's "zeal" to see through a solution in northeastern China.[43] Honjo took great offense at the suggestion that the army was plotting independence, saying that belief in such views signaled a "lack of confidence in myself and my staff" and was therefore "greatly deplored."[44] While Honjo admitted that the Kwantung Army generally acted "positively and arbitrarily," it nonetheless was acting in the national interest, and requested that the AGS "retrieve" the previous message it had sent. The Kwantung Army staff was even more outraged in its reply saying it could not allow the "glorious Kwantung Army [to] be regarded with ineffaceable suspicion," and that anyone who violated military discipline would be subject to "drastic measures."[45] These outraged replies suggest that the rumors of the Kwantung Army declaring independence may not have been completely accurate. Nonetheless, the Kwantung Army's protests did not stop the high command from using the independence issue to continue censoring Ishiwara and Itagaki.

Army Minister Minami sent another message to the Kwantung Army on October 18th, directing it to stop any talk of independence or absorbing the whole of northeastern China. He also sent Imamura

Hitoshi to speak directly to Ishiwara and explain his rationale for these directives.[46] Meeting at a restaurant in Mukden, Imamura attempted to explain the complexities of Japan's international situation and the bind in which Ishiwara's unilateral actions put the Army Ministry. Ishiwara did not accept his arguments and accused the central headquarters of lacking a "backbone." Imamura shot back that nothing could be accomplished through the "arbitrary decisions of field elements," and that if such arbitrary decisions continued, they could create a crisis.[47] After Imamura made this statement, Ishiwara declared that he was going to sleep, to which the former furiously rebuked him for insubordination. But tempers had cooled enough by the next day that Imamura was able to admit that he agreed with Ishiwara on the issue of regime change in northeastern China. Nonetheless, Itagaki and Ishiwara refused to budge, and told Imamura that such change had to come in the form of an independent state.[48] Unable to convince Ishiwara and Itagaki, Imamura departed, and soon thereafter the Kwantung Army began making plans for a new "Manchurian" state. These plans were put into writing in documents titled "Draft Principles for a New Manchurian-Mongolian Republic" and "Basic Principles Concerning the Proposed Establishment of the Free State of Manchuria and Mongolia," which provided the blueprint for a Kwantung Army puppet state in the Three Eastern Provinces.[49]

The October Incident

Back in Tokyo, the Cherry Society was once again plotting a coup to achieve the long-desired "national reconstruction." From the point of view of Hashimoto and his co-conspirators, October 1931 was the perfect opportunity to strike, as the situation in northeastern China represented a moment of national crisis that required a military government.[50] But the plotters also believed that Ishiwara, Itagaki, and the rest of the Kwantung Army plotters would benefit from having the anti-expansionist Wakatsuki government out of the way.[51] As Cho Isamu, one of the Cherry Society plotters, put it, the plotters sought to "set up a new government, and with the power of the new government to rally the support of the entire population toward the settlement of the Manchurian incident."[52]

Hashimoto, Cho, and the rest of the Cherry Society learned from their previous failures in the March Plot, where civilian demonstrators played a key role in undermining the plotters' plans. In the October Plot, Hashimoto opted to exclusively utilize IJA forces to overthrow the government, with the IJA's 23rd Infantry Regiment selected for the task.[53] The gist of their plan was to bomb Wakatsuki's residence with naval aircraft on October 24th while the cabinet was engrossed in a meeting. The army would then occupy key buildings in Tokyo including the AGS Headquarters and Army Ministry and assassinate figures who opposed the Cherry Society, including Major General Katsuki Kiyoshi, Director of the War College, and Major Nakano Naozo of the Military Academy.[54] In the event that the cabinet could not be bombed while in session, each of the key ministers would be assassinated individually, with Wakatsuki, Shidehara, and Minami receiving particular emphasis.[55] A new national government would be formed in their place, with General Araki Sadao as Prime Minister and Army Minister, Admiral Kobayashi as Navy Minister, General Tatekawa as Foreign Minister, Colonel Hashimoto as Home Minister, Okawa Shumei as Finance Minister, and Colonel Nemoto Hiroshi as head of the military police.[56] The new national government would oversee continued operations in northeastern China and root out the system of *zaibatsu*-funded parliamentarianism that defined Japanese politics up to that point.

Unfortunately for Hashimoto, his Cherry Society could not agree on much outside of broad dissatisfaction with the status quo. The disagreement within the Cherry Society took the form of a divide between radicals and moderates. The radicals, spearheaded by Hashimoto, believed that the old society, government, and economic system needed to be destroyed in order to implement their program of national reconstruction. The moderates opposed this view, advocating the minimization of destruction during the coup, and possibly eschewing a violent insurrection all together. This fissure would go on to undermine the unity of the coup plotters and the ultimate implementation of the envisioned coup.

The fissure between the moderates and radicals within the organization led many officers to get cold feet on the coup plot. Muranaka Koji and Isobe Asaichi were two officers who became disenchanted with Hashimoto and the Cherry Society, as they did not believe they had a refined ideology or post-coup plan. Tanaka Kiyoshi was a more high-profile defector from the coup plot, as he believed that it would destroy the army and prevent it from implementing its program of national reconstruction. Higuchi Kiichiro, founding member of the Cherry Society, even went so far as trying to convince Hashimoto to give up the plot. The sudden opposition of these and other figures to the October plot led Cho Isamu to complain that "members of the Sakura-kai began to hesitate, and therefore [I] tried to forcibly drag them toward execution of the plan."[57]

By mid-October, Hashimoto, Cho, and the other diehards dedicated to the coup plot were growing desperate. Up to that point, they had tried to cultivate support among young junior and field-grade officers. But the defection of junior officers *en masse* from the ranks of the coup plotters led Hashimoto to try and court the attention of more senior officers. On October 15th Hashimoto had a conversation with Vice-Minister of War Sugiyama, where he tried to secure his backing for the plan. The next day Cho and some of the other coup plotters had a conversation with Araki Sadao, getting him to promise that he would support the overthrow of the Wakatsuki cabinet. These conversations greatly frightened many in the AGS and Army Ministry, who pondered what should be done about the coup plotters.[58] Faced with turmoil in their own backyard, the AGS and Army Ministry needed answers, fast.

Fortunately for the AGS and Army Ministry, the coup plot was coming apart at the seams. In part, the sloppiness of the plotters was to blame, as they repeatedly discussed their plans while drinking with figures outside the inner circle of conspirators.[59] Some of the defectors also probably informed their superiors of the plot, and thereby further compromised its secrecy.[60] Whatever the case, these incidents were enough for Army Minister Minami to meet with General Araki Sadao on the night of October 16th to discuss what should be done about the coup plotters. After an all-night meeting with Araki and a wide variety of figures within the Army Ministry, Minami made his final decision: to arrest the conspirators and nix the coup. By 4 a.m. on October 17th Hashimoto, Cho, Nemoto, and nine other figures were arrested, and taken to jail.[61] But once again, the army failed to set an example of the plotters, opting to give them light sentences in a private trial. The IJA's prioritization of its "honor" and prestige triumphed over the necessity of military discipline, just as it had in prior months and years. This pattern would continue for the foreseeable future and lead Japan down the path of war.

Chapter 14

To the North

As October gave way to November, the Kwantung Army prepared to resume operations in northeastern China. Zhang Haipeng's forces continued to advance, and the Kwantung Army began deploying its own forces further north. By late October, the Kwantung Army had found the perfect pretext to attack the Chinese again: Ma Zhanshan's destruction of a Japanese-owned railway bridge in Heilongjiang province. As Ma hesitated to allow the Japanese to repair this bridge, fighting broke out once again, and the Kwantung Army advanced towards the city of Qiqihar. While the AGS intervened to prevent the army from immediately absorbing Heilongjiang province, the collapse of the Wakatsuki government in December facilitated another Kwantung Army advance towards Jinzhou and Harbin. Ultimately, the Kwantung Army would continue advancing until it had conquered all of northeastern China, which it finally achieved by February. The conquest of northeastern China was capped off by the formation of the puppet state of Manchukuo on February 14th, 1932, with the former Qing Emperor, Puyi at its head. Manchukuo represented the culmination of the Kwantung Army's activities in northeastern China over the previous two decades, and would serve as the avatar of Japanese imperialism in the coming years.

Action at the Nen River Bridges

While Tokyo was in flux due to the October Incident, Kwantung Army collaborationist forces under the control of Zhang Haipeng continued to advance against Ma Zhanshan's forces.[1] To hinder the advance of

Zhang's forces, Ma decided to blow up bridges of the Taonan-Angangchi Railway over the Nen River, which the SMR had built and provided the capital for.[2] On October 20th, 1931, the Japanese Consul General at Qiqihar requested Ma Zhanshan to allow the bridges to be repaired, which Ma declined. Subsequently, the SMR sent five engineers to inspect the damage to the bridges, but these came under fire from Ma's forces.[3] In light of this hostile reception, on October 28th the Kwantung Army directly intervened, with Major Hayashi—the Kwantung Army's representative at Qiqihar—informing Ma Zhanshan's clique that they had until November 3rd to complete the repair of the Nen River bridges. If they did not assent to Hayashi's request, the SMR would repair the bridges with Kwantung Army protection.[4]

In Tokyo, the AGS and Army Ministry viewed the situation in Heilongjiang with great concern. The AGS and Army Ministry feared that if the Kwantung Army continued its advance into Heilongjiang province, Japan would be risking war with the Soviet Union. Consequently, upon hearing Major Hayashi's ultimatum to Ma Zhanshan, the Army Ministry and AGS both sent the Kwantung Army telegraphs on October 30th telling it to stand down and not take action against Ma's forces.[5]

Despite the high command's concerns, Ishiwara and others on the Kwantung Army staff were not particularly worried about Soviet intervention in northeastern China. To a great extent, the Kwantung Army's previous intuition about the Soviets' intentions was borne out by Consul General Ohashi in Harbin and the Harbin Special Service agency, who both confirmed that the Soviets would not send military forces in the event of Sino-Japanese conflict in Heilongjiang.[6] Armed with this new information, the Kwantung Army responded to the AGS and Army Ministry's concerns with a telegram of its own denying the likelihood of Soviet intervention and putting forth the necessity of occupying Jinzhou and Qiqihar.[7]

On November 2nd, Ma Zhanshan still had not allowed the Kwantung Army to repair the Taonan-Angnagnchi railway. Even though the Chinese apparently requested an extension of Hayashi's November 2nd deadline, no reply was forthcoming from the Kwantung Army, which was not interested in Ma's response. The Kwantung Army's lack of in-

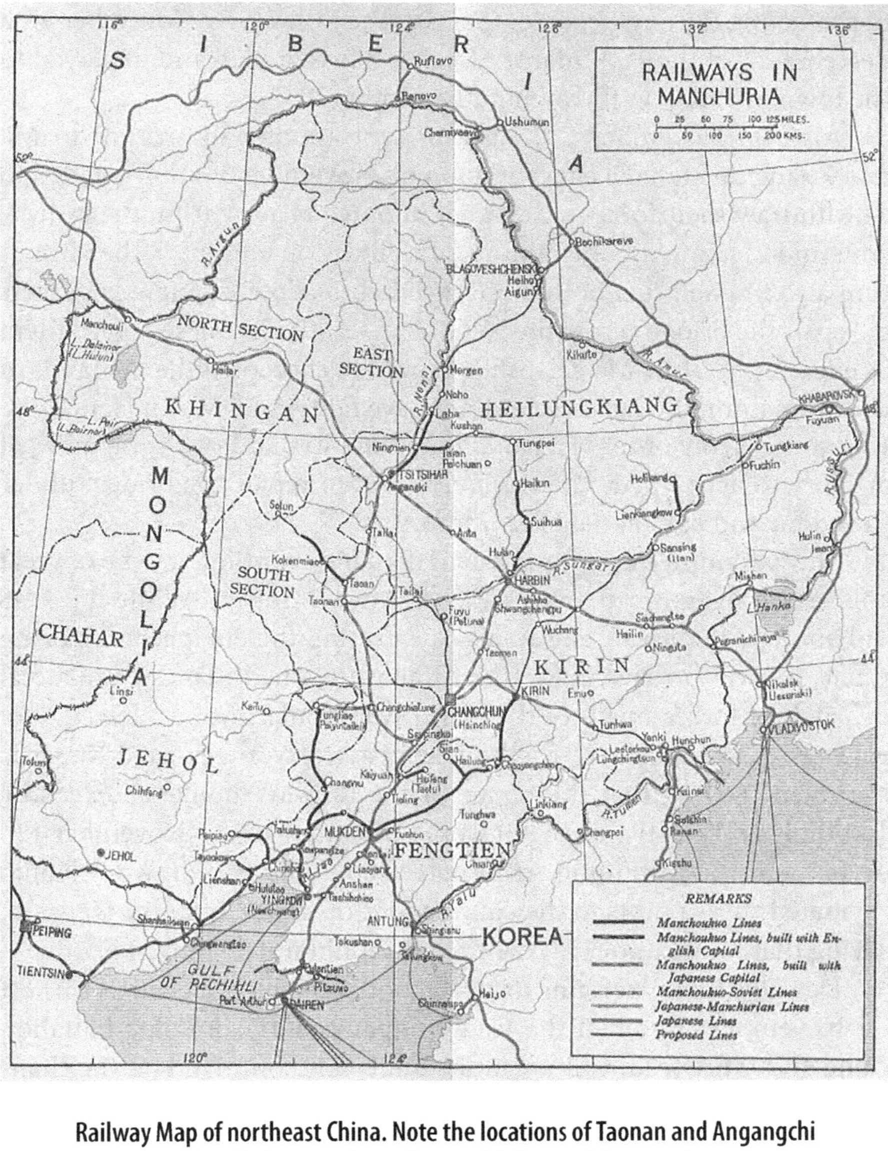

Railway Map of northeast China. Note the locations of Taonan and Angangchi (spelled Angangki on the map) (Library of Congress).

terest is evidenced in its unilateral decision to repair the railway on October 30th, three days before the deadline.[8] That day, the 16th Infantry Regiment was given orders to be ready for action, with the mission of protecting the engineers set to repair the railway. By November 1st, a detachment of the 16th Infantry Regiment assembled and deployed to the town of Tailai on the night of November 2nd.[9]

Also on November 2nd, Hayashi sent another ultimatum to Ma and Zhang due to take effect at noon on November 3rd: they needed to 1) withdraw their forces at least ten kilometers away from the railway zone and 2) not to use the railway for military operations.[10] The ultimatum also implied that both forces needed to allow Japanese engineers to repair the bridge across the Nen River, lest the Japanese regard them as enemies.[11] While Ma Zhanshan nominally accepted the demands of the ultimatum, the Japanese did not have faith in his word. Hence, the Kwantung Army ordered a battalion of infantry and two companies of artillery to accompany the engineers sent to repair the bridge, under the command of Colonel Hashimoto.[12]

The Kwantung Army's persistent demands continued to be of great concern to the government, AGS and Army Ministry. While the AGS did approve the Kwantung Army's protecting the engineers repairing the Nen River bridges, AGS Chief Kanaya warned that any unilateral dispatch of forces north of the Nen River would not be allowed, "regardless of the reason."[13] The Vice Minister of War repeated Kanaya's sentiment, saying that the repair of the railway should be executed "skillfully and swiftly" so that it would be completed by November 13th or 14th and the Kwantung Army forces could be withdrawn.[14] Honjo reiterated these points to the commander of the 2nd Infantry Division, saying that forces under his command should not be moved north.[15]

Despite Honjo's refrains to avoid conflict, fighting began in earnest on November 4th, when the lead company of Hashimoto's battalion came under fire from Chinese units under the command of Ma Zhanshan.[16] Hashimoto quickly responded to this turn of events, and reconnoitered the battlefield. Upon review, Hashimoto ruled out a frontal assault over swampy ground in the center of his line; instead, he opted for a concentrated blow on the Chinese left flank, which rested on a critical hill on the far left. Hashimoto deployed his reserve companies to seize this hill, which they did by 8:30 p.m. on November 4th.[17] However, these companies could not advance further due to their minimal numbers and lack of artillery support; any further assaults had to wait until the following morning.

On the morning of November 5, the Kwantung Army headquarters dispatched another infantry battalion to reinforce Hashimoto's forces. Upon receiving these reinforcements, Hashimoto's forces went over to the offensive once again. Even though Hashimoto's forces—roughly 1.5 battalions at this point—were able to take the first enemy position at 10 a.m., the Chinese were able to retreat to a prepared second line of defense some 700-800 meters to the rear. This line of defense consisted of strong field fortifications (mainly trenches) with roughly 70 machine guns and other automatic guns according to Hashimoto's recollection.[18] Hashimoto's forces made little progress in the face of these strong positions.

Suddenly, Hashimoto's forces were threatened by 1000 Chinese cavalry attacking the right flank and rear of the Japanese forces. This outflanking maneuver nearly enveloped the stunned Japanese units. But the Japanese soon came to their senses and called for reinforcements, which came in the form of two infantry battalions and three batteries of artillery on the night of November 5th. After receiving these reinforcements, Hashimoto's detachment was able to go over to the offensive once more, finally forcing the Chinese to retreat by November 6th.[19]

Despite the Kwantung Army's success in the battles around the Nen River bridges, the AGS still looked on the situation with great unease. Fear of war with the USSR was still the primary consideration for the general staff, which did not want to entrust the Kwantung Army with adhering to its previous orders. Therefore, on November 5th, Kanaya and the AGS asked the Emperor to assume the "right of command" in northeastern China.[20] The Kwantung Army staff did not take kindly to this measure, protesting that it was impossible for Tokyo to direct operations at the tactical level and a violation of the Emperor's right to Supreme Command.[21] The Kwantung Army proceeded to ask the AGS if its command over the Kwantung Army applied to aims, the sphere of operations, troop numbers, tactics, or all of the above; the AGS responded that its control over the Kwantung Army corresponded not only to these categories but also to "that part of Kwantung Army action, that, in the judgement of the chief of staff, needs to be directed from the center."[22]

The AGS issued its first order to the Kwantung Army on November 5th mandating that no further "positive action" in northeastern China was authorized.[23] This order caused a great stir in Kwantung Army headquarters, where Katakura, Itagaki, Miyake and Honjo all agreed that the AGS's seizure of supreme command authority was outrageous. Honjo took matters upon himself to send a telegram to Kanaya asking how long the AGS's stewardship over Kwantung Army affairs would continue. Miyake also sent a telegram to Ninomiya stating that the AGS could not make decisions regarding the pursuit of enemy forces in the field or the occupation of given areas. The AGS did not dignify their telegrams with a response, likely because there was little need to reiterate its previous positions.[24]

During the Battle for the Nen River Bridges, Tokyo authorized the assembly of a Kwantung Army corps to crush Chinese forces near the bridge. However, Tokyo also restricted the pursuit of these forces to the Xinmin-Tangchi-Daputai line, with a withdrawal in order after the operation was completed.[25] Honjo abided by the AGS guidelines and dispatched Major Hayashi to begin another round of negotiations between Kwantung Army headquarters and Ma Zhanshan on November 6th. During these negotiations, Hayashi made two demands to Ma: 1) that he resign as governor of Heilongjiang in favor of Zhang Haipeng and 2) that he form a "public safety committee" to protect Japanese interests in the region.[26] Ma's failure to give a prompt reply led to a temporary renewal of the Kwantung Army offensive near Sanchienfeng on November 7th.[27] The next day, both Hayashi and Honjo repeated their demand that Ma Zhanshan resign in favor of Zhang Haipeng, but Ma replied that he could not see Zhang due to the presence of Japanese troops.[28] Ma's rebuff led Honjo to request reinforcement from Tokyo by another infantry division on November 9th.[29] The next day, November 10th, the Kwantung Army resolved to launch an offensive against Ma and began making plans to do so.[30]

To justify their renewed offensive against Ma, the Kwantung Army decided to renew their demands knowing full well that he was not able to accept. These demands included that 1) Ma resign in favor of Zhang Haipeng, 2) Ma withdraw his forces from Qiqihar and 3) allow Kwantung Army forces to advance to Angangchi station.[31] Honjo told Hayashi to deliver these demands to Ma on November 11th.[32] But Ma Zhanshan rebuffed these terms, which gave the Kwantung Army the justification it needed to attack. On November 11th, the army issued orders to the 2nd Infantry Division to begin preparing for a new offensive

The Kwantung Army had already requested permission from Tokyo to undertake "unilateral action for self-defense" against Ma's forces on the night of November 10th.[33] Kanaya and Ninomiya once again expressed their opposition to this request, saying that "military action based solely on local military intelligence and judgment will complicate the overall picture, damage national unity, and even ruin the great achievements of the Kwantung Army since the beginning of the conflict. The chief of staff is absolutely opposed, in particular, to offensive action in the Angangchi area as contrary to national policy."[34] But others within the Japanese foreign policy establishment, including Consul General Hayashi and Uchida Yasuya, began to embrace the Kwantung Army's adventurism in Heilongjiang. After Kanaya's initial telegram, the AGS also began to come around to the Kwantung Army's views, sending instructions that Ma's army was to withdraw north of the CER, inclusive of Qiqihar. Even Shidehara began to moderate his opposition to the Kwantung Army's proposals, likely due to the dangerous atmosphere of Japanese politics in the aftermath of the October Incident. Because the new consensus in Tokyo embraced "positive action" for the Kwantung Army, it was able to renew its offensive in the region on November 14th and 15th, after negotiations with Ma broke down.[35] By persistently pushing the envelope, the Kwantung Army finally got the government and supreme command to back its program.

Despite the government and supreme command's acquiescence to the Kwantung Army's offensive, many within the Kwantung Army believed that the approved list of demands to Ma was still insufficient.[36] Therefore, on November 15th, the Kwantung Army altered the AGS's approved instructions, saying Ma's forces should withdraw north of the CER and from Qiqihar altogether. It also added a deadline of November

16th—the next day—for Ma to accept the proposals, knowing full well that this amount of time was insufficient for him to deliberate. These measures were designed to justify a further Kwantung Army advance on Qiqihar and eventually the rest of Heilongjiang.[37]

Ma once again rebuffed the Kwantung Army's demands, which led Honjo to renew his offensive on November 18th, 1931.[38] The Kwantung Army was well-prepared for this new offensive, as it had assembled the 2nd Infantry Division in the area and received ample reinforcements from the AGS, including three air regiments.[39] While vastly outnumbered (the Japanese only had roughly 3,000-5,000 men to Ma's 25,000), the Kwantung Army forces had a significant advantage in supporting weapons such as artillery and armored cars. Furthermore, Ma's forces were starting to run out of ammunition after fighting the Japanese around the Nen River Bridges for the better part of two weeks. Consequently, after two days, Ma Zhanshan's forces had to withdraw from the vicinity of the Nen River area towards Qiqihar.[40] Even though Ma asked Chiang for reinforcements, his request went unanswered as Chiang continued adhering to his "first pacification, then resistance" policy.[41]

After Ma's troops retreated, they joined a force numbering roughly 20,000 men, 24 field guns, and 100 machine guns in the outskirts of Qiqihar.[42] Here these forces had dug in, creating fortifications that stretched four miles.[43] These fortifications included machine gun positions with clear fields of fire and shrapnel-proof bunkers that were sited every eighty yards. The Chinese also had good telephone lines at multiple levels of command including the battalion and brigade levels. The advantages such prepared positions afforded should have made them almost impregnable to the mere 4,500 Japanese who would assault them, especially considering the fact they now had less fire support than the Chinese did.[44]

On November 17th, the Japanese began their attack on Qiqihar with a two-hour bombardment of Chinese lines, which did little damage to the well-prepared Chinese defenses. An advance of light tanks commenced at 8:30 a.m., but all except for one broke down in the course of the assault. Soon after, the Japanese infantry and cavalry joined the fray on a five-mile frontage.[45] At first Ma's right flank managed to hold firm against the Japanese assault, which led Ma to send cavalry to attack the Japanese right flank; but Japanese field guns and close air support stopped these attacks.[46] After the defeat of Ma's attacks, the Japanese resumed their advance, and got within 800 yards of the Chinese positions.[47] At this point, the Chinese defenders broke ranks and ran, throwing away "rifles, coats, hats, canteens, valuable pairs of officers' boots to run faster."[48] The Japanese pursued the retreating Chinese along both sides of the Qiqihar-Koshan railway line.[49] By midnight on November 19th, the Japanese entered an abandoned Qiqihar.[50]

Crossing the Nen River on the way to Qiqihar (Author's Collection, taken at the Museum of War of Chinese People's Resistance Against Japanese Aggression)

Qiqihar's capture represented a great paradox for the Japanese government and supreme command. On the one hand, they had greenlit the Kwantung Army's advance against Ma's forces in Heilongjiang on the pretext they threatened Japanese lives, property, and interests. On the other hand, elements within the government and supreme command worried about the international optics of occupying Qiqihar and still wished to avoid antagonizing the Soviet Union. The AGS made its final decision on November 16th, invoking its temporary right of command over the Kwantung Army to prevent it from permanently occupying Qiqihar. Most of the Cabinet and Wakatsuki agreed with this measure, given their tendency towards moderation.[51] While Minami initially advocated on behalf of occupying Qiqihar, he soon came around to the consensus of the Cabinet, sending a telegram to the Kwantung Army forbidding any long-term occupation of Qiqihar on November 19th.[52]

The telegrams the AGS and Army Ministry sent to the Kwantung Army were not sufficient to stop it from trying to remain in Qiqihar and continue advancing in Heilongjiang. In light of this reality, the AGS sent Vice Chief of Staff Ninomiya to Mukden to tell the Kwantung Army in person that it had to halt further offensive operations and withdraw. But Ninomiya soon found himself convinced by the Kwantung Army staff officers' determination to continue their operations. Therefore, Ninomiya sent a telegram to Tokyo expressing his belief that a complete withdrawal of the military forces in Heilongjiang could only occur in two weeks.[53]

Unfortunately for Ninomiya and the belligerent Kwantung Army staff officers, the AGS would not allow for this delay. The AGS's reply to Ninomiya's telegram demanded that the Kwantung Army immediately withdraw its forces from Qiqihar save for one regiment, "without regard to other considerations and as quickly as possible."[54] Ishiwara and some other staff officers did not want to accept the AGS's orders, and cabled Kanaya asking him to allow for some discretion in the Kwantung Army's withdrawal timetable from Qiqihar. But Kanaya shot back with another telegram repeating his order that the Kwantung Army submit to his previous order "without delay."[55] Realizing that little was to be gained from further argument with Kanaya, the Kwantung Army opted to withdraw its forces, save for a single brigade.[56] This withdrawal marked the second instance since September of the high command preventing the Kwantung Army from expanding operations.

While Ma Zhanshan's forces resisted the Kwantung Army and its Chinese collaborators in Heilongjiang, Colonel Kenji Doihara attempted to coax Puyi out of Tianjin and install him as the leader of a "Manchurian" puppet state, for which he had been advocating since September. To achieve this aim, Doihara went to Tianjin and attempted to persuade Puyi to come with him to northeastern China.[57] His pitch began with justifications for the Kwantung Army's intervention in northeastern China, which he attributed to Zhang Xueliang's bad government. Doihara went on to say that Japan had no territorial designs on northeastern China, and merely wanted to "protect the Manchurian people and enable them to set up their own independent state." He concluded by saying that Japan would be a treaty-bound ally of the new "Manchurian" state and that the Kwantung Army would protect its sovereignty.[58] Doihara hoped that this pitch would persuade Puyi to come to northeastern China and serve the Kwantung Army as a puppet Emperor.

Even though Doihara's pitch was clever and persuasive, Puyi wanted to know the answer to one vital question: whether or not the new state would officially be a monarchy or a republic.[59] At first, Doihara was noncommittal, saying that the new state would be "independent and autonomous" and headed by Puyi. When Puyi pressed again, Doihara dodged the question, saying that the matter would be decided when Puyi arrived in Mukden. But Puyi insisted on being regarded as Emperor, stating that he would not go if the new state were merely a republic. Realizing Puyi's adamance on this question, Doihara quickly changed his tune, saying of the new state "[o]f course it will be a monarchy; there will be no question of that." Thanks to Doihara's reassurance about the nature of the new state, Puyi agreed to go to northeastern China and be "restored" as Emperor.

After Doihara went back to northeastern China, the press exposed his meeting with Puyi and its purpose, leading many figures around Puyi to counsel against leaving Tianjin. Puyi's wife, the Empress Wanrong, led the way in voicing displeasure with her husband's decision to go to northeastern China.[60] Another member of Puyi's family also implored Puyi not to go to the Three Eastern Provinces, saying that he should not "acknowledge the enemy as my father." Additionally, Puyi's advisor Zhen Baoshen was "horrified" by his meeting with Doihara and made clear his desire for him to remain in Tianjin. But as Puyi later put it, he was "too carried away by my desire for restoration" to heed their warnings.[61] Come what may, Puyi intended to go to northeastern China and resume his position as "Emperor."

Despite Puyi's determination to go to the region, he did not make his feelings known to the public. Consequently, Doihara was still not certain of Puyi's intentions, despite their earlier conversation. Furthermore, because the port of Yingkou would freeze over in November, Doihara wanted to hasten Puyi's decision and compel him to come to northeastern China.[62] He therefore took a variety of steps to ensure Puyi would leave Tianjin by November 16th. First, he sent a Japanese agent who happened to be a relative of Puyi named Kawashima Yoshiko to convince him that he should go.[63] Then, Doihara organized, and armed Chinese gangsters and Japanese plainclothes policemen to riot in the Chinese section of Tianjin. The first riot occurred on November 8th, 1931, and saw roughly 2,000 rioters clash with Chinese police.[64] Puyi also received threatening letters, and a bomb placed in a basket of fruit at Doihara's behest.[65] The combination of these events was enough for Puyi to expedite his decision to leave Tianjin. On November 10th, Puyi left the city and boarded a boat to the city of Yingkou.[66]

While Doihara induced Puyi to leave Tianjin, the Japanese garrison in the city sought a pretext to advance into North China. As the garrison commander Lieutenant General Kashii-Kohei put it in a telegraph to Honjo in Mukden, "it will not be easy to let Chang Hsueh-liang fall by himself. To facilitate the solution of the Manchurian question, it is necessary not only to wipe out the old Northeastern warlords in Manchuria but also to destroy Hsueh-liang's influence in north China."[67] But in order to destroy Zhang Xueliang's forces in the north, the weak Tianjin garrison needed more manpower, as it was only 583 men strong. These realities forced the Tianjin garrison to seek reinforcements from the AGS.

Unfortunately for the Tianjin garrison, the AGS was not keen on reinforcing them. The General Staff made clear its unwillingness to reinforce Tianjin in a November 12th telegram, saying that "overall considerations make it impossible to send you reinforcements. You are instructed to inspire the morale of your men, maintain close contact with related organs, and be prepared to deal with all developments with your existing strength."[68] The AGS did not believe that sending reinforcements to Tianjin would resolve the situation in the northeast. But the Kwantung Army believed that Doihara and the Tianjin garrison could fabricate a pretext for facilitating a relief expedition for Tianjin via Jinzhou, where the Kwantung Army could also engage Zhang Xueliang's recently reestablished forces. Therefore, the Kwantung Army indicated its willingness to support Doihara and the Tianjin garrison in late November.

The Kwantung Army's pretext to move on Jinzhou came in the form of renewed riots in Tianjin on November 26. These riots, bankrolled by Doihara, were characterized by a great explosion and plain-clothes men emerging from the Japanese concession of the city brandishing small arms.[69] In the name of suppressing the "riots" in the city, Ishiwara and other Kwantung Army staff officers persuaded Honjo to reinforce the embattled Japanese concession at Tianjin.[70] The Kwantung Army also asked the Korea Army for reinforcements, and sent a telegram to Tokyo asking permission for their dispatch upon the request of Korea Army commander Hayashi Senjuro.[71]

After receiving Honjo's orders on the morning of November 27th, the 4th Mixed Brigade and 2nd Infantry Battalion began moving south along the Beijing-Mukden railway towards Jinzhou and Shanhaiguan.[72] By mid-morning, one armored train, a troop train, and some supporting aircraft crossed the Liao River. Soon thereafter, Japanese troops attacked the remnants of Chinese outposts with support from the armored train and aircraft, hastening the withdrawal of Zhang's forces.[73] It appeared by that evening that the Kwantung Army would be well on its way to Jinzhou, Shanhaiguan, and Tianjin.

The AGS learned of the second Tianjin incident at midnight on November 27th.[74] The AGS discussed what should be done the whole night, with some advocating the cancellation of the Kwantung Army's operations and others, spearheaded by Tatekawa, advocating to allow the army some leeway in the event that the situation in Tianjin spiraled out of control. By 11 a.m., after it was established that the situation in Tianjin was under control, the AGS finally decided to send a series of telegrams to the Kwantung Army clarifying its position. The telegrams directed the army not to cross the Liao River, not attack Jinzhou, and refused reinforcement for the Kwantung Army from Korea.[75]

Unbeknownst to the AGS, the Kwantung Army's forces had already crossed the Liao River, rendering their first attempt to prevent the Kwantung Army's advance moot. Indeed, the Kwantung Army cited the 4th Brigade's deployment and engagement with Chinese forces as the reason why it could not immediately withdraw back across the Liao River. The AGS did not take kindly to this suggestion, replying that all of its forces needed to be withdrawn back across the Liao "regardless of the existing circumstances."[76] Because the Kwantung Army did not respond to this telegram, the AGS received imperial sanction to invoke its right of supreme command, and delivered a message to the Kwantung Army castigating its unilateral actions at Tianjin.[77] This message from the AGS combined with the restoration of order in Tianjin led the Kwantung Army to finally halt its assault on November 29.[78]

To a great extent, the AGS's adamance on the Kwantung Army's withdrawal back across the Liao River stemmed from diplomatic negotiations abroad, which began when Chinese Foreign Minister Wellington Koo promised to the US, UK and French Foreign Ministers that China was prepared to withdraw from Jinzhou to Shanhaiguan on November 24th.[79] This proposal was conveyed to Shidehara by the French ambassador on November 26th. After Shidehara told the Army Minister and Navy Minister of this Chinese proposal, all three ministers were in favor of withdrawing the Kwantung Army, which the AGS finally succeeded in accomplishing on November 29th.[80]

Unfortunately for Shidehara and other advocates of restraint in the Three Eastern Provinces, China's KMT government nixed Koo's offer of unilateral withdrawal on November 25th. Instead, the administration proposed the creation of a neutral zone at the League of Nations.[81] Soon thereafter, negotiations between China and Japan fell through because the KMT government suddenly changed its stance, instructing its representative to the League, Alfred Sze, to oppose the creation of a neutral zone. The sudden turn against the formation of a neutral zone led the Kwantung Army to request one division reinforcement from the AGS. The AGS agreed to send one mixed brigade to the Kwantung Army from Korea, in large part because of their resentment of China's sudden shift in diplomatic posture.[82] With the AGS's approval, the Kwantung Army once again began preparing for operations against Zhang's forces near Jinzhou.

In Japan, the Wakatsuki government was fracturing due to the Kwantung Army's actions in northeastern China. Back in October, Home Secretary Adachi Kenzo began the process of fracturing the Wakatsuki Cabinet, declaring that he was boycotting all further Cabinet meetings due to Shidehara's "spineless" China policy.[83] Adachi's stated views were more in line with the Seiyukai's claims about Wakatsuki's government, which it accused of being too conciliatory to the Chinese and not hawkish enough on the question of northeastern China. While Adachi did not resign his post, his public opposition to Shidehara's foreign policy undermined the integrity of the Minseito party, which was unprepared for a public boycott from one of its key members. Many figures in the party feared the loss of Home Ministry resources for the next general elections, set to take place in the spring of 1932, forcing them to try and find a solution to the situation at home and abroad.[84]

On Adachi's suggestion, Wakatsuki tried to form a coalition with the Seiyukai party in a desperate bid to prevent his government from fracturing. Wakatsuki hoped that this move would force the Kwantung Army to obey a united civilian government, and remove overt political opposition from the Seiyukai party.[85] But because Shidehara and Inoue opposed this measure, Wakatsuki dropped the idea on October 31st.[86] Nonetheless, later in November, Adachi betrayed Wakatsuki's trust, informing the press of Wakatsuki's intention to form a coalition government.[87] Adachi's statement did great damage to his former Prime Minister's government as it undermined its legitimacy in the eyes of the army and public. By late November, the downfall of the Wakatsuki Cabinet was only a matter of time.

Adachi also tried to unilaterally form a coalition government with the Seiyukai behind Wakatsuki's back.[88] However, when word of Adachi's negotiations reached the Wakatsuki Cabinet, they vehemently opposed Adachi's gambit and requested his resignation. Adachi in turn played hardball, saying he would only resign once the Wakatsuki cabinet resigned.[89] Adachi's betrayal signaled the final nail in the coffin for Wakatsuki's cabinet, all of whose members resigned on December 11th.[90] A new Seiyukai cabinet formed on December 13th, with Inukai Tsuyoshi as Prime Minister and Lieutenant General Araki Sadao as Army Minister.[91] This new government was to be more amenable to the Kwantung Army's initiatives and activities than its predecessor.

The League of Nations assembled again on December 10th and passed a new resolution on the situation in northeastern China, mandating that the Japanese withdraw back to the railway zone.[92] The Japanese accepted this resolution, but only on the condition that Japan could utilize its military forces against bandits and elements threatening Japanese lives and property. The League agreed to this Japanese condition, saying that "circumstances might arrive [in Manchuria] causing damage to Japanese lives and property and in such an emergency, it might be inevitable Japanese forces in the neighborhood take action."[93] The League's agreement to accept Japan's demands represented, in effect, the capitulation of liberal internationalism to Japan's aggression, despite the League's charter, China's appeals, and Japan's pre-existing treaty commitments. The League's December 10th resolution was the first step towards its being dis-credited as a peace-keeping institution.

The Kwantung Army interpreted the League's December 10th resolution as allowing Japan the "right to maintain her troops" in northeastern China and continue operations towards Jinzhou.[94] To prepare for operations against Jinzhou, the Kwantung Army began conducting reconnaissance in the last half of November and first half of December. These missions and the resulting skirmishes were aptly summarized in a Japanese report on army activities near Jinzhou:[95]

> Based on reconnaissance by our forces stationed around Huludao, west of the Tiger River line and its vicinity, a significant enemy force has been identified, with fortified positions and defensive arrangements established across several areas, including Jinzhou. Our troops are conducting movements to probe enemy defenses near these key points.

> Along the railway lines stretching from Jinzhou toward Shanhaiguan and further along the coast, multiple groups of scattered enemy units have been observed. Our units stationed in nearby areas are maintaining defensive positions and monitoring enemy activity.
>
> Enemy forces on the north bank of the Luan River, north of Fengtian, are being closely watched. Although only a few dozen enemy troops have been sighted, they are likely part of a reconnaissance or diversionary operation.
>
> Due to the terrain, our response has been cautious, and reinforcements are being considered to support the operations near Jinzhou.
>
> From November onward, there have been a total of 29 confirmed incidents, including 17 in November and 12 in December, involving skirmishes, shelling, and firefights along various sectors. Among these, five involved large-scale confrontations.
>
> Currently, the regular army in this district is fully engaged in combat operations. Additional forces are being deployed to address the situation along the western section of the Changchun Railway line, especially in the area around Maoling.
>
> Enemy units in the region are increasingly active, and our command is now formulating countermeasures to secure the rail network and strategic points.
>
> Enemy retreats have been observed in multiple sectors, including the western front of Manchuria, suggesting a planned regrouping.
>
> Our forces remain on alert and are pursuing scattered enemy units while maintaining control of occupied positions.

The successful probes of existing strongpoints, combined with the withdrawal of Chinese troops, gave the Kwantung Army the confidence necessary to begin preparing for offensive operations. To these ends, the army assembled the 2nd Infantry Division in the vicinity of Mukden.[96] In mid-December, the 4th Brigade of the 8th Infantry Division reinforced the 2nd Infantry Division. Later that same month, on December 27th, the Kwantung Army gained imperial approval for the dispatch of the 20th Infantry Division and another brigade from the Korea Army.[97] The arrival of these reinforcements increased the Kwantung Army's odds of success in the coming offensive on Jinzhou.

On December 28th, the 2nd Infantry Division began its offensive against Jinzhou.[98] Even though the Kwantung Army feared a protracted fight in the event the Chinese decided to resist, Zhang's forces in fact continued retreating.[99] By the end of the month, the Kwantung Army realized Zhang had little intention of resisting their advance on Jinzhou and ordered the 20th Infantry Division to rapidly advance on the city. On January 3rd, at 10:40 a.m., the 20th Infantry Division entered Jinzhou.[100] Soon thereafter, the Kwantung Army reached the Great Wall at Shanhaiguan, thereby threatening the gateway to the North China plain and the capital, Beijing.[101]

Soldiers of the Kwantung Army celebrating their conquest of Jinzhou (Author's Collection, taken at the Museum of War of Chinese People's Resistance Against Japanese Aggression)

After the Kwantung Army seized Jinzhou, it turned its sights north towards Harbin. Here, Ma Zhanshan and his forces continued to resist in coalition with General Ting Chao. Crucially, both Ma and Ting maintained communication with their commanders, Chiang Kai-shek and Zhang Xueliang, meaning that the Chinese command and control structure remained intact.[102] Consequently, Doihara gave Xi Qia the task of attacking Harbin with his Jilin Self-Defense Army on the pretext of flushing out troops loyal to the former governor. Xi obliged, seizing the town of Shuangcheng in late January. But Ting Chao's forces soon checked the advance of the Jilin Self Defense Army at the end of January.[103] Ting's resistance temporarily frustrated Doihara's desire to take Harbin and eliminate Ma's forces.

To aid Xi Qia's forces and advance on Harbin, the Kwantung Army needed a justification. Doihara thus utilized the connections he made as an intelligence officer to help instigate riots that led to the deaths of a Japanese and three Koreans.[104] The Kwantung Army then used these riots to justify deploying the 3rd Mixed Brigade of the 2nd Infantry Division north to Harbin to "protect" Japanese residents and property in late January. Honjo also asked the AGS for a further two battalions of reinforcements.[105] Unlike in past months where Tokyo had opposed the Kwantung Army's unilateral activities, the supreme command approved this advance on Harbin because of the replacement of the Wakatsuki cabinet with the new Seiyukai government and the success of the Qiqihar operation in November, which did not incur the wrath of the Soviet Union.[106] The Kwantung Army was now finally able to advance on Harbin, which it had intended to do since mid-September.

The Kwantung Army initially had a difficult time deploying forces to Harbin, in large part because the Soviets denied the Kwantung Army passage along the Chinese Eastern Railway.[107] But Japanese officers soon forced the transport of some of the remaining forces and managed to get three troop trains moving towards Harbin on the night of January 28th.[108] By January 30th, the Soviet manager of the CER finally agreed to transport the remaining Japanese units to Harbin, thereby unleashing a full-scale assault on the city.[109]

For seven days, the Japanese converged on Harbin in multiple columns, enduring repeated Chinese raids and frigid temperatures. But on February 1st, the 2nd Infantry Division's main forces began to arrive in the Harbin area, concentrating at Shuangcheng by February 3rd.[110] That same day Ting Chao's roughly 14,000 strong anti-Jilin Army took up entrenched positions south of Harbin. Of these 14,000 men, 1,000 with six field guns were thought to be on the right flank, 3,000 with two field guns in the center, 7,000 with four field guns on the left, and a further 2,000 to 3,000 in reserve.[111]

On February 3rd, the 2nd Infantry Division began advancing on the Chinese positions, reaching the city of Weitangkou, twenty miles north of Shuangcheng on the night of February 3-4th. The next day at 3:30 p.m., the 2nd Infantry Division began its assault in earnest, concentrating its main effort on the center of Ting Chao's line.[112] A supporting offensive occurred on the right flank, where the Japanese advanced to within 500 meters of the Chinese positions. By nightfall on February 4th, the Japanese had partially seized the Chinese positions; by 12 p.m. the next day, the Japanese had completely routed the Chinese from their positions, before marching into Harbin in the afternoon.[113]

Immediately after the seizure of Harbin, the Kwantung Army did not exploit its success by pursuing the Chinese forces. But since Ma's forces still controlled the railroads north and east of Harbin, as well as the Sungari River on the border of the Soviet Union, they were still a formidable threat. The Kwantung Army also wanted to seize these objectives because they would gain the advantage of interior lines and establish defensible borders in the event of a conflict with the USSR. Therefore, at the beginning of March, the 2nd Infantry Division moved towards the town of Fangcheng to suppress the remnants of Ma Zhanshan's forces. They advanced as far as the town of Sanxing before returning to Harbin in early April. The 10th Infantry Division arrived in the city and relieved the 2nd Infantry Division that same month.[114] The 10th Infantry Division proceeded to engage the anti-Jilin Army near Sanxing and near the Chinese Eastern Railway at Hailin in the spring of 1931.

Meanwhile, the 14th Infantry Division reinforced the 10th Infantry Division's mopping up campaign of Ma Zhanshan's forces that May.

The 14th Infantry Division's main element engaged Ma's army north of Harbin late in the month, along the Halun-Hailun Railway. Supporting elements attacked east of Koshan near the junction of the Qiqihar-Koshan Railway. The 14th Infantry Division effectively routed Ma's forces, and forced them to become irregular by the autumn of 1932.[115] By this point, the Kwantung Army's campaign in northeastern China was effectively over, as Ma's forces were forced to the margins of relevancy. The campaign's termination marked the end of one of the most efficient exercises in limited warfare of the twentieth century, netting Japan a vast tract of land the size of Germany and France put together.

Chapter 15

Japan and the World Reacts to Kwantung Army Aggression

The Mukden Incident represented a great blow to the regime of pacifism, disarmament, and respect for territorial sovereignty that defined the liberal international order after World War I. For the first time since the war's end, Japan—a great power—had invaded a neighboring country, despite signing treaties that respected China's sovereignty and renounced war as a means of national policy. Chiang Kai-shek and the KMT did little to resist the Kwantung Army's invasion of its Three Eastern Provinces, instead misplacing trust in the League of Nations, the United States, and the existing system of pacts and treaties to rescue China. In retrospect, the failed promise of peace in a postwar liberal international order raises some interesting questions. First, how did people in China react to Japan's invasion of northeastern China and Chiang's "first pacification, then resistance" policy? Second, how did the Japanese public react to the Kwantung Army's invasion? Third, how did other foreign powers, namely the United States, react to Japan's invasion of the Three Eastern Provinces? Finally, what measures did the League of Nations and the liberal international system take to respond to Japan's invasion, and why were they ultimately inadequate? While the answers may be complex, to answer these questions get at the reasons the Mukden Incident succeeded. Therefore, the final substantive chapter of this book will be a summary and analysis of reactions to the incident in China, Japan, and other nations around the globe. Only by examining these reactions can one fully grasp the Mukden Incident's place in history.

China's Reaction to the Mukden Incident

In China, the incident and its aftermath triggered indignation from the KMT government, articulated in an article titled "Japan's Violation of International Law." In it, the KMT condemned Japan's invasion of north-eastern China and called on both the signatories of the Kellogg-Briand Pact and the League of Nations to take action against Japan. Excerpts from this article can be seen below:[1]

> The sudden invasion of South Manchuria by Japanese troops... is in effect a gross violation of a number of treaties and agreements to which Japan is a party...
>
> In the first place, Japan has violated the "Regulations Respecting the Laws and Customs of War on Land" of both the 1899 and 1907 Hague conventions. By shelling the city of Mukden and the North Camp [the Beidaying barracks] without warning, the Japanese military authorities have violated Article 26 Section II of the Annex of the Convention relating to land warfare: 'The officer in command of an attacking force must, before commencing a bombardment, except in cases of assault, do all in his power to warn the authorities...'
>
> Again, in attaching her signature to anti-war Kellogg Pact of August 27, 1928, which binds the contracting parties to declare "in the names of their respective peoples that they condemn recourse to war for the solution of international controversies, and renounce it as an instrument of national policy in their relations with one another," Japan declared in her note to the United States on May 26th, 1928: 'The Government of Japan sympathize warmly with the high and beneficent aims of the proposal now made by the United States which they take to imply the entire abolition of the institution of war, and that they will be glad to render their most cordial cooperation towards the attainment of that end...'
>
> In becoming a member of the League of Nations 'by the acceptance of obligations not to resort to war' Japan, as a permanent member of the Council, must respect Article X of the League Covenant which says 'The members of the League undertake to respect and

> preserve as against external aggression the territorial integrity and existing political independence of all members of the League.'
>
> Japan's obligations *to maintain peace and refrain from making aggressions upon China* are further put down in black and white in the Nine Power Treaty signed by the United States, Great Britain, Belgium, China, France, Italy, Japan, the Netherlands, and Portugal at the Washington Conference, binding these nations to adopt "a policy designed to stabilize conditions in the Far East, to safeguard rights and interests of China, and to promote intercourse between China and the other powers on the basis of equal opportunity...'
>
> Japan has violated the law of nations by invading Chinese territory without previous warning and has committed acts of barbarity which should brand her as a criminal in the eyes of the world. Who is the aggressor is only too self-evident—when Japan slapped one cheek, China turned the other. Will the signatories of the anti-War Kellogg Pact who attached their signatures to high-sounding phrases outlawing aggressions upon weak and unoffending states continue to fold their hand while the Chinese people are being massacred by the thousands and their land forcibly taken from them by the Japanese military brigands? Will the League of Nations do nothing and advance a few technical excuses for inaction? How long will these treaties respecting the integrity of China be treated as mere scraps of paper by Japan? These are the questions which every intelligent Chinese and foreigner in this country are asking themselves.

The KMT made the case that Japan had violated multiple treaties by invading northeastern China. This case was relatively straightforward, as Japan had previously committed to renouncing war as an instrument of policy and respecting China's territorial integrity. But the question of whether the League of Nations would sanction Japan remained to be answered. While Chiang put forth a request to the League to "take immediate steps to prevent the further development of a situation endangering the peace of nations," the League's mechanisms for enforcing its decisions remained relatively untested, despite its widespread membership.[2] Therefore, as Chiang Kai-shek himself put it in a speech on September 25th, the Mukden Incident would "test whether or not there is a such thing as international morality or justice between and among nations, and whether or not the Powers will sanction the punishment of international commitments and to preserve the peace of the world." Chiang nonetheless had faith in the League, as evidenced by his statement that he believed "the League of Nations and the signatories of the Kellogg-Briand Pact will take appropriate steps to bring justice to the disgraceful conduct of the Japanese which is obviously contradictory to the Covenant of the League and the spirit of the Pact."[3] From this quote, one can conclude that Chiang hoped the League and existing treaty infrastructure would be enough to force Japan to abandon northeastern China. In this way, he would avoid committing his forces to fighting the Japanese while his internal enemies again grew stronger.

On September 23rd, Chiang Kai-shek implored his fellow Chinese to keep calm and place their faith in international institutions centered around the League of Nations in Geneva. If these entities did not come to the country's aid, Chiang promised to lead an army himself to take the Northeast back. Chiang's exact words are excerpted below: [4]

> For the moment we await judgment of the world and we must suppress our indignation and remain calm but if the League of Nations and signatories of Kellogg Pact fail to uphold justice the National Government is prepared for a final and supreme struggle. I shall lead the Army to fight for preservation of our race and to uphold dignity of our people. I shall go to the front and if need be fall with other patriots.

While Chiang paid lip service to fighting the Japanese if necessary, many Chinese did not regard his response to be sufficient in the face of a blatant attack on China's sovereign territory. Their attitudes were summarized in a telegram from the US Consul General in Mukden:[5]

> The first thing to be recorded about this terrible affair is the utter and complete humiliation that is felt by everyone. One cannot witness bitter weeping by strong men in financial and educational circles without

> realizing that the rancor of this experience will last a lifetime. Secondly, one is moved by the confession of personal failures and the general condemnation of the folly and worse, committed by official leaders in the face of terrible suffering within the country and invasion from without. What as educationalists and business men have we done to prepare our people to meet these disasters, and where are the guardians of our homes and country?"—such are some of the questions that are now being asked of individual hearts and concerning the public officers at present in Peiping, in Nanking and in Canton.

The condemnation of Chiang and the KMT came into the open on September 28th, when a mob of students invaded the Foreign Ministry and attacked its head, Wang Zhengting.[6] Later in November and December, patriotic students and merchants triggered boycotts of Japanese goods and violent protests of Chiang's "first pacification, then resistance" policy.[7] The Chinese press promoted these activities, publishing polemics that relentlessly criticized Chiang's timidness. While the KMT attempted to crack down on the pro-resistance press, enforcement was relatively lax until 1934. Therefore, journals such as Zou Tao-fen's *Shenghuo Zhoukan* were able to publish anti-Chiang articles with impunity.[8]

The success of the KMT's "first pacification, then resistance" policy was predicated on the effectiveness of the untested liberal international system. But even though Chiang Kai-shek hoped Geneva could rein in the Japanese, the League was actually in no position to pull Chiang's proverbial chestnuts from the fire. The League of Nations did not have an armed force at its disposal for the settlement of disputes. The United States and the other western powers were in the midst of the Great Depression and disarmament. In light of such shortcomings, Chiang should have recognized that the agreements of the 1920s were indeed "mere scraps of paper," and taken measures to put together a coalition that could have resisted Japan. That he did not immediately do so indicates an ill-advised prioritization of his own power at the expense of a strong defense of the Northeast.

Japan's Reaction to the Mukden Incident

In Japan, the public by and large supported the Kwantung Army's actions in northeastern China. Mass fundraisers were held, goods were donated to the army, and large rallies organized to convey support for the Kwantung Army's actions. The mood was summarized by a foreign observer:[9]

> war maps and vivid pictures attract crowds to the show windows where they are displayed; at night electric ideographs flash across the faces of the great newspaper offices telling the day's happenings in Manchuria to the populace of Osaka and Tokyo. Young middle-school students, emphasizing their importance with pennants and arm bands, patrol the downtown streets, haranguing the crowds for funds to help the 'heroes' up in the bitter northern territory. Girls make appeals for nursing volunteers. Newspaper headlines clamor about 'the enemy', 'the latest engagement', the 'shifting front', 'our troops' progress westward'. It is difficult to reconcile all this activity, all this deliberate stirring of patriotism, with the solemn official pronouncements of 'no war in Manchuria'.

From this excerpt, it is obvious that the Japanese media played a significant role in the public's jingoistic embrace of aggression in the Three Eastern Provinces. For years beforehand, the press had written about how the Chinese were endangering Japanese citizens in the region, and disrespecting Japan's extraterritorial rights. By 1931, the supposed "attack" on the SMR represented the proverbial "final straw" in the views of many Japanese, who believed it was necessary to teach the insolent Chinese a lesson once and for all. Once war broke out, the press capitalized on the Kwantung Army's advances to "sell" the war to the public through sensationalism and jingoism; this reporting enthralled many Japanese, as evidenced by the fact that maps of northeastern China in Osaka sold out within five days of the Mukden Incident.[10] Such sensationalism continued as the Kwantung Army advanced further into northeastern China and whipped up more support for the war, a trend that would continue in future years as Japan moved down the war path.

Importantly, the Kwantung Army explained its invasion of northeastern China in terms of protecting the empire's subjects and their rights in the region. Honjo justified the Kwantung Army's invasion on the basis that Zhang Xueliang's Northeastern Army had "butchered" Koreans (who were Japanese subjects) "like dogs." The Japanese government corroborated Honjo's declaration, stating that only once it felt confident in the ability of Chinese to "protect life and property of the Japanese residents" would it withdraw its forces.[11] Indeed, even Shidehara stated that the Kwantung Army had intervened as a matter of "police protection" for Japanese and Korean citizens in northeastern China.[12] The emphasis these government officials placed on protecting the lives and property of Japanese in the region picked up a press theme from previous years which emphasized the supposed danger posed to Japanese citizens by the unruly Chinese. Official delegation of responsibility to protect Japanese citizens abroad further legitimized the Kwantung Army's intervention in the eyes of many Japanese. Therefore, the government and army played just as great a role in "selling" the invasion of northeastern China as the press, despite the real opposition of many officials to the Kwantung Army's unilateral activities. Such officials opposed the form, but not the content, of empire in the Three Eastern Provinces.

To be sure, there were some who opposed the invasion of northeastern China. Committed communists were at the vanguard on the basis of their anti-imperialist principles.[13] Certain liberal intellectuals also opposed the invasion as pacifists who wanted to maintain good relations with the west.[14] Even certain members of the press were skeptical of the Kwantung Army's official narrative, and managed to communicate that skepticism in their columns.[15] However, these elements were isolated, making little impact on mainstream Japanese society as it got swept up in war fever. In part, the ineffectiveness of these elements was due to army censorship, which largely prevented reporting that was critical of its initiatives.[16] But the "war fever" that gripped Japan took hold mostly with the complicity and support its press, academia, and mainstream politicians. All were willing to go along with the invasion once it became clear that the Kwantung Army's operations would succeed. The willingness of liberals to stay silent or even endorse the invasion was a stark contrast from previous years, when these same liberals had allied with other societal elements to check the army's power. The new state of affairs indicated the extent to which army radicals had captured politics in Japan and augured the army's future political ascendancy.

The West Responds to the Mukden Incident

In light of the disarmament treaties and pacifist atmosphere of the 1920s, Japan's invasion of northeastern China came as a surprise to many in the west. But even without the element of surprise, the west's most powerful countries, the United States and United Kingdom, were ill-prepared to immediately take action against Japan because their economies were mired in the Great Depression. These countries instead called for both sides to make efforts to end the fighting and restore peaceful relations in East Asia. To these ends, the Americans sent both the Chinese and Japanese Foreign Ministries the following message:[17]

> The Government and people of the United States have observed with regret and with great concern events of the past few days in Manchuria. In view of the sincere desire of the people of this country that principles and methods of peace shall prevail in international relations, and of the existence of treaties, to several of which the United States is a party, the provisions of which are intended to regulate the adjustment of controversies between nations without resort to use of force, the American Government feels warranted in expressing to the Chinese and the Japanese Governments its hope that they will cause their military forces to refrain from any further hostilities, will so dispose respectively of their armed forces as to satisfy the requirements of international law and international agreements, and will refrain from activities which may prejudice the attainment by amicable methods of an adjustment of their differences.

The British, French, German, Italian, and Spanish Foreign Ministries soon followed America's lead. In October, these countries' Foreign Ministries each sent China and Japan telegrams "calling the attention" of both governments to Article II of the Kellogg-Briand Pact (in which signatories foreswore war as a means of resolving disputes).[18] The attribution of responsibility to both sides for the Mukden Incident indicates the Western powers fecklessness in the face of Japan's aggression.

While the Western powers called for both China and Japan to adhere to the Kellogg-Briand pact and maintain peace in the Asia-Pacific region, the private assessments of the US State Department were entirely different. Vice Consul to Mukden Lynch went so far as to say in a report to Stimson on September 20th that he "found it difficult to give credence to the Japanese explanation of the Incident" because the Japanese foray into northeastern China was "too sudden and too concerted to have been caused by an explosion on the South Manchuria Railway Line at 10:30 p.m."[19] These views were corroborated by other US diplomatic and intelligence assessments from the region, which held the Kwantung Army responsible for the Mukden Incident and invading northeastern China. Three of these assessments are excerpted below, with one each from the period before, during, and after the incident:[20]

> Dr. Ferguson called today and after some conversation about various matters and before leaving he said that he felt very much concerned about the situation that was growing up between Japan and China. He suggested that I might not wish to go south to resume negotiations in regard to extraterritoriality in view of the fact that Japan was bound to take drastic action vis-a-vis China very shortly. I asked him what he meant by drastic action. He said his information was that Japan would occupy Manchuria within the next three months. He said that a high Japanese official had made a tour in China for the purpose of investigating the situation here and had reported to his Government that the opportunity for taking this action had now arrived and he had recommended it.
>
> Mr. Donald stated that he had received a personal message to the effect that firing was continuing at one o'clock this morning and that

> Japanese soldiers had been seen marching in the direction of the west gate of the city, the inference being that the Japanese were making a move to occupy the city of Mukden. Mr. Donald stated that their information was that the Japanese military had gotten completely out of hand at Mukden, that the Japanese civilian authorities, namely the consul general, were powerless to do anything.
>
> Nobody here, not even the Japanese themselves, appears to credit the Japanese claims that the Chinese provoked hostilities and that the Japanese forces acted in self-defence. The very nature of the Japanese measures, punitive from the outset, precludes belief in their self-defensive character. Outside of stereotyped statements for foreign consumption, the local Japanese authorities, residents, and press seem to have forgotten the original charge of Chinese aggression, and look upon the occupation of South Manchuria as a political measure to settle international disputes of long standing.

US Secretary of State Henry Stimson likely read reports such as those above and incorporated their analysis into his assessment of events. The content of these reports was reflected in Stimson's September 19th diary entry that stated the Japanese "made a coup" in northeastern China, with an added caveat that the Kwantung Army might have acted unilaterally.[21] As Stimson learned more, he became even more convinced that the invasion was premeditated, because the "orderliness, precision, coordination of time, and comprehensiveness" could not have been executed without thorough staff planning beforehand.[22] He also incorporated the factor of *gekokujo* into his analysis, stating his belief that the Kwantung Army had acted "against the opposition of their government representatives," in an off the record conversation with US media members.[23] But Stimson still wanted to give Shidehara and his Foreign Ministry a chance to get the Kwantung Army back in line, as he knew that these elements were a force for moderation in the Japanese government.[24] Therefore, in the first months after the Mukden Incident, his State Department refrained from making any public statement explicitly condemning Japan, despite its opposition to the invasion of northeastern China.

By November, it became clear that Shidehara and his fellow moderates were losing ground in their effort to stop the Kwantung Army's advances in northeastern China. Consequently, Stimson and State Department officials began making public statements that undermined the legitimacy of the Kwantung Army's presence in northeastern China. These statements began on November 1st, when Undersecretary Castle said that "the United States would feel compelled under the Treaties guaranteeing China's integrity to disapprove of permanent Japanese occupation of Manchuria."[25] On November 6th and 18th, Stimson reiterated the official position of the United States government, which centered around cooperating with other nations to ensure a peaceful resolution to the situation in northeastern China.[26] But after rumors of renewed operations at Jinzhou came to the surface in late November, Stimson held a public press conference where he condemned Japan's duplicitous diplomacy and its actions in northeastern China.[27] Even though he tried to mitigate the fallout of this press conference, the Japanese press ran with his purportedly incendiary statements, and helped contribute to a deterioration of US-Japan relations.[28] The fallout from Stimson's statements in Japan was so severe that the Army and Navy Minister's called for Shidehara's resignation.[29] Such an outcome would have been contrary to what Stimson intended, even though he was correct about Japan's duplicitousness.

After the Japanese renewed their advance in December, Stimson took a harder tack against Japan's foreign policy in northeastern China. Japan clearly had no intention of withdrawing back to the SMR zone, and would likely attempt to install a puppet government to administer the Three Eastern Provinces. Consequently, the Secretary of State formed a new doctrine known as the "Stimson Doctrine," stipulating the US not recognize states that were created in the course of an aggressive war. The following text makes clear the Americans' new foreign policy doctrine:

> [T]he American Government deems it to be its duty to notify both the Imperial Japanese Government and the Government of the Chinese Republic that it cannot admit the legality of any situation de facto nor does it intend to recognize any treaty or agreement entered

> into between those Governments, or agents thereof, which may impair the treaty rights of the United States or its citizens in China, including those that relate to the sovereignty, the independence, or the territorial and administrative integrity of the Republic of China, or to the international policy relative to China, commonly known as the open door policy; and that it does not intend to recognize any situation, treaty, or agreement which may be brought about by means contrary to the covenants and obligations of the Pact of Paris of August 27th, 1928, to which treaty both China and Japan as well as the United States are parties.[30]

Much like the previous note to the Chinese and Japanese governments, the American government nominally stated its desire for both sides to adhere to the Kellogg-Briand Pact. The United States also stated that it would not recognize any situation, treaty, or agreement that contradicts this pact, whether brought about by China or Japan. But the context of this note was different from the one sent in 1931, because Japan had widened the scale and scope of its invasion despite maintaining the appearance of desiring a negotiated withdrawal. Furthermore, the US also caught wind of the fact that Japan sought to establish a puppet government in Manchukuo, using the last Qing Dynasty Emperor Puyi as a figurehead.[31] In light of the threat these actions posed to America's "open door policy" in China, the US refused to acknowledge the state of Manchukuo or the legitimacy of Japan's conquest.

The Lytton Commission

The uncertain nature of events surrounding the Mukden Incident meant that few conclusive assessments could be made about the nature of the conflict between China and Japan. While military attachés and fact-finding missions had been proposed—and, in some cases, sent to certain areas of northeastern China—they did not have the time or access to key figures to make a comprehensive assessment. But in November 1931, Japan proposed that a commission be sent from the League of Nations to investigate the situation in the region.[32] The League of Nations Council endorsed this Japanese proposal, and shortly thereafter, assembled a team of five people—one each from Britain, the US, Germany, Italy, and France—to go to northeastern China, evaluate the circumstances behind the Mukden Incident, and make recommendations to resolve the conflict. This Commission, dubbed the "Lytton Commission" after its head Lord Robert Lytton, began its investigation in the spring of 1932.

Once in northeastern China, the Lytton Commission went about interviewing the key figures on both sides to try and put together a comprehensive picture of the situation. The Kwantung Army, for its part, stuck to its preexisting line that the Chinese had blown up the South Manchuria Railway. The army also alleged that Japanese and Korean nationals were threatened by the rising current of anti-Japanese nationalism. The Kwantung Army further accused the Chinese of consistently violating the treaty rights Japan had acquired in northeastern China since the beginning of the twentieth century. Indeed, the Kwantung Army cited 300 incidents in the run up to the Mukden Incident as evidence of increased Chinese hostility to Japan, which in turn supposedly proved that China was "not an organized state" and "in a condition of complete chaos and incredible anarchy."[33]

The Lytton Commission, in their assessment of the situation in Manchukuo, noted that both the restoration of the situation before September 18th and the maintenance of the newfound puppet government of Manchukuo were "unsatisfactory."[34] The Commission reasoned that restoration of the previous situation was unsatisfactory due to the weakness of the Chinese Central Government and its inability to safeguard Japanese treaty privileges, citizens, and property. The Commission saw the Manchukuo government as unsatisfactory because most Chinese did not approve of Puyi's Manchukuo puppet government, which assumed power in February, 1932. The Commission promulgated the following proposals and principles as a means of achieving a satisfactory solution in northeastern China for all parties:[35]

1. Compatibility with the interests of both China and Japan
2. Consideration for the interests of the U.S.S.R.

3. Conformity with existing multilateral treaties
4. Recognition of Japan's interests in Manchuria
5. The establishment of new treaty relations between China and Japan
6. Effective provision for the settlement of future disputes
7. Manchurian autonomy
8. Internal order and security against external aggression
9. Encouragement for an economic rapprochement between China and Japan
10. International cooperation in Chinese reconstruction

These League of Nations suggestions explicitly attempt to placate both China and Japan on the issue of northeastern China. But the Lytton Commission never questioned the basis of Japan's presence in the region; rather, the Commission assumed that Japan's extraterritorial rights and interests were legitimate, and that China—an ancient civilization with historical ties to the region—could not govern the province effectively. In particular, the Lytton Commission's proposal for increased "Manchurian autonomy" acquiesced to Japan's imperialist designs, especially when one considers the fact that Manchukuo's "autonomy" was to be complete with foreign advisors—including Japanese ones—in the "autonomous" government. Such an arrangement would have been anathema to most Chinese, who already opposed the imperialist and extraterritorial system of "unequal treaties" that had been imposed on China since the 19th century. However, contrary to the hopes of Chiang Kai-shek, the Commission gave only minimal consideration to the national and civilizational aspirations of most Chinese. Instead, the Commission prioritized Japan's strategic and economic interests in northeastern China. Therefore, the Kwantung Army's belief that no outside intervention would forestall the Mukden Incident was confirmed. By contrast, Chiang Kai-shek's pursuit of the "first pacification, then resistance" policy made his government seem weak and ineffectual. This weakness would only further embolden Kwantung Army officers to attempt Mukden-style gambits in future years, further eroding China's sovereignty.

Conclusion

From 1906 to 1931, the Kwantung Army's embrace of *gekokujo* influenced the course of history in northeastern China, contributing substantially to Japan's warlike and imperial posture of the period. This influence was evident from the 1910s, when Kawashima Naniwa attempted to foment Manchu and Mongol separatism. After the official formation of the Kwantung Army in 1919, incidents of *gekokujo* continued to increase, as the army no longer answered directly to the civilian government in Tokyo. In the early part of the 1920s, the Kwantung Army repeatedly provided support to Chinese warlord Zhang Zuolin to keep the northeast separated from the rest of China. The Kwantung Army usurped strategic decision-making when it bribed Feng Yuxiang to switch his allegiance to Zhang in 1924 and created a buffer zone to halt the advance of Guo Songling's forces in 1925. These actions helped shore up Zhang's position in northeastern China and thereby enhanced Japan's influence in the region.

While Japan and the Kwantung Army sponsored their client Zhang Zuolin throughout the 1920s, the rising tide of Chinese nationalism began to threaten Japanese interests in the Three Eastern Provinces. This threat was perhaps most evident in the Northern Expedition of 1926, which saw the nationalist Kuomintang advance against the Japan-backed warlords and establish control over a large swath of China's territory. But even in the Northeast, Zhang began taking measures that threatened Japan's hegemony, including supporting Chinese-owned railways and industries, and allowing nationalist, anti-imperialist political movements. Furthermore, Zhang began to take a firmer line with the Japanese, denying their requests to build railways in northeastern China. Zhang's unwillingness to function as a complete lackey for Japan led many in the Kwantung Army to begin flirting with the idea of removing the "Old Marshal" once and for all.

By 1928, Komoto Daisaku plotted Zhang's assassination. His plan called for blowing up the warlord's train at Huanggutun, where the Chinese-owned PMR intersected the Japanese-owned SMR. The assassination would then be blamed on the Chinese and supply the excuse for the Kwantung Army to take control of the Three Eastern Provinces. Komoto did not receive approval from Tokyo but nonetheless carried out the assassination, killing the "Old Marshal," on June 4th. Komoto was, however, unable to trigger an invasion of northeastern China, as he had not coordinated effectively with other figures in the Kwantung Army. The government in Tokyo was unable to hold Komoto accountable for his egregious act of *gekokujo* because the IJA prevented him from being put on trial in public. The failure to discipline Komoto emboldened many young Japanese officers, who were growing increasingly radical amidst years of economic crisis and political corruption.

The Mukden Incident and invasion of northeastern China in 1931 were defining acts in the Kwantung Army's long history of *gekokujo* in China. Not only did the Kwantung Army act without orders, it also defied government and AGS policies which sought to limit the scope of the incident to the SMR zone. This time, its brazen acts of *gekokujo* were well planned by its main perpetrators, Ishiwara and Itagaki, who had been preparing the attacks for two years. Their in-depth reconnaissance, operational planning, and coordination with other officers allowed them to rapidly implement an offensive starting on the night of September 18th, and to continue launching attacks throughout the rest of the fall and winter. By February 1932, the Kwantung Army effectively established control over northeastern China, culminating in the formation of the puppet state of "Manchukuo." Manchukuo would be the medium through which the Kwantung Army would exert its authority over northeastern China and implement its various programs.

There are many historical takeaways from both the Mukden Incident and the Kwantung Army's subsequent invasion of northeastern China. First, the invasion was a criminal endeavor that contradicted multiple international treaties to which Japan was a party. The most notable were the Nine-Power Treaty, which guaranteed China's sovereignty and territorial integrity, and the Kellogg-Briand Pact, in which signatories swore not to use war as a means of national policy. The Kwantung Army's violation of these treaties provided the basis for the prosecution of figures such as Itagaki in the Tokyo trials for crimes against peace in 1945. Additionally, the Kwantung Army's conspiracy to wage aggressive war set the stage for Japan's future war crimes, including mass atrocities committed in northeastern China. The criminal invasion of China after a false flag attack must be considered one of the most serious Japanese imperial offenses, because it set the stage for total war in the Asia-Pacific region.

Second, the Mukden Incident revealed the feebleness of the post-World War I liberal international order, the roots of which lay in Woodrow Wilson's 14-point plan for peace. During the 1920s, pacifism and disarmament treaties were promoted to such a degree that they seemed to represent the consensus of a new cosmopolitan class of modern politicians, businessmen, and elites. Wilson also promoted the "Open Door Policy" in China, national self-determination, and the League of Nations as means of preventing another inter-imperial war. But to many in Japan, the liberal international order was hypocritical and contradictory, especially after the United States refused Japan's racial equality clause at the Treaty of Versailles, stopped Japanese immigration to the US, and cut off Japan's silk export markets. Such disregard for Japan's interests led many in Japan to conclude that the US's liberal internationalism was simply an attempt to entrench Anglo-American hegemony over lesser powers. Therefore, some came to see an invasion of northeastern China as the only solution to Japan's post-World War I economic problems and sought to establish a self-sufficient empire there. When Ishiwara and Itagaki finally did invade in 1931, many Japanese supported them, especially after it became apparent that the west lacked the desire and capability to stop Japan's invasion. The liberal international order's lack of an enforcement mechanism encouraged the Kwantung Army to continue pushing the envelope of aggression in future years, consistent with a system still reliant on self-help and "realist diplomacy."

Third, even though the Kwantung Army's actions undermined the post-World War I liberal international order, the roots of its aggression against northeastern China lay in previous world historical patterns of imperialist expansion. In East Asia, a defining instance of liberal imperialism was the First Opium War, when the United Kingdom forced China to open up several ports for unfettered trade and took control of ports and customs. Free trade, freedom of religion, and the rule of law were causes the west championed as it repeatedly subordinated China in the decades before Japan's imperial rise. Japan followed suit by establishing its extraterritorial "rights" in the Liaodong Peninsula after the Russo-Japanese War of 1905 and forming the Kwantung Army to protect these claims. As Chinese nationalism grew the Kwantung Army moved to secure and expand its economic and strategic dominance in northeastern China. The Kwantung Army invaded in 1931, on the pretext that patriotic Chinese nationalists were somehow threatening Japanese citizens and their property. Though the Kwantung Army nominally turned its back on the post-World War I international system, its actions resembled previous liberal iterations of empire in East Asia, which had asserted dominion over China and its people pursuant to a set of universal principles that conveniently aligned with western interests.

Fourth, the Kwantung Army's junior and field officers took control over national policy by means of a highly coordinated conspiracy, effectively undermining the authority of the center in Tokyo. While the capital repeatedly attempted (sometimes successfully) to rein in their unruly subordinates, the Kwantung Army always remained a step ahead of its nominal superiors, forcing them to react to events. This predicament was rooted in Tokyo's failure to discipline its rogue subordinates in the 1920s, which in turn fueled an institutional culture of *gekokujo* that only grew worse after the Great Depression. The Kwantung Army's success in overrunning northeastern China gave further credibility to the strategists behind the Mukden Incident and the culture they helped create, leading to further Kwantung Army pursuit of aggressive, unilateral expansion in future years. Even though the conquest of northeastern China was operationally successful, in the end it was a strategic failure, because it forced Japan down a path of inevitable, self-destructive conflict.

Finally, the Kwantung Army's invasion of northeastern China made future expansion strategically unavoidable and increased the likelihood of war with China and other powers. The Army's first collateral target was the province of Rehe, because it threatened the flank of Manchukuo, and many of Zhang Xueliang's forces were garrisoned here. The threat these forces presented led the Kwantung Army to invade the province in 1933. But the Kwantung Army could not stop at Rehe province, as other provinces to the South and West had more raw materials and territory to be exploited. Consequently, the Kwantung Army compelled the KMT to sign treaties that effectively stripped China of its sovereignty over Hebei and Chahar provinces in 1935.

Later in 1935 and 1936, the Kwantung Army continued to support regional separatism in Hebei and Suiyuan, even sending a Mongolian separatist army to take control of Suiyuan. The defeat of this Mongolian army in turn precipitated the so-called "Xian Incident," in which Zhang Xueliang kidnapped Chiang Kai-shek, forced him to end the "first pacification, then resistance" policy and, instead, form a "United Front" to resist Japan. Soon thereafter, the Second Sino-Japanese War broke out in 1937. Because Japan could not inflict a decisive defeat on China in the first three years of fighting, it began running out of key resources, most notably rubber, tin and oil. The Japanese thus decided to "strike south" into southeast Asia against British, American, and Dutch colonies to obtain key raw materials. But to prevent American intervention, the Japanese bombed the US Pacific Fleet at Pearl Harbor on December 7th, 1941, in a surprise attack that brought the US into the war in the Asia-Pacific. While the bombing of Pearl Harbor seems unrelated to the Mukden Incident of 1931, it was the Mukden Incident ten years earlier that precipitated Japan's continuous imperial expansion in the 1930s, leading to the outbreak of World War II in the Asia-Pacific region. Absent the Mukden Incident, World War II in the Asia-Pacific region would certainly have followed quite a different course, if it had even broken out at all.

The Mukden Incident represented the culmination of decades of Japanese imperialism in northeastern China. This process began in the 1890s when Japan ventured onto the Asian mainland in search of security and a "line of advantage." But what Japan found in decades on the continent was insecurity and uncertainty, as both China and Russia collapsed, and with them the assumptions of stability that defined the geopolitics of the old order. While the Kwantung Army attempted to secure Japan's economic and strategic interests in northeastern China through the person of Zhang Zuolin, he proved to be an unreliable client, and after the Kwantung Army assassinated him, there was no replacement. The Kwantung Army took matters into its own hands in 1931 with the Mukden Incident and wholesale conquest of northeastern China. Puyi's puppet state of "Manchukuo," provided a thin veneer of legitimacy to the Kwantung Army's heavy-handed rule. But the puppet regime existed to fulfill the Kwantung Army's objectives: the economic exploitation of the region, the suppression of resistance, and further expansion into north China. As such, the Kwantung Army began a reign of terror in its newly occupied territory and thereby came to embody the depredations of fascism in East Asia.

Notes

Introduction

1. Shimada, Toshihiko, "The Extension of Hostilities," in *Japan's Road to the Pacific War, Volume I: Japan Erupts the London Naval Conference and the Manchurian Incident*, (Columbia University Press, 1984), 228.
2. Ibid.
3. Ogata, Sadako, *Defiance in Manchuria: Making of Japanese Foreign Policy, 1931-32*, (University of California Press, 1964), 63.
4. Coox, Alvin, *Nomonhan: Japan Against Russia, 1939*, (Stanford University Press, 1985), 30.
5. Otterstedt, Charles, "The Kwantung Army and the Nomonhan Incident: Its Impact on National Security," Defense Technical Information Center, 10.
6. Peattie, Mark, *Ishiwara Kanji and Japan's Confrontation with the West*, (Princeton University Press, 1975), 120.
7. Goldman, Stuart, *Nomonhan, 1939: The Red Army's Victory That Shaped World War II*, (Naval Institute Press, 2013), 15.
8. Cameron, Craig, M., "Race and Identity: The Culture of Combat in the Pacific War," *The International History Review 27,* no. 3 (2005): 560, https://doi.org/10.1080/07075332.2005.9641072.

Chapter 1: European Modernity Arrives in East Asia

1. For more on China's iron and steel industry, see Wagner, Donald, "The Traditional Chinese Iron Industry and its Modern Fate," *East Asian Science Technology and Medicine*, 12 no. 1 (July 1995): 138-161.

2. For facts and figures on the China's share of global GDP in the 1800s, see Angus, Maddison, *Development Centre Studies: The World Economy Historical Statistics,* (OECD Publications, 2003), 261.
3. For more on the Tributary system in East Asia, see Kang, David, "International Order in Historical East Asia," *International Organization,* 74, no. 1, (September 2019): 71, https://doi.org/10.1017/s0020818319000274.
4. Ibid.
5. Ibid.
6. Ibid.
7. For more on how the European empires in the New World helped give rise to the Industrial Revolution, see Darity, William Jr, "A Model of Original Sin: Rise of the West and Lag of the Rest," *American Economic Review 82,* no. 2 (May 1992): 162-167.
8. Platt, Stephen, "New Domestic and Global Challenges," in *The Oxford Illustrated History of Modern China* ed. by Wasserstrom, Jeffrey, (Oxford University Press, 2016), 46-48.
9. Spence, Jonathan, *The Search for Modern China* (W.W. Norton, 1999), 150-152.
10. Ibid.
11. For more details on how the war-like elements in the UK argued for war with China, see Su, Christine, "Justifiers of the British Opium Trade: Arguments by Parliament, Traders, and the Times in the Leading Up to the Opium War," *Stanford Undergraduate Research Journal,* 45-51.
12. For more on British steamships on the First Opium War, see Stine Jeffrey K. & Headrick, Daniel, "The Tools of Empire: Technology and European Imperialism in the Nineteenth Century," *The Journal of Modern History,* 51 no. 2. (June 1979), 48-49.
13. While Lin Zexu and the Chinese Navy did attempt to adapt British cannon and some of the larger wooden ships, these ships by and large did not participate in the fighting according to Spence, *The Search,* 158.
14. For more information on Chinese weapons systems in the first half of the Nineteenth Century see Hacker, Barton, "The Weapons of the West: Military Technology and Modernization in 19th-Century China and Japan," *Technology and Culture,* 18 no. 1 (January 1977), 47.
15. See Makeham, John, *China: The World's Oldest Living Civilization Revealed* (Thames & Hudson, 2009), 331, for more details

16. For more on the Treaty of Nanjing, see Spence, *The Search*, 158-160.
17. Ibid.
18. Kang, "International Order," 66.
19. Fairbank, John, Liu, Kwang-ching, Twitchett, Denis, *The Cambridge History of China, Volume 11 Part 2*, (Cambridge University Press, 1980), 202.
20. Spence, *The Search*, 179-180.
21. For more on the burning of the Summer Palace, see Hibbert, Christopher, *The Dragon Wakes: China and the West 1793-1911* (Penguin, 1988).
22. For more on the Treaty of Tianjin, see the World and Japan Database, "The Treaty of Tientsin," https://worldjpn.net/documents/texts/pw/18580626.T1E.html.
23. For more on the self-strengthening movement, see Spence, *The Search*, 199-224.
24. Hsu, Wilbur, *Art of War Papers: Survival Through Adaptation, The Chinese Red Army and the Extermination Campaigns, 1927-1936* (Combat Institute Studies Press, 2012), 8.
25. Esposito, Gabriele, & Rava, Giuseppe, *Armies of the First Sino-Japanese War*, (Bloomsbury Publishing, 2022), 33.
26. Fairbank, Liu, & Twitchett, *The Cambridge History of China, Volume 11 Part 2*, 203.
27. For more on the problems of the Qing forces, see *Art of War Papers*, 8-9.
28. Fairbank, Liu, & Twitchett, *The Cambridge History of China, Volume 11 Part 2*, 204.
29. Ibid., 205-207.
30. Hsu, Wilbur, *Art of War Papers*, 8-10.
31. Fairbank, Liu, & Twitchett, *The Cambridge History of China, Volume 11 Part 2*, 243-244.
32. Ibid., 244-245.
33. Spector, Stanley, *Li Hung-chang and the Huai Army* (University of Washington Press, 1964), 42-45.
34. Fairbank, Liu, & Twitchett, *The Cambridge History of China, Volume 11 Part 2*, 245.
35. Esposito & Rava, *Armies of the Sino-Japanese War*, 36.
36. For more on the Manchurian Army, see James, Henry, *A Long White Mountain… Or a Journey in Manchuria with Illustrations and a Map* (Longmans, Green, & Co, 1888), 119-121.

37. Hsu, *Art of War*, 11.
38. Ibid.
39. Liu, Kwang-Ching, and Chu, Samuel, *Li Hung-chang and China's Early Modernization* (Routledge, 2015), 9-12, 67-71.
40. Beasley, W.G., *The Meiji Restoration*, (Stanford University Press, 1972), 78.
41. Ibid.
42. For more on the unequal treaties Japan signed with western countries, see Auslin, Michael, *Negotiating With Imperialism: The Unequal Treaties and the Culture of Japanese Diplomacy* (Harvard University Press, 2022), 44.
43. Norman, Herbert, & Lawrence, Timothy, *Japan's Emergence as a Modern State* (UBC Press, 2007), 7.
44. Drea, Edward, *Japan's Imperial Army: Its Rise and Fall, 1853-1945* (University of Kansas Press, 2016), 3.
45. Ibid., 4.
46. Harries, Meirion & Harries, Susie, *Soldiers of the Sun; The Rise and Fall of the Imperial Japanese Army* (Random House, 1991), 14.
47. Jaundrill, Collin D., *Samurai to Soldier: Remaking Military Service in Nineteenth Century Japan* (Cornell University Press, 2017), 65.
48. Whitehurst, William G., *The China Incident: Igniting the Second Sino-Japanese War* (McFarland & Company Publishers, 2021), 13-14.
49. Jaundrill, *Samurai to Soldier*, 66.
50. Ibid.
51. Drea, *Japan's Imperial Army*, 5.
52. Ibid., 5.
53. Ibid. 5-6.
54. Beasley, W.G., *The History of Modern Japan*, 95-96.
55. Ibid., 96.
56. Drea, *Japan's Imperial Army*, 6-7.
57. Nish, Ian, *The Iwakura Mission to America and Europe: A New Assessment*, (Routledge, 2008), 35.
58. Read Nish, *The Iwakura Mission* for more in-depth treatment of the Japanese delegation's time in each place
59. Allen, G.C., *A Short Economic History of Modern Japan*, (G Allen and Unwin, 1950), 28-31.

60. Kawamoto, Bondon, "How Old Regimes Help New Economic Growth" (Honors Thesis, Portland State University, 2013), 8.
61. Norman & Lawrence, *Japan's Emergence*, 138-141.
62. Jaundrill, *Samurai to Soldier*, 95.
63. Drea, Edward, "The Imperial Japanese Army (1868-1945): Origins, Evolution, and Legacy," in *War in the Modern World Since 1815*, ed. Jeremy Black (Taylor & Francis, 2003), 23.
64. Jaundrill, *Samurai to Soldier*, 95.
65. Harries & Harries, *Soldiers of the Sun*, 21.
66. Drea, *Japan's Imperial Army* 29.
67. Jaundrill, *Samurai to Soldier*, 96-97.
68. Drea, *Japan's Imperial Army*, 30.
69. Jaundrill, *Samurai to Soldier*, 108.
70. Drea, *Japan's Imperial Army*, 29.
71. Whitehurst, *The China Incident*, 17.
72. Drea, *Japan's Imperial Army*, 29.
73. Ibid. 30
74. Kublin, Hyman, "The Modern Army of Early Meiji Japan," *The Far Eastern Quarterly 9*, no.1 (November 1949), 32, https://doi.org/10.2307/2049123.
75. Drea, *Japan's Imperial Army*, 30.
76. Kublin, "The Modern Army," 28-29.
77. Drea, *Japan's Imperial Army*, 25.
78. McClain, Kendra, "Explaining Meiji Japan's Top-Down Revolution," (Master's Thesis, Naval Postgraduate School, 2017), 47.
79. Drea, *Japan's Imperial Army*, 26.
80. Ibid., 26.
81. Harries & Harries, *Soldiers of the Sun*, 24.
82. Drea, *Japan's Imperial Army*, 57-58.
83. Harries & Harries, *Soldiers of the Sun*, 34.
84. Drea, *Japan's Imperial Army*, 49.
85. Ibid.
86. Hendrix, Henry, "The Roots of Japanese Militarism," (Master's Thesis, Naval Postgraduate School, 1994), 108-109.
87. Ito, Miyoji, "The Constitution of the Empire of Japan."
88. Harries & Harries, *Soldiers of the Sun*, 48.
89. Ibid., 58-59.

90. Drea, *Japan's Imperial Army*, 59.
91. Whitehurst, *The China Incident*, 26.
92. Ibid.
93. Drea, *Japan's Imperial Army*, 59.
94. Ibid.
95. Harries & Harries, *Soldiers of the Sun*, 49.
96. Esposito & Rava, *Armies of the First Sino-Japanese War*, 19-20.
97. Ibid.
98. Drea, *Japan's Imperial Army*, 59.
99. Orbach, Danny, *Curse on this Country: The Rebellious Army of Imperial Japan* (Cornell University Press, 2017), 135.
100. Ibid., 135
101. Humphreys, Leonard, *The Way of the Heavenly Sword: The Japanese Army in the 1920s* (Stanford University Press), 9.
102. Ibid., 9-10.
103. Kitaoka, Shinichi, "The Army as a Bureaucracy: Japanese Militarism Revisited," *The Journal of Military History*, 57 no. 5, (October 1993), 67-68: https://doi.org/10.2307/2951803.
104. Drea, *Japan's Imperial Army*, 22.
105. For more on Han-batsu within the IJA, see Kitaoka, "The Army as a Bureaucracy: Japanese Militarism Revisited," 68-70.
106. For more on Japan's view of the Korean Peninsula as a "dagger pointed at the heart of Japan," see Cha, Victor, "Japan's Grand Strategy on the Korean Peninsula: Optimistic Realism," *Japanese Journal of Political Science* 1, no. 2, (March 2001), 250-254, https://doi.org/10.1017/S1468109900002048.
107. Whitehurst, *The China Incident*, 26.
108. Harries & Harries, *Soldiers of the Sun*, 49-50.
109. McClain, "Explaining Meiji Japan's Top Down Revolution," 27.
110. End, Nicholas, "Japan During the Interwar Period: From Monetary Restraint to Fiscal Abandon," *Debt and Entanglements Between the Wars* ed. by Norris, Era, (International Monetary Fund, 2019), 245-246.
111. Metzler, Mark, *Lever of Empire: The International Gold Standard and the Crisis of Liberalism in pre-War Japan*, (University of California Press, 2006), 40.

112. For more details on these Korean reformists and their relations with Japan see Duus, Peter, *The Abacus and the Sword: The Japanese Penetration of Korea, 1895-1910*, (University of California Press, 1995), 54-55.
113. Ibid., 55.
114. Ibid., 58.
115. Among those who supported Toyama's foundation of the Genyosha were Hirakoa Kotaro, a coal mine owner, Yamamoto Jotaro, a former Mitsui director and later one of the presidents of the South Manchuria Railway corporation, and Kuhara Fusanosuke according to Norman, E.H., "The Genyosha: A Study in the Origins of Japanese Imperialism," *Pacific Affairs 17*, no. 3, (September 1944), 270.
116. Ibid., 267.
117. Conroy, Hilary, "Government Versus Patriot: The Background of Japan's Asiatic Expansion," *Pacific Historical Review*, 20, no. 1, (February 1951), 39, https://doi.org/10.2307/3634474.
118. Hall, "Blinded by the Rising Sun: Japanese Military Intelligence From the First Sino-Japanese War to the End of World War II," PhD Thesis, University of Adelaide (2016), 8.
119. Orbach, Danny, "The Military-Adventurous Complex: Officers, Adventurers, and Japanese Expansion in East Asia, 1884-1937," *Modern Asian Studies*, 53, no. 2 (August 2018), 344. https://doi.org/10.1017/s0026749x17000543.
120. Hall, "Blinded by the Rising Sun," 24.
121. Orbach, Danny, "The Military-Adventurous Complex," 346-347.
122. Norman, "The Genyosha," 279-280.
123. Hall, "Blinded by the Rising Sun," 24-35.
124. See Dwight Tat Wai Kok's Master's Thesis "A Translation of Datsu A-Ron: Decoding A Prewar Japanese Nationalistic Theory," 15-20, for a complete translation of Fukuzawa's work
125. Hendrix, "The Roots," 50-51.
126. Jansen, Marius, "Japanese Imperialism: Late Meiji Perspectives," in *Japan's Colonial Empire, 1895-1945*, ed. by Ramon Myers and Peter Duus, (Princeton University Press, 1984), 67.
127. Hendrix, "The Roots," 51-52

Chapter 2: Japan's Two Splendid Little Wars

1. "Convention of Tientsin, 1885," WorldJpn.net.
2. The Treaty of Tianjin stripped China of its traditional role as the unilateral protector of Korea's territorial integrity. Instead, Japan and China agreed to notify one another if either dispatched forces to the Korean peninsula according to *The Imperial Japanese Army in Manchuria, 1894-1945*. Japanese Research Division, Japanese Studies on Manchuria 2, (1959), 3.
3. Olender, Piotr, *Sino-Japanese Naval War 1894-95*, (Stratus, 2014), 42.
4. Eastlake, Warrington F., & Yamada, Yoshiyaki, *Heroic Japan: A History of the War Between China and Japan* (University Publications of America, 1979), III.
5. Ibid., IV
6. Lone, Stewart, *Japan's First Modern War: Army and Society in the Conflict with China* (Palgrave Macmillan, 1994), 22-23.
7. Drea, *Japan's Imperial Army*, 79.
8. Ibid.
9. To see translations of the dispatches the Japanese sent to China, see Eastlake & Yamada, *Heroic Japan*, IV-VI.
10. Harries & Harries, *Soldiers of the Sun*, 55.
11. Ibid.
12. Olender, *Sino-Japanese Naval War*, 44.
13. Volpicelli, Zenone, *The China-Japan War Compiled from Japanese, Chinese, and Foreign Sources* (Charles Scriber's Sons, 1896), 81.
14. Esposito, & Rava, *Armies of the First Sino-Japanese War*, 20.
15. Drea, *Japan's Imperial Army*, 73.
16. Lone, *Japan's First Modern War*, 28.
17. Drea, *Japan's Imperial Army*, 74.
18. Ibid., 72.
19. Lone, *Japan's First Modern War*, 29.
20. *The Imperial Japanese Army in Manchuria*, 3.
21. Drea, *Japan's Imperial Army*, 73-74.
22. Jowett, Philip, *China's Wars: Rousing the Dragon, 1894-1949* (Osprey Publishing, 2013), 24.
23. Esposito & Rava, *Armies of the Sino-Japanese War*, 33.
24. Ibid.

25. Ibid., 34.
26. Ibid.
27. Ibid.
28. Ibid.
29. For more on the armaments of Chinese armies in the run-up to the First Sino-Japanese War, see Hacker, "The Weapons of the West," 43-55.
30. Esposito & Rava, *Armies of the Sino-Japanese War*, 37.
31. Olender, Piotr, *Sino-Japanese Naval War*, 1894-95, 20.
32. Ibid.
33. Evans, David C, Peattie, Mark, *Kaigun: Strategy, Tactics, and Technology in the Imperial Japanese Navy, 1887-1941* (Naval Institute Press, 2015), 5-7.
34. Ibid., 7-9.
35. Ibid., 13-15.
36. Ibid., 14.
37. Olender, Piotr, *Sino-Japanese Naval War*, 27.
38. Ibid.
39. For more on the Imperial Chinese Navy's foreign procurement, see Roberts, Stephen S., "The Imperial Chinese Navy 1862-1895," *Warship International,* 11, no. 1, (1974): 31, 44.
40. Olender, *Sino-Japanese Naval War,* 43.
41. Seth, Michael, J., *A History of Korea: From Antiquity to the Present* (Rowman & Littlefield, 2011), 262.
42. Paine, S.C.M., *The Sino-Japanese War of 1894-95: Power, Perceptions, and Primacy* (Cambridge University Press), 165-166.
43. For more on the landing of these forces, see *The Imperial Japanese Army in Manchuria*, 7-11.
44. For more on the IJA's advance and assault on Pyongyang, see Volpicelli, *The China-Japan War, 120-163.*
45. Olender, *Sino-Japanese Naval War*, 61-68.
46. Ibid., 60.
47. Evans, & Peattie, *Kaigun,* 40-41.
48. Perry, John Curtis, "The Battle off the Tayang, 17 September, 1894," *The Mariner's Mirror, 50,* no. 4, (January 1964): 243-259, https://doi.org/10.1080/00253359.1964.10657787.
49. Evans & Peattie, *Kaigun*, 42-43.

50. Perry, "The Battle off the Tayang," 243-259.
51. Evans & Peattie, *Kaigun*, 48.
52. *The Imperial Japanese Army in Manchuria,* 13.
53. Paine, S.C.M, *The Sino-Japanese War,* 197-213.
54. *The Imperial Japanese Army in Manchuria,* 14.
55. Ibid., 16-18.
56. Esposito & Rava, *Armies of the Sino-Japanese War*, 15-16.
57. Morse, Hosea Ballou, *The International Relations of the Chinese Empire*, (Longman Greens & Co, 1910), 40-42.
58. Olender, *Sino-Japanese Naval War*, 274.
59. For more on Weihaiwei's coastal defenses, see Volpicelli, *The China-Japan War,* 273-276.
60. Evans and Peattie, *Kaigun*, 46.
61. To see the whole Treaty of Shimonoseki, see Queen's University Belfast's "Treaty of Shimonoseki," 1-2
62. Zachmann, Mathias, Urs, "Imperialism in a Nutshell: Conflict and the 'Concert of Powers,' in the Tripartite Intervention, 1895," *Japanstudien,* 17, no. 1, (January 2006): 63, https://doi.org/10.1080/09386491.2006.11826924.
63. Ibid., 65.
64. Ibid.
65. Ibid., 67.
66. Ikle, Frank, "The Triple Intervention: Japan's Lesson in the Diplomacy of Imperialism," *Monumenta Nipponica,* 22, no 1/2, (1967): 128, https://doi.org/10.2307/2383226.
67. Nish, Ian, *The Origins of the Russo-Japanese War*, (Longman Group, 1985), 31.
68. For more on the joint Sino-Russian bank, see Kowner, *Historical Dictionary* (Rowman & Littlefield, 2017), 209-210.
69. To see the whole lease treaty, see "Convention for the Lease of the Liaotung Peninsula," in *Treaties and Agreements with and Concerning China, 1894-1919,* (Oxford University Press, 1921), 119-121.
70. Matsusaka, Yoshihisa Tak, *The Making of Japanese Manchuria, 1904-1932,* (Harvard University Asia Center, 2001), 28-30.
71. Ibid., 30.

72. For more on the assassination of Queen Min, see Kim, Youngsoo, "Two Perspectives on the 1895 Assassination of Queen Min," *Korea Journal,* 48 no. 2 (June 2008): 160-185.
73. Nish, *The Origins*, 32-33.
74. Matsusaka, *The Making of Japanese Manchuria*, 30.
75. Nish, *The Origins*, 33-34.
76. Matsusaka, *The Making of Japanese Manchuria*, 32.
77. For more on the Nishi-Rosen agreement, see Nish, *The Origins,* 47.
78. For more on the Kogetsukai, see *The Making of Japanese Manchuria,* 33-37.
79. For more on the Kogetsukai, see ibid., 33-37.
80. Iguchi's views are elaborated on in ibid.
81. Koda, Yoji, "The Russo-Japanese War: Primary Causes of Japanese Success," *Naval War College Review,* 58, no. 2, (2005): 58.
82. Connaughton, Richard, *Rising Sun and Tumbling Bear: Russia's War with Japan* (Orion, 2007), 17.
83. Given the size of the Russian Army, precise estimates are difficult to find; some estimates indicate the Russians had roughly 1.2 million men in their army, while others suggest over the Russian Army may have had as many as two million men under arms.
84. Connaughton, *Rising Sun,* 20.
85. Tyler, Sydney, *The Russo-Japanese War: An Illustrated History of the War in the Far East*, (Sharp Ink, 2022), 17.
86. Ibid.
87. Drea, *Japan's Imperial Army*, 102.
88. Connaughton, *Rising Sun*, 17.
89. Drea, *Japan's Imperial Army*, 102.
90. Information about the Japanese linear logistical system can be found in Connaughton, *Rising Sun*, 14-15.
91. Ibid., 19.
92. Ibid.
93. *The Imperial Japanese Army in Manchuria*, 24-25.
94. Connaughton, *Rising Sun*, 21.
95. *The Imperial Japanese Army in Manchuria*, 25.
96. Drea, *Japan's Imperial Army*, 60.
97. Connaughton, *Rising Sun*, 16.

98. Harries and Harries, *Soldiers of the Sun*, 82.
99. Connaughton, *Rising Sun*, 23.
100. Ibid.
101. Harries & Harries, *Soldiers of the Sun*, 82.
102. Hall, "Blinded by the Rising Sun," 21-22.
103. Jacob, Frank, "Secret Societies in Japan and Preparation for the Russo-Japanese War," *Diacronie* 28, no. 4 (December 2016): 6, https://doi.org/10.4000/diacronie.4738.
104. Norman, Herbert, "The Genyosha: A Study in the Origins of Japanese Imperialism," 266.
105. Sabey, John W., *The Genyosha, the Kokuryukai and Japanese Expansionism*, PhD Thesis, University of Michigan, 185
106. Ibid., 192.
107. Jacob, "Secret Societies," 9.
108. Drea, *Japan's Imperial Army*, 50.
109. Harries & Harries, *Soldiers of the Sun*, 80-81.
110. Ibid., 79.
111. Tikowara, Hesbio, & Grant, Robert F.S., *Before Port Arthur in A Destroyer: The Personal Diary of a Japanese Naval Officer* (John Murray, 1907), 11.
112. Jukes, Richard, *The Russo-Japanese War, 1904-1905* (Osprey, 2022), 5.
113. Ibid., 2.
114. Ibid, 5-6.
115. Warner Denis & Warner Peggy, *The Tide at Sunrise: A History of the Russo-Japanese War, 1904-1905*, (Charterhouse, 1974), 249.
116. Harries & Harries, *Soldiers of the Sun*, 82-83.
117. Ibid., 84.
118. *The Imperial Japanese Army in Manchuria*, 33.
119. Warner & Warner, *The Tide at Sunrise*, 288.
120. Kowner, *Historical Dictionary of the Russo-Japanese War*, 250-251.
121. Harries & Harries, *Soldiers of the Sun*, 85.
122. Warner & Warner, *The Tide at Sunrise*, 294-295.
123. Jukes, *The Russo-Japanese War*, 14.
124. Ibid., 22.
125. *The Imperial Japanese Army in Manchuria*, 49.
126. Warner & Warner, The Tide at Sunrise, 376-378.

127. To read more about the last stages of the assault on Port Arthur, see ibid., 373-383.
128. Connaughton, *Rising Sun,* 230-246.
129. Clodfelter, Michael, *Warfare and Armed Conflicts, Statistical Reference Volume II* (McFarland & Company Publishers, 2017), 648.
130. Harries & Harries, *Soldiers of the Sun*, 89.
131. Jukes, *The Russo-Japanese War*, 49-52.
132. *The Imperial Japanese Army in Manchuria,* 42.
133. Kowner, *Historical Dictionary*, 205-208.
134. *The Imperial Japanese Army in Manchuria*, 59.
135. Harries & Harries, *Soldiers of the Sun*, 90.
136. *The Imperial Japanese Army in Manchuria*, 63.
137. Warner & Warner, *The Tide at Sunrise*, 469.
138. *The Imperial Japanese Army in Manchuria*, 59.
139. Ibid., 62
140. Jukes, *The Russo-Japanese War*, 40.
141. For more on the disadvantages of the Russians' ships, see Boot, *War Made New, Technology, Warfare, and the Course of History, 1500 to Today* (Gotham Books, 2006), 183-187.
142. Ibid., 183-184.
143. Watts, Anthony, *The Imperial Russian Navy*, (Arms and Armour, 1990), 22.
144. Boot, *War Made New*, 188.
145. Mainardi, Benjamin, "The Russo-Japanese War: Origins and Implications," *James Madison Undergraduate Research Journal* 7 no. 1, 2019-2020: 10.
146. To see the full version of the Treaty of Portsmouth, see The World and Japan Database's "Russo-Japanese Peace Treaty," https://worldjpn.net/documents/texts/pw/19050905.T1E.html.
147. Noguchi, Masayoshi, "The South Manchuria Railway Company: An Accounting and Financial History, 1907-1943," Kobe University Research Institute for Economics and Business Administration, March 2013, 3.
148. Lytton, Robert, "Appeal by the Chinese Government: Report of the Commission of Enquiry," 44.

Chapter 3: Japan Consolidates its Empire

1. Coox, *Nomonhan*, 3.
2. Japanese "dual use initiatives" in the Liaodong Peninsula included the construction of meatpacking plants in Kwantung, the afforestation of large areas, and construction of hospitals, among other measures, in Matsusaka, *The Making of Japanese Manchuria*, 74-75.
3. Coox, *Nomonhan*, 3.
4. Ibid., 3-4.
5. Coox, Alvin, "The Kwantung Army Dimension," in *The Japanese Informal Empire in China, 1895*-1937 edited by edited by Peter Duus, Mark Peattie, and Ramon Myers (Princeton University Press, 2014), 398.
6. Coox, *Nomonhan*, 4.
7. Matsusaka, *The Making of Japanese Manchuria*, 88.
8. Ibid.
9. Ibid.
10. Ibid.
11. Ibid.
12. Ibid., 88-89
13. Coox, "The Kwantung Army Dimension," 398.
14. Ibid.
15. Fogel, Joshua & Ito, Takeo, *Life Along the South Manchuria Railway: The Memoirs of Ito Takeo* (Routledge, 2015), 9.
16. Young, Louise, *Japan's Total Empire: Manchuria and the Culture of Wartime Imperialism* (University of California Press, 1999), 31.
17. Myers, Ramon, "Japanese Imperialism in Manchuria: The South Manchuria Railway Company," in *The Japanese Informal Empire in China* edited by Duus, Peattie and Myers, (Princeton University Press, 2014), 102.
18. Ibid., 104.
19. Noguchi, "The South Manchuria Railway," 3.
20. Myers, "Japanese Imperialism in Manchuria," 104.
21. Matsusaka, *The Making of Japanese Manchuria*, 90.
22. Myers, "Japanese Imperialism in Manchuria," 105.
23. When the Russians built the CER, they used broad-gauge Russian tracks (five feet wide), which the Japanese changed to a narrow gauge during

the Russo-Japanese war, according to *Railway and Locomotive Engineering*, 91

24. Ibid.
25. Matsusaka, *The Making of Japanese Manchuria*, 105.
26. Myers, "Japanese Imperialism in Manchuria," 105.
27. *Railway and Locomotive Engineering*, 91.
28. Matsusaka, *The Making of Japanese Manchuria*, 105.
29. Young, *Japan's Total Empire*, 31-32.
30. Matsusaka, *The Making of Japanese Manchuria*, 132.
31. Ibid., 133.
32. Myers, "Japanese Imperialism in Manchuria," 104-108.
33. Bix, Herbert, "Japanese Imperialism and the Manchurian Economy, 1900-1901," *The China Quarterly*, 51 (July 1972): 429, https://doi.org/10.1017/s0305741000052231.
34. Fogel & Ito, *Life Along the South Manchuria Railway*, 5.
35. Ibid.
36. Matsusaka, *The Making of Japanese Manchuria*, 148.
37. For more on measures designed to attract settlers, see ibid., 161.
38. Ibid.
39. Coox, *Nomonhan*, 3.
40. Iyenaga, Tokoyichi, "Japan in South Manchuria," *The Journal of Race Development* 2, no. 4 (1912): 384, https://doi.org/10.2307/29737926.
41. Matsusaka, *The Making of Japanese Manchuria*, 161.
42. Lattimore, Owen, *Manchuria: Cradle of Conflict*, (Macmillan, 1932), 237.
43. Drea, *Japan's Imperial Army*, 125.
44. Ibid., 125-126.
45. Matsusaka, *The Making of Japanese Manchuria*, 98.
46. Drea, *Japan's Imperial Army*, 127.
47. Harries & Harries, *Soldiers of the Sun*, 101.
48. Drea, *Japan's Imperial Army*, 127.
49. Matsusaka, *The Making of Japanese Manchuria, 98.*
50. Ibid., 98.
51. For more on the IJA's operational plans, see ibid., 104-106.
52. Ibid., 104.
53. Ibid., 141.
54. Ibid., 106.

55. Davis, Malcolm, "Railway Strategy in Manchuria," *Foreign Affairs 4,* no. 3, (1926): 499-500, https://doi.org/10.2307/20028472.
56. Matsusaka, *The Making of Japanese Manchuria*, 114-115.
57. Ibid., 115.
58. Ibid.
59. Ibid., 141-142.
60. Ibid.
61. Ibid., 142.
62. Ibid.
63. For more on the great powers' "scramble for China," see Bickers, Robert, *The Scramble for China: Foreign Devils in the Qing Empire, 1832-1914*, (Penguin Books, 2016).
64. Mallory, H. Walter, "The Open Door in China: A Reappraisal," *Foreign Affairs 26,* no. 1, (October 1947): 156-157, https://doi.org/10.2307/20030095.
65. For more on the Open Door Policy, see the US Secretary of State's article "Secretary of State John Hay and the Open Door in China, 1899-1900," United States State Department Office of the Historian, https://history.state.gov/milestones/1899-1913/hay-and-china.
66. For more on why the US implemented its Open Door Policy, see Hu, Shizhang, *Stanley K. Hornbeck and the Open Door Policy, 1919-1937,* (Greenwood Press, 1965).
67. Mallory, "The Open Door in China," 157.
68. For more on the origins of the Boxer Rebellion, see Esherick, Joseph, *The Origins of the Boxer Uprising*, (University of California Press, 1987), xiii-xix.
69. For more on the assault on the Dagu Forts, see Silbey, David, *The Boxer Rebellion and the Great Game in China,* (Hill and Wang, 2013), 91-100.
70. For more on the Battle of Tianjin and eventual seizure of Beijing see ibid., 137-164, 199-214.
71. To see the complete Boxer Protocol, see Queen's University Belfast's "Boxer Protocol, Peking 7.1901." https://reparations.qub.ac.uk/assets/uploads/Boxer-Protocol.pdf.
72. Spence, Jonathan, *The Search*, 252-253.
73. Zheng, Xiaowei, *The Politics of Rights and the 1911 Revolution in China*, (Stanford University Press, 2020), 3.

74. For a more detailed treatment of the Wuchang Uprising, see Esherwick, Joseph, and C.X. George Wei, *China: How the Empire Fell*, 107-155.
75. For more on the Tongmenghui's influence on the New Army, see Fung, Edmund S.K., "Military Subversion in the Chinese Revolution of 1911," 114-120
76. Esherwick & Wei, *How the Empire Fell* (Routledge, 2014), 184.
77. Zarrow, Peter, "Felling a Dynasty and Founding a Republic," in *The Oxford Illustrated History of Modern China,* edited by Jeffrey Wasserstrom (Oxford University Press, 2016), 106.
78. Spence, *The Search*, 267.
79. For a more detailed treatment of Yuan Shikai's betrayal of the Qing Dynasty, see Esherwick & Wei, *How the Empire Fell*, 195-145.
80. Spence, *The Search*, 28.0
81. Carter, James, "The Rise of Nationalism and Revolutionary Parties, 1919-1937," in *The Oxford Illustrated History of Modern China* edited by Jeffrey, Wasserstrom, (Oxford University Press, 2016), 130.
82. Spence, *The Search*, 281.
83. Weale, Putnam, and Lenox, Bertam, *The Fight for the Republic in China* (Dodd, Mead, and Co, 1917), 50.
84. Ibid., 188-189.
85. For more on Cai's National Protection Army, see Buttner, Klemens, "The National Protection War and the Intellectual Foundations of Chinese Warlordism," *Modern China,* 49, no. 6, (April 2023): 679-708, https://doi.org/10.1177/00977004231153331.
86. Ibid., 689.
87. Ibid., 692-693.
88. Chi, Hsi-sheng, *The Chinese Warlord System* (American University: Center for Research in Social Systems, 1969), 16.
89. Matsusaka, *The Making of Japanese Manchuria*, 150-152.
90. Orbach, Danny, *Curse on this Country*, 135.
91. Ibid.
92. "5 December 22, 1911 to December 31, 1911" Japan Center for Asian Historical Records (JACAR) Ref.B03050659300, Overseas miscellaneous news about the revolution conflicts in Qing China/Resident Generals Office of Kwantung Report Volume 4 (1.6.1.51-1_004) (Diplomatic Archives of the Ministry of Foreign Affairs), Translated by Ms. Fatima Janaghir.

93. For more on the various interests at play in Japanese politics between 1911 and 1914, see Matsusaka, *The Making of Japanese Manchuria*, 155-158.
94. Ibid.
95. Valiant, Robert B., "Japanese Involvement in Mongol Independence Movements," *The Mongolian Society Bulletin 11,* no. 2 (1972): 2.
96. For more details, see Matsusaka, *The Making of Japanese Manchuria*, 155-158.
97. For more on the Kokuryukai's activities in China during this period, see Hall, "Blinded by the Rising Sun," 79-83.
98. Valiant, "Japanese Involvement," 23.
99. Ibid., 3-4.
100. Ibid., 3.
101. Hall, "Blinded by the Rising Sun," 104.
102. Orbach, "The Military-Adventurous Complex," 353.
103. Weland, "Misguided Intelligence: Japanese Military Intelligence Officers in the Mukden Incident, September 1931," *The Journal of Military History,* 58 no. 3, (July 1994): 450-451, https://doi.org/10.2307//2944134.
104. Coox, "The Kwantung Army Dimension," 401.
105. Valiant, "Japanese Involvement," 4.
106. Weland, "Misguided Intelligence," 450.
107. Orbach, "The Military-Adventurous Complex," 355.
108. Valiant, "Japanese Involvement," 5.
109. Weland, "Misguided Intelligence," 450.
110. Orbach, "The Military-Adventurous Complex," 356.
111. Valiant, "Japanese Involvement," 4-5.
112. Orbach, "The Military-Adventurous Complex," 356.
113. Valiant, "Japanese Involvement," 5.
114. Orbach, "Curse on this Country," 140.
115. For more on the active duty rule and its effect on the Saionji government, see ibid., 132-141.
116. Ibid., 143.
117. For more on this controversy, see ibid., 143-145.
118. Harries & Harries, "Soldiers of the Sun," 107.
119. Orbach, "Curse on this Country," 144-146.
120. For more on the issues of Manchurian expansion Yamamoto and his cabinet supported, see Matsusaka, *The Making of Japanese Manchuria*, 160-171.

121. The Japanese Foreign Ministry made this offer after Sun Yat-sen proposed the construction of Chinese-owned railways from Mukden to Jilin and Jilin to Taonan (through Changchun)
122. Ibid., 171.
123. Orbach, "The Military-Adventurous Complex," 357-358.
124. Valiant, "Japanese Involvement," 8.
125. Orbach, "The Military-Adventurous Complex," 360.
126. For more details on Kawashima's plans, see Valiant, "Japanese Involvement," 9-10.
127. Weland, "Misguided Intelligence," 450.
128. Orbach, "The Military-Adventurous Complex," 361-362.
129. Ibid., 362.
130. Coox, "The Kwantung Army Dimension," 401.
131. Weland, "Misguided Intelligence," 451.
132. Kwong, *War and Geopolitics in Interwar Manchuria: Zhang Zuolin and the Fengtian Clique During the Northern Expedition,* (Brill, 2017), 58-59.
133. McCormack, *Chang Tso-lin in Northeast China*, (Stanford University Press, 1977), 19.
134. Ibid., 26-27.
135. For more on Zhang's connections and courting of local elites, see Kwong, *War and Geopolitics*, 60-65.
136. Valiant, "Japanese Involvement," 10.
137. Ibid.
138. Ibid., 10-11.
139. Orbach, "The Military-Adventurous Complex," 365.
140. Ibid.

Chapter 4: World War I, Japan's Imperial Opportunism, and the Rise of the Liberal International Order

1. Hsu, Immanuel, *The Rise of Modern China*, (Oxford University Press, 1983), 495.
2. Schlitz, Michael, "Separating the Roots of the Chrysanthemum: Nishihara Kamezo and the Abortive China Loans," Thesis, Catholic University of Leuven, Munich Personal REpEC Archive, 2007, 2.

3. For more on the Qingdao campaign, see Wilmott, H.P., *First World War* (De Gruyter, 2022,) 91.
4. For more analysis on the European power vacuum in China during World War I, see Matsusaka, *The Making of Japanese Manchuria*, 187-188.
5. Jacob, Frank, *East Asia and the First World War*, (De Gruyter, 2022), 15.
6. Matsusaka, *The Making of Japanese Manchuria*, 188.
7. To see the Twenty One Demands listed in full and a comparison to what such demands on the US would be like, see Millard, Thomas, F, "The ABC's of the Twenty-One Demands" (Chinese Students' Committee on Washington Conference, 1921), 5-15.
8. The further privileges for Japanese nationals in northeast China included the right to lease land, travel, reside, and conduct business.
9. Jacob, *East Asia and the First World War*, 20-21
10. For more on the boycotts during the Twenty-One Demands, see Luo, Zhitian, "National Humiliation and National Assertion: The Chinese Response to the Twenty-One Demands," *Modern Asian Studies*, 27, no. 2, (May 1993): 303-304, https://doi.org/10.1017/s0026749x00011501.
11. Ibid., 302.
12. Jacob, *East Asia and the First World War*, 20.
13. Japanese Government, "Twenty-One Demands," April 26, 1915, USC US-China Institute.
14. Weale, Putnam & Lenox, B.L., *The Fight for the Republic*, 271.
15. Gray, Jack, *Rebellions and Revolutions: China from 1868-2000*, (Oxford University Press, 2003), 171-172.
16. Ibid.
17. For more on Zhang Xun's attempted Qing Restoration, see Weale, Putnam, *The Fight for the Republic*, 355.
18. McCormack, *Chang Tso-lin in Northeast China*, 46.
19. For more on the question of China entering World War I, see Putnam, *The Fight for the Republic*, 309-370.
20. McCormack, *Chang Tso-lin*, 47.
21. Ibid.
22. For more on this Incident and Zhang's demands, see ibid., 47-48.
23. Ibid., 48.
24. Schlitz, "Separating the Roots," 19-20.
25. Ibid., 21.

26. Chi, Madeleine, "Bureaucratic Capitalists in Operation: Ts'ao Ju-lin and His New Communications Clique," *The Journal of Asian Studies* 34, no. 3, (May 1975): 680-681, https://doi.org/10.2307/2052549.
27. The "Financial Cutting Edge of the Partition of China," *Pacific Affairs* 9, no. 2 (June 1936): 172, https://doi.org/10.2307/2751404.
28. Matsusaka, *The Making of Japanese Imperialism*, 210.
29. Schlitz, "Separating the Roots," 33-34.
30. "The Financial Cutting Edge," 172.
31. Wasserstrom, Jeffrey, *Twentieth Century China: New Approaches*, (Taylor & Francis, 2013), 49.
32. Humphreys, *The Way of the Heavenly Sword*, 31-32.
33. Harries & Harries, *Soldiers of the Sun*, 120.
34. Drea, *Japan's Imperial Army*, 133.
35. Harries & Harries, *Soldiers of the Sun*, 120.
36. Drea, *Japan's Imperial Army*, 138.
37. Ibid.
38. Harries & Harries, *Soldiers of the Sun*, 121.
39. Barnhart, Michael, "The Origins of the Second World War in Asia and the Pacific: Synthesis Impossible," *Diplomatic History 20,* no. 2, (April 1996): 254, https://doi.org//10.1111/j.1467-7709.1996.tb00624.x.
40. Drea, *Japan's Imperial Army*, 138.
41. Harries & Harries, *Soldiers of the Sun*, 121.
42. Yamamuro & Fogel, *Manchuria Under Japanese Dominion,* (University of Pennsylvania Press, 2006), 18.
43. Ibid.
44. "Preface," (JACAR), Ref.C12121558200, Japanese national defense resources August, 1917 (National Institute for Defense Studies), 1-2, Translated by Ms. Fatima Janaghir.
45. Ibid.
46. Yamamuro & Fogel, *Manchuria Under Japanese Dominion*, 19.
47. Matsusaka, *The Making of Japanese Manchuria,* 216.
48. Barnhart, Michael, *Japan Prepares for Total War: The Search for Economic Security, 1919-1941*, 23.
49. Matsusaka, *The Making of Japanese Manchuria*, 220-221.
50. Drea, *Japan's Imperial Army*, 142.
51. Ibid.

52. Dunscomb, Paul, "A Great Disobedience Against the People: Popular Press Criticism of Japan's Siberian Intervention 1918-1922," *Journal of Japanese Studies 32,* No. 1, (Winter 2006): 57-58, https://doi.org/10.5771/9780739146026.
53. Hosoya Chihiro, "Origin of the Siberian Intervention, 1917-1918," *The Annals of the Hitotsubashi Academy,* 9, no. 1, (October 1958): 96-97
54. Ibid.
55. Dunscomb, "A Great Disobedience Against the People," 58.
56. Ibid.
57. Hosoya, "Origins of the Siberian Intervention," 102.
58. Drea, *Japan's Imperial Army*, 142.
59. Humphreys, *Way of the Heavenly Sword*, 26.
60. Hosoya, "Origins of the Siberian Intervention," 1917-1918, 103.
61. For more on the Rice Riots of 1919, see Lewis, Michael, *Rioters and Citizens: Mass Protest in Imperial Japan,* (University of California Press, 1990), 1-32.
62. Drea, *Japan's Imperial Army*, 143.
63. "Wilson's Fourteen Points," United States State Department Office of the Historian, Accessed August 19th, 2025, https://history.state.gov/milestones/1914-1920/fourteen-points.
64. For more on the New Diplomacy, see Sadao, Asada, "Between the Old Diplomacy and the New, 1918-1922: The Washington System and the Origins of Japanese-American Rapprochement," *Diplomatic History* 30, no. 2, (April 2006): 212-213.
65. Matsusaka, *The Making of Japanese Manchuria*, 236.
66. To read more about Japanese officials' resentment of Wilson's liberal internationalism, see Sadao, "Between the Old Diplomacy and the New," 213.
67. Kawamura, Noriko, "Wilsonian Idealism and Japanese Claims at the Paris Peace Conference," *Pacific Historical Review,* no. 4, (November 1997): 513, https://doi.org/10.2307/3642235.
68. Ibid., 515-516.
69. Matsusaka, *The Making of Japanese Manchuria*, 236-237.
70. Ibid.
71. Pesek, Pauline, "The Second Chinese Banking Consortium: The Open Door and Finance Diplomacy," (Master's Thesis, University of Nebraska, 1976), 69-70.

72. Matsusaka, *The Making of Japanese Manchuria*, 237.
73. Ibid., 239.
74. Kawamura, "Wilsonian Idealism," 518.
75. Ibid., 519-520.
76. Lauren, Paul Gordon, *Power and Prejudice: The Politics and Diplomacy of Racial Discrimination*, (Routledge, 2019), 188.
77. Temperley, H.W.V., *A History of the Paris Peace Conference*, (Oxford University Press, 1920), 352.
78. Kawamura, "Wilsonian Idealism," 521.
79. MacMillan, Margaret, *Paris 1919: Six Months that Changed the World*, (John Murray Publishers, 2019), 331.
80. Kawamura, "Wilsonian Idealism," 521.
81. MacMillan, *Paris 1919*, 331.
82. Ibid., 524.
83. Ibid.
84. For more on the views of Chinese intellectuals and students on Wilson, see Manela, Erez, "Imagining Woodrow Wilson in Asia: Dreams of East-West Harmony and the Revolt Against Empire in 1919," *The American Historical Review,* 111 no. 5 (December 2006): 1334-1348, https://doi.org/10.1086/ahr.111.5.1327.
85. Ibid.
86. For more on the ideas of "Mr. Science" and "Mr. Democracy," in the May 4th Movement, see Xu, Edward, "Who Was Mr. Democracy? The May Fourth Discourse of Popularity Democracy and the Radicalization of Chinese Intellectuals (1915-1922)," *Modern Asian Studies,* 35 no. 3, (July 2001): 589-590, https://doi.org/10.1017/s0026749x01003043.

Chapter 5: The Foundation of an Army

1. Coox, "The Kwantung Army Dimension," 399-400.
2. "Revision of Kwantung Province defense structure zones ordinance," (JACAR), Ref.A03034065900, Resolution at Privy Council, / Government organization of Kwantung Territory Government / Abolishing 1917 Imperial ordinance No. 87 concerning appointment and ranks of Secretary-General of the Security Department, police superinten-

dents, and police inspectors at Resident Generals Office of Kwantung / Amending Kwantung Province defense structure zones ordinance / Resolution on 1919, 2-3, Translated by Ms. Fatima Janaghir.

3. Coox, "The Kwantung Army Dimension" 399.
4. Coox, *Nomonhan*, 7.
5. Bix, Herbert, *Hirohito and the Making of Modern Japan*, (HarperCollins Publishers, 2000), 120-122.
6. For more details on the views of Japanese officers on China and the Soviet Union, see Yamamuro & Fogel, *Manchuria Under Japanese Dominion*, 21-28.
7. Coox, *Nomonhan*, 7-8.
8. Myers, "Japanese Imperialism," 124.
9. Drea, *Japan's Imperial Army*, 146-147.
10. Ibid., 147.
11. Kitaoka, "The Army as a Bureaucracy: Japanese Militarism Revisited," 76.
12. Drea, *Japan's Imperial Army*, 146-147.
13. Ibid., 147.
14. Seki, Hiroharu, "The Manchurian Incident, 1931," in *Japan's Road to the Pacific War Volume I, 1928-1932,* edited by James Morley (University of Columbia Press, 1984), 145.
15. Ibid.
16. Ibid.
17. Drea, *Japan's Imperial Army*, 155.
18. Ogata, *Defiance in Manchuria*, 26.
19. Whitehurst, *The China Incident*, 60.
20. Ogata, *Defiance in Manchuria*, 26-27.
21. Humphreys, *Way of the Heavenly Sword*, 34.
22. Ibid., 35.
23. Ogata, *Defiance in Manchuria*, 26.
24. Seki, "The Manchurian Incident," 145.
25. Crowley, James, *Japan's Quest for Autonomy: National Security and Foreign Policy, 1930-1938* (Princeton University Press, 1966), 85.
26. Seki, "The Manchurian Incident," 146.
27. On his death bed, Taro apparently drew lines on a map of the world at the sixty and one-hundred and seventy degree markers, and said "all

of that is Japan's!" "All of that" included Siberia, China, all of Southeast Asia, India Australia, and New Zealand, Ibid., 146

28. Humphreys, *The Way of the Heavenly Sword*, 37.
29. Ibid., 37.
30. Ibid., 40.
31. Ibid.
32. Ibid.
33. McCormack, *Chang Tso-lin*, 50.
34. Young, John, "The Hara Cabinet and Chang Tso-lin," 1920-21, *Monumenta Nipponica*, 27, no. 2 (1972): 125, https://doi.org/10.2307/2383715.
35. Ibid.
36. Jowett, Philip, *Chinese Warlord Armies, 1911-30*, (Osprey, 2010), 10-11.
37. Ibid.
38. McCormack, *Chang Tso-lin*, 50.
39. For more details on the formation of the anti-Zhang alliance, see Chi, *Warlord Politics in China,* 30-35.
40. McCormack, *Chang Tso-lin*, 50.
41. Kwong, *War and Geopolitics*, 90.
42. Young, "The Hara Cabinet," 127.
43. McCormack, *Chang Tso-lin*, 51.
44. Matsusaka, *The Making of Japanese Manchuria*, 260.
45. Ibid.
46. For more details about Zhang's dealings with Japanese political and military figures in 1920, see Young, "The Hara Cabinet," 129-134.
47. For more on Sato's meeting with Zhang, see ibid., 133.
48. Matsusaka, *The Making of Japanese Manchuria*, 261.
49. Young, "The Hara Cabinet," 134.
50. Ibid., 138.
51. Ibid., 139.
52. For more on the specific means of Japanese support for Zhang, including the provision of workers, operation of railways and warehouses, and the dispatch of military advisers see "Invitation to Chang Tso-lin as a Japanese Officer" (JACAR), Ref.C03022654100, (National Institute for Defense Studies), Translated by Ms. Fatima Janaghir.
53. Wright, Quincy, "The Washington Conference," *The American Political Science Review* 16, no. 2, (May 1922): 291

54. Ibid.
55. Ibid.
56. Sadao, Asada, "Japan's 'Special Interests' and the Washington Conference," *The American Historical Review* 67 no. 1, (October 1961): 64, https://doi.org/10.2307/1846262.
57. Hattori, Ryuji, *Japan at War and Peace: Shidehara Kijuro and the Making of Modern Diplomacy*, (Australian National University Press, 2021), 70.
58. Sadao, "Japan's Special Interests," 64.
59. Ibid.
60. Matsusaka, *The Making of Japanese Manchuria*, 265.
61. Wright, "The Washington Conference," 295.
62. Hattori, *Japan at War and Peace*, 73.
63. To read more about the IJA's views on the "lifeline" of northeast China, see Matsusaka, *The Making of Japanese Manchuria*, 269-270.
64. "Japan in Manchuria, 1930," Box 2, Organizational History Files, Record Group 59, National Archives at College Park, 14.
65. Hattori, *Japan at War and Peace*, 79-81.
66. "Foreign Relations of the United States: Treaty Between the United States of America, Belgium, the British Empire, China, France, Italy, Japan, the Netherlands, and Portugal," *The Avalon Project—Documents in Law, History and Diplomacy*, Accessed June 26, 2025. https://avalon.law.yale.edu/20th_century/tr22-01.asp.

Chapter 6: Zhang's Adventurous Wars

1. Akira, Iriye, *After Imperialism: The Search for a New Order in the Far East, 1921-1931*, (Harvard University Press, 1965), 27-28.
2. For more on Japan's support for Zhang, see "Invitation to Chang Tso-lin as a Japanese Officer," Translated by Ms. Fatima Janaghir.
3. Chi, *The Chinese Warlord System*, 36.
4. Matsusaka, *The Making of Japanese Manchuria*, 271.
5. McCormack, *Chang Tso-lin*, 62.
6. "Union OF ALL CHINA IS SOUGHT IN PEKING; Chang Tso-lin, Manchurian Chief Who Caused Cabinet's Resignation, Takes Control. Delegation Here Hopeful Statement on Rehabilitation Move at Home Says They Still," *The New York Times*, December 20, 1921, https://www.nytimes.com/

1921/12/20/archives/union-of-all-china-is-sought-in-peking-chang-tsolin-manchurian.html.

7. "CHANG TO GIVE CHINA NEW CONSTITUTION; Unity is Possible Under the Right Kind of Government, General Declares," *The New York Times*, December 30, 1921, https://www.nytimes.com/1921/12/30/archives/chang-to-give-china-new-constitution-unity-is-possible-under-the.html.
8. Ibid.
9. Chi, *The Chinese Warlord System*, 36.
10. McCormack, *Chang Tso-lin*, 62.
11. Chan, Anthony, Arming the Chinese: *The Western Armaments Trade in Warlord China, 1920-1928* (University of British Columbia Press, 2010), 69.
12. "Ultimatum Orders Peking Cabinet Out, Wu Pei-fu, with Troops Already Moving, Gives Liang Shih-yi Three Days to Resign, Civil War Believed Near, Rival Threatens Attack Unless Chang Tso-lin Dictatorship Ends Within a Week," *The New York Times,* January 16, 1922, https://www.nytimes.com/1922/01/16/archives/ultimatum-orders-peking-cabinet-out-wu-peifu-with-troops-already.html.
13. Ibid.
14. McCormack, *Chang Tso-lin*, 62.
15. Ibid., 64.
16. Ibid., 65.
17. Ibid.
18. Matsusaka, *The Making of Japanese Manchuria*, 271.
19. McCormack, *Chang Tso-lin*, 65-66.
20. Ibid., 66-67.
21. Ibid.
22. Matsusaka, *The Making of Japanese Manchuria*, 271.
23. Chi, *The Chinese Warlord System*, 36.
24. McCormack, *Chang Tso-lin*, 70-71.
25. Ibid.
26. Ibid.
27. Chi, *The Chinese Warlord System*, 36.
28. Hall, "Blinded by the Rising Sun," 108.
29. Matsusaka, *The Making of Japanese Manchuria*, 271-272.

30. Ibid.
31. McCormack, *Chang Tso-lin*, 73.
32. Ibid.
33. Ibid., 88
34. For more on Zhang's "incorrigible" opposition to Chinese nationalism, see ibid., 88-89.
35. Matsusaka, *The Making of Japanese Manchuria*, 273.
36. Ibid.
37. Ibid., 278
38. Ibid., 279
39. Ibid., 291
40. Ibid., 299-303
41. McCormack, *Chang Tso-lin*, 94.
42. Ibid.
43. Ibid., 95
44. Matsusaka, *The Making of Japanese Manchuria*, 272.
45. McCormack, *Chang Tso-lin*, 95.
46. Myers, "Japanese Imperialism," 111.
47. Ibid.
48. Ibid.
49. "2nd Report Manchuria on Progress in Manchuria to 1930," Box 261, Organizational History Files, Record Group 550, National Archives at College Park, 4.
50. Myers, "Japanese Imperialism," 117, in Duus et al., *The Japanese Informal Empire in China.*
51. "2nd Report Manchuria on Progress in Manchuria to 1930," 4.
52. Myers, "Japanese Imperialism," 117.
53. Young, *Japan's Total Empire*, 32.
54. For more on why the SMR's monopolization of the soy trade was detrimental to Chinese peasants in the northeast, see Bix, "Japanese Imperialism," 429-434,
55. Ibid.
56. Ibid.
57. Eckstein, Alexander, Chao, Kang, and Chang, John, "The Economic Development of Manchuria: The Rise of a Frontier Economy," *The Jour-*

nal of Economic History, 34, no. 1, (March 1974): 255. https://doi.org/10.1017/s0022050700079729.

58. Bix, "Japanese Imperialism," 436.
59. Ibid., 437.
60. "2nd Report Manchuria on Progress in Manchuria to 1930," 171.
61. Bix, "Japanese Imperialism," 437.
62. For more on the plans and intentions of Japanese policymakers regarding the industrialization of northeast China, see Matsusaka, *The Making of Japanese Manchuria,* 250-258.
63. Young, *Japan's Total Empire,* 31-32.
64. Eckstein, Chao, and Chang, "The Economic Development of Manchuria," 253.
65. "2nd Report Manchuria on Progress in Manchuria to 1930," 171-172.
66. Stewart, John, R., "Manchuria: The Land and its Economy," *Economic Geography* 8, no. 2, (April 1932): 149, https://doi.org/10.2307/140247.
67. Ibid.
68. "2nd Report Manchuria on Progress in Manchuria to 1930," 123.
69. Matsusaka, *The Making of Japanese Manchuria,* 305.
70. "2nd Report Manchuria on Progress in Manchuria to 1930," 124.
71. Ibid., 175.
72. Ibid., 179.
73. Ibid., 180.
74. For more on the SMR's spinoff ventures and their profitability, see Young *Japan's Total Empire* 32-33.
75. 2nd Report Manchuria on Progress in Manchuria to 1930," 124.
76. Myers, 123 "Japanese Imperialism," in Duus et al., *The Japanese Informal Empire in China.*
77. Ibid.
78. 2nd Report Manchuria on Progress in Manchuria to 1930," 124.
79. Matsusaka, *The Making of Japanese Manchuria,* 256.
80. "2nd Report Manchuria on Progress in Manchuria to 1930," 84.
81. Young, *Japan's Total Empire,* 34.
82. Fogel & Ito, *Life Along the South Manchuria Railway,* 22.
83. Young, *Japan's Total Empire,* 32.
84. Harries & Harries, *Soldiers of the Sun,* 133.
85. Hattori, *Japan at War and Peace,* 115.

86. Ibid.
87. Seki, "The Manchurian Incident," 123-124.
88. Ogata, *Defiance in Manchuria*, 8.
89. Ibid.
90. Ibid.
91. McCormack, *Chang Tso-lin*, 127.
92. Ibid.
93. Ibid.
94. Matsusaka, *The Making of Japanese Manchuria*, 295.
95. McCormack, *Chang Tso-lin*, 128-129.
96. Ibid.
97. Ibid.
98. Ibid., 123.
99. Hall, "Blinded by the Rising Sun," 120-121.
100. Hattori, *Japan at War and Peace*, 117-118.
101. McCormack, *Chang Tso-lin*, 141.
102. Ibid.
103. Hattori, *Japan at War and Peace*, 118.
104. Matsusaka, *The Making of Japanese Manchuria*, 295.
105. To learn more about the circumstances surrounding the Japanese Cabinet's views on the Second Fengtian-Zhili war, see McCormack, *Chang Tso-lin*, 141-142.
106. Coox, "The Kwantung Army Dimension," 402.
107. Ibid.
108. "The Fengtien-Chihli war, No. 3/ 4 From October 9, 1914 to October 24, 1914," (JACAR), Ref.B03050758900, Confusional relationships of Kiangsu-Chekang and Fengtian-Chihli/ A strategic situation Vol.7 (1.6.1.85-2_007) (Diplomatic Archives of the Ministry of Foreign Affairs), 2-5, Translated by Ms. Fatima Janaghir.
109. McCormack, *Chang Tso-lin*, 144.
110. Ibid.
111. Ibid.
112. "Message to be conveyed to Chang Tsolin," (JACAR), Ref.C03022725300, Mitsu Dainikki, 4 of 6, 1925 (National Institute for Defense Studies), Translated by Ms. Fatima Janaghir.
113. McCormack, *Chang Tso-lin*, 189.

114. Ibid., 157-158.
115. "War and political situations in general (from October 21, 1925 to November 24, 1925)" (JACAR), Ref.B03050778100, Occurrence of confusion in the anti-Mukden (Shenyang) groups, Vol. 1 (1.6.1.86_001) (Diplomatic Archives of the Ministry of Foreign Affairs), 2, Translated by Ms. Fatima Janaghir.
116. For more on this dynamic, see Matsusaka, *The Making of Japanese Manchuria*, 313-315.
117. McCormack, *Chang Tso-lin*, 159.
118. For more on Sun's attack and its aftermath, see ibid., 160-161.
119. Ibid., 161.
120. For more on Guo Songling and his decision to rebel, see ibid., 162-163.
121. "War and political situations in general (from October 21, 1925 to November 24, 1925)" 1-2, Translated by Ms. Fatima Janaghir.
122. McCormack, *Chang Tso-lin* 162.
123. Ibid., 163.
124. For more on Guo's war against Zhang, see ibid., 164-165.
125. Ibid.
126. Coox, "The Kwantung Army Dimension," 404.
127. For more on the Kwantung Army's views of Guo, see McCormack, *Chang Tso-lin*, 170-172.
128. Matsusaka, *The Making of Japanese Manchuria*, 301.
129. Kitaoka, "The Army's China Experts," in *The Japanese Informal Empire in China*, edited by Duus, Peattie, and Myers (University of Princeton Press, 2014), 363.
130. McCormack, *Chang Tso-lin*, 177.
131. Ibid.
132. Ibid.
133. Ibid.
134. Coox, *Nomonhan*, 13.
135. Ibid., 179
136. Coox, "The Kwantung Army Dimension," 403.
137. Coox, *Nomonhan*, 13.
138. Weale, *The Vanished Empire*, 271.
139. Kwong, *War and Geopolitics*, 104-105.
140. Weale, *The Vanished Empire*, 271.

141. "General Condition of War and Politics/6 November 28, 1925 to December 8, 1925," (JACAR), Ref.B03050778500, Occurrence of confusion in the anti-Mukden (Shenyang) groups, Vol. 1 (1.6.1.86_001) (Diplomatic Archives of the Ministry of Foreign Affairs), Translated by Ms. Fatima Janaghir.
142. McCormack, *Chang Tso-lin*, 183.

Chapter 7: Reform in the IJA and the Northern Expedition

1. Crowley, *Japan's Quest for Autonomy*, 87.
2. Humphreys, *The Way of the Heavenly Sword*, 84.
3. Ibid.
4. Ibid., 85.
5. Ibid.
6. Drea, *Japan's Imperial Army*, 146.
7. Ibid., 147.
8. Ibid., 153.
9. Humphreys, *The Way of the Heavenly Sword*, 91.
10. Drea, *Japan's Imperial Army*, 153.
11. Humphreys, *The Way of the Heavenly Sword*, 92.
12. Drea, *Japan's Imperial Army*, 153.
13. Ibid., 154.
14. Ibid.
15. Humphreys, *The Way of the Heavenly Sword*, 92.
16. For more on this dynamic, see ibid., 95-99.
17. Ibid., 97-98.
18. Ibid.
19. Jordan, Donald, A., *The Northern Expedition: China's National Revolution of 1926-1928*, (University of Hawaii Press, 2019), 4.
20. Ibid., 4-5.
21. Jowett, *China's Warlord Armies*, 17.
22. Iriye, *After Imperialism*, 41.
23. Crean, Jeffery, *The Fear of Chinese Power: An International History*, (Bloomsbury Publishers, 2024), 55.
24. Spence, *The Search*, 336.

25. For more details on foreign support for warlords, see Chan, *Arming the Chinese: The Western Armaments Trade in Warlord China, 1920-1928.*
26. For more on these protests, see Smith, Stephen Anthony, *A Road is Made: Communism in Shanghai,* (Curzon, 2000), 89-108.
27. Iriye, *After Imperialism*, 73.
28. Ibid.
29. Ibid.
30. Hattori, *Japan at War and Peace*, 131.
31. Iriye, *After Imperialism*, 82.
32. Fenby, *Jonathan, Generalissimo: Chiang Kai-shek and the China he Lost*, (Simon & Schuster, 2003), 115.
33. Iriye, *After Imperialism*, 91.
34. Jordan, *The Northern Expedition*, 39-40.
35. Jowett, *China's Warlord Armies*, 17.
36. Fenby, *Generalissimo*, 116-117.
37. Ibid., 115.
38. Ibid.
39. Chi, *China's Warlord Armies*, 225.
40. Jordan, *The Northern Expedition*, 79-84.
41. Ibid., 85-86.
42. Ibid., 86.
43. Iriye, *After Imperialism*, 90.
44. Jordan, *The Northern Expedition*, 88-89.
45. Fenby, *The Generalissimo*, 118.
46. Jordan, *The Northern Expedition*, 89.
47. Fenby, *The Generalissimo*, 119.
48. For more on the NRA's Wuhan campaigns during the Northern Expedition see Jordan, *The Northern Expedition*, 90-95.
49. Ibid., 92.
50. For more on the invasion of Jiangxi, see ibid., 96-98.
51. Ibid., 96.
52. Wilbur, Clarence Martin & How, Julie Lien-ying, *Missionaries of Revolution: Soviet Advisors and Nationalist China 1920-1927*, (Harvard University Press, 1989), 331-332.
53. Smith, *A Road is Made*, 147-148.
54. Fenby, *The Generalissimo*, 122.

55. Jordan, *The Northern Expedition*, 106.
56. For more on the NRA's exploits in Fujian province, see *ibid.*, 107-110.
57. Ibid., 110.
58. Ibid., 110-111.
59. Iriye, *After Imperialism*, 90.
60. McCormack, *Chang Tso-lin*, 209-210.
61. Jordan, *The Northern Expedition*, 118.
62. Ibid., 119.
63. Ibid.
64. Worthing, Peter, *General He Yingqin: The Rise and Fall of Nationalist China*, (Cambridge University Press, 2018), 75.
65. Ibid.
66. For more on these strikes and uprisings, see Smith, A Road is Made, 178-184.
67. For more on how Chiang Kai-shek reassured the foreign financial and business community of Shanghai that he would safeguard their position, see ibid., 192.
68. Jordan, *The Northern Expedition*, 139.
69. Ibid., 137.
70. For more on how the Green Gang helped Chiang kill Communists, see Smith, *A Road is Made*, 190-191.
71. Rebecca, Karl E., *Mao Zedong and China in the Twentieth Century World: A Concise History*, (Duke University Press, 2010), 33.
72. Coble, Parks, M., *Facing Japan: Chinese Politics and Japanese Imperialism*, (Council on East Asian Studies, Harvard University, 1991), 18-19.

Chapter 8: The Positive Policy and its Effects

1. For more on the 1927 Japanese financial crisis, see Metzler, *Lever of Empire*, 175-200
2. Ogata, *Defiance in Manchuria*, 9.
3. McCormack, *Chang Tso-lin*, 233.
4. Ibid.
5. "Japan in Manchuria, 1930," 15.
6. Iriye, *After Imperialism*, 183.
7. Yoshihashi, *Conspiracy at Mukden*, 14.

8. Ibid., 14-16.
9. Ibid., 16.
10. Yoshihashi, *Conspiracy at Mukden*, 16.
11. Koo, Wellington, "Memorandum on Japan's Plots and Schemes Against the Unification of China" in *Memoranda nos. 1-29 Submitted by the Chinese Assessor to the Commission of Enquiry into the Sino-Japanese Dispute*, From Library of Congress, Main Reading Room, 24.
12. Yoshihashi, *Conspiracy at Mukden*, 18-19.
13. Iriye, *After Imperialism*, 146.
14. "Japan in Manchuria, 1930," 15.
15. Koo, Wellington, "Memorandum on Japan's Plots and Schemes," 25.
16. Ibid.
17. Yoshihashi, *Conspiracy at Mukden*, 22.
18. Ibid.
19. Ibid.
20. Matsusaka, *The Making of Japanese Manchuria*, 328.
21. Ibid., 329.
22. McCormack, *Chang Tso-lin*, 236.
23. "Japan in Manchuria, 1930," 17-18.
24. Humphreys, *The Way of the Heavenly Sword*, 139.
25. Koo, Wellington, "Memorandum on Japan's Plots and Schemes," 25.
26. Humphreys, *The Way of the Heavenly Sword*, 140.
27. Matsusaka, *The Making of Japanese Manchuria*, 320-321.
28. Ibid.
29. Yoshihashi, *Conspiracy at Mukden*, 30.
30. Ibid.
31. "Japan in Manchuria, 1930," 21.
32. McCormack, *Chang Tso-lin*, 238.
33. Iriye, *After Imperialism*, 178.
34. Ibid., 179.
35. Ibid.
36. Humphreys, *The Way of the Heavenly Sword*, 144.
37. Iriye, *After Imperialism*, 180.
38. Ibid., 180-181.
39. Humphreys, *The Way of the Heavenly Sword*, 144.
40. Iriye, *After Imperialism*, 182-183.

41. Yoshihashi, *Conspiracy at Mukden*, 29-30.
42. Ibid., 31.
43. For more analysis on why Zhang accepted Yamamoto's proposals, see ibid., 32.
44. Ibid., 31.
45. Jordan, *The Northern Expedition*, 156-157.
46. Humphreys, *The Way of the Heavenly Sword*, 145.
47. Matsusaka, *The Making of Japanese Manchuria*, 342.
48. Jordan, *The Northern Expedition*, 178.
49. For more details on the details of the fighting between the NRA and other forces, see ibid., 178-179.
50. Iriye, *After Imperialism*, 194-196.
51. Ibid., 195.
52. Yoshihashi, *Conspiracy at Mukden*, 34.
53. Iriye, *After Imperialism*, 195.
54. Humphreys, *The Way of the Heavenly Sword*, 148.
55. Ibid., 147.
56. Yoshihashi, *Conspiracy at Mukden*, 34.
57. Ibid.
58. Humphreys, *The Way of the Heavenly Sword*, 148.
59. Jordan, *The Northern Expedition*, 179-180.
60. Humphreys, *The Way of the Heavenly Sword*, 148.
61. Jordan, *The Northern Expedition*, 179-180.
62. Ibid.
63. Koo, Wellington, "Memorandum on Japan's Plots and Schemes," 27.
64. Jordan, *The Northern Expedition*, 179-180.
65. Ibid.
66. Fairbank, John King, *Republican China, 1912-1949 Volume 12, Part I* (Cambridge University Press, 1985), 179.
67. Jordan, *The Northern Expedition*, 179.
68. Ibid.
69. Fairbank, *Republican China Volume 12, Part I*, 179.
70. Ibid.
71. Koo, Wellington, "Memorandum on Japan's Plots and Schemes," 27.
72. Ibid., 26
73. Abend, Hallett, "Japan Wins Tsinan; Drives out Chinese in Week's Fight-

ing," *The New York Times,* May 11, 1928. https://www.nytimes.com/1928/05/11/archives/japan-wins-tsinan-drives-out-chinese-in-weeks-fighting-nippons.html.

74. Koo, Wellington, "Memorandum on Japan's Plots and Schemes," 26.
75. "Telegrams on Jinan Incident," (JACAR) Ref.C04021700000, 1928 Rikushi Mitsu dai nikki Vol. 2 (National Institute for Defense Studies), 2-3, Translated by Ms. Fatima Janaghir.
76. For more on the final stages of the Northern Expedition, see Jordan, *The Northern Expedition*, 186-194.
77. Yoshihashi, *Conspiracy at Mukden*, 31.
78. "Japan in Manchuria, 1930," 22.
79. Humphreys, *The Way of the Heavenly Sword*, 154.
80. Ibid., 155.
81. Iriye, *After Imperialism*, 211-212.
82. For more on the Kwantung Army's advocacy of these solutions, see Humphreys, *The Way of the Heavenly Sword,* 153-156.
83. Dull, Paul S., "The Assassination of Chang Tso-lin," *The Far Eastern Quarterly,* 11, no. 4, (August 1952): 454, https://doi.org/10.2307/2942022.
84. Humphreys, *The Way of the Heavenly Sword*, 155.
85. Iriye, *After Imperialism*, 211.
86. Yoshihashi, *Conspiracy at Mukden*, 37-38.
87. Humphreys, *The Way of the Heavenly Sword*, 155.
88. Ibid., 155-156.
89. Yoshihashi, *Conspiracy at Mukden*, 38.
90. For more on these debates see ibid., 38-39.
91. Ibid.
92. Humphreys, Way of the Heavenly Sword, 158.
93. Yoshihashi, *Conspiracy at Mukden*, 39-40.
94. Ibid., 40.
95. Ibid.
96. Ibid.
97. Ibid.
98. Ibid., 141.
99. Ibid.
100. Ibid.
101. Iriye, *After Imperialism*, 206.

Chapter 9: They Behave Like Children

1. Yoshihashi, *Conspiracy at Mukden*, 43.
2. Ibid.
3. Matsusaka, *The Making of Japanese Manchuria*, 345.
4. Dull, "The Assassination of Chang Tso-lin," 453-454.
5. Orbach, *Curse on this Country*, 174.
6. Ibid., 175.
7. Ibid.
8. Matsusaka, *The Making of Japanese Manchuria*, 347.
9. Ibid., 347.
10. Yoshihashi, *Conspiracy at Mukden*, 43.
11. Orbach, *Curse on this Country*, 176.
12. Yoshihashi, *Conspiracy at Mukden*, 45.
13. Ibid.
14. Orbach, *Curse on this Country*, 176.
15. For more on Hata and Doihara's plans to assassinate Zhang Zuolin, see ibid., 176-177.
16. Ibid., 177.
17. Yoshihashi, *Conspiracy at Mukden*, 45.
18. Orbach, *Curse on this Country*, 178.
19. Yoshihashi, *Conspiracy at Mukden*, 46.
20. Ibid.
21. For more on Tomiya's meeting with Komoto, see ibid.
22. Ibid.
23. Ibid., 48.
24. Orbach, *Curse on this Country*, 178.
25. Ibid.
26. Yoshihashi, *Conspiracy at Mukden*, 46-48.
27. Orbach, *Curse on this Country*, 179.
28. Ibid.
29. Yoshihashi, *Conspiracy at Mukden*, 48.
30. Orbach, *Curse on this Country*, 178.
31. Yoshihashi, *Conspiracy at Mukden*, 48.
32. Jordan, *The Northern Expedition*, 191.
33. Iriye, *After Imperialism*, 213.

34. Jordan, *The Northern Expedition*, 191.
35. Iriye, *After Imperialism*, 213.
36. For more on these security measures, see Dull, "The Assassination of Chang Tso-lin," 455.
37. Ibid.
38. Yoshihashi, *Conspiracy at Mukden*, 49.
39. Dull, "The Assassination of Chang Tso-lin," 455.
40. Yoshihashi, *Conspiracy at Mukden*, 49.
41. Ibid.
42. Dull, "The Assassination of Chang Tso-lin," 456.
43. Humphreys, *The Way of the Heavenly Sword*, 161.
44. Matsusaka, *The Making of Japanese Manchuria*, 347.
45. Seki "The Manchurian Incident," 134.
46. Yoshihashi, *Conspiracy at Mukden*, 50.
47. Orbach, *Curse on this Country*, 181.
48. Seki "The Manchurian Incident," 134.
49. Yoshihashi, *Conspiracy at Mukden*, 51.
50. Ibid.
51. Orbach, *Curse on this Country*, 181.
52. Yoshihashi, *Conspiracy at Mukden*, 50.
53. Orbach, *Curse on this Country*, 182.
54. Ibid.
55. Ibid., 183.
56. Humphreys, *The Way of the Heavenly Sword*, 165.
57. Coox, "The Kwantung Army Dimension," 405.
58. Dull, "The Assassination of Chang Tso-lin," 457.
59. For more on the activities of the committee, see Orbach, *Curse on this Country*, 183.
60. Ibid.
61. Orbach, *Curse on this Country*, 183.
62. Humphreys, *The Way of the Heavenly Sword*, 166.
63. Orbach, *Curse on this Country*, 184.
64. Ibid.
65. Humphreys, *The Way of the Heavenly Sword*, 166.
66. Bix, *Hirohito and the Making of Modern Japan*, 200.
67. Ibid., 200-201.

68. Orbach, *Curse on this Country*, 186.
69. Bix, *Hirohito and the Making of Modern Japan*, 200-201.
70. Ibid.
71. Humphreys, *The Way of the Heavenly Sword*, 168.
72. Ibid., 168-169.
73. Bix, *Hirohito and the Making of Modern Japan*, 200-201.
74. Humphreys, *The Way of the Heavenly Sword*, 169.
75. Bix, *Hirohito and the Making of Modern Japan*, 200-201.
76. Yoshihashi, *Conspiracy at Mukden*, 51.
77. Iriye, Akira, "Chang Hsueh-liang and the Japanese," *The Journal of Asian Studies* 20 no. 1, (November 1960): 34, https://doi.org/10.2307/2050070.
78. Yoshihashi, *Conspiracy at Mukden*, 51-52.
79. For more on this divide, see Iriye, Chang Hsueh-liang, 35-36.
80. Iriye, *After Imperialism*, 233.
81. Matsusaka, *The Making of Japanese Manchuria*, 350.
82. Ibid., 350-351.
83. Ibid.
84. Iriye, *After Imperialism*, 235.
85. Yoshihashi, *Conspiracy at Mukden*, 53.
86. For more on these talks, see Mitter, Rana, *The Manchurian Myth: Nationalism, Resistance, and Collaboration in Modern China*, (University of California Press, 2000), 51-53.
87. Iriye, "Chang Hsueh-liang," 37.
88. Seki, "The Manchurian Incident," 135.
89. Iriye, "Chang Hsueh-liang," 37-38.
90. Ogata, *Defiance in Manchuria*, 16.
91. Ibid., 17
92. Yamamuro, *Manchuria Under Japanese Dominion*, 13.

Chapter 10: Ishiwara's Final War and Japan's Preparations

1. End, "Japan During the Interwar Period," 255.
2. Maeda, Kiyotaka, "Colonial Impact on Japan's Economy: Dynamics of the Rice Market During the Interwar Period," (Thesis, Keio University, 2022), 5
3. Yoshihashi, *Conspiracy at Mukden*, 110.

4. Ibid., 109-110.
5. Lockwood, William, Jr. "Japanese Silk and the American Market," *Far Eastern Survey* 5, no. 4, (February 1936): 32, https://doi.org/10.2307/3021472.
6. Harada, Yutaka, "Why Was Wilsonian-Taisho Moment Lost in Japan in Spite of its Economic Success," *Japanese Journal of Political Science 19,* no. 4, (December 2018): 572, https://doi.org/10.1017/s1468109918000348.
7. For more on the Hamaguchi government's return to the gold standard, see Metzler, *Lever of Empire*, 199-216.
8. For more on Hamaguchi's austerity policies and their effects, see ibid., 217-239.
9. Bix, *Hirohito and the Making of Modern Japan,* 210.
10. For more on Tachibana and Gondo's views, see Havens, Thomas R.H., *Farm and Nation in Modern Japan: Agrarian Nationalism 1870-1940,* (Princeton University Press, 1974), 163-274
11. For more on the Tachibana's patriotic reform movement, see ibid., 254-274
12. Kita, Ikki, *An Outline Plan for the Reorganization of Japan*, (Adept Press, 2024), 3, Translated by Jake Glanor.
13. Ibid., 1.
14. Ibid., 1-2.
15. For more on Kita's proposed economic policies, see ibid., 17-52.
16. Ogata, *Defiance in Manchuria*, 24.
17. For more on Kita Ikki's views on external expansion see Kita, *An Outline Plan for the Reorganization of Japan.*
18. For more on the Tenkento and Kozakurakai, see Ogata, *Defiance in Manchuria*, 29-30.
19. Ibid.
20. Humphreys, *The Way of the Heavenly Sword*, 111-112.
21. Ibid.
22. Orbach, *Curse on this Country*, 202.
23. Ibid., 202.
24. "Brief-Hashimoto Kingoro" Box 8, Files on Defendants: Hashimoto-Hata, 1946, Tavenner Papers & IMFTE Official Records, University of Virginia Law Library, 1.
25. Orbach, Curse on this Country, 197.
26. Ibid.

27. Ibid., 197-198.
28. Ibid.
29. "Hashimoto, Kingoro-Miscellaneous Notes," Box 8, Files on Defendants: Hashimoto-Hata, 1946, Tavenner Papers & IMFTE Official Records, University of Virginia Law Library, 1.
30. Ogata, *Defiance in Manchuria*, 30.
31. Yoshihashi, *Conspiracy at Mukden*, 99.
32. Ibid., 97-100.
33. "Amplified Report on Dr. Okawa Shumei," Box 8, Files on Defendants: Okawa, Oshima, 1946, Tavenner Papers and IMFTE Official Records, University of Virginia Law Library, 4.
34. Ibid., 5.
35. Yoshihashi, *Conspiracy at Mukden*, 139.
36. "Addition to Doc 684, Asia, Europe, Japan, by Okawa, Shumei," Box 10, IPS Documents 0-999, Tavenner Papers and IMFTE Official Records, University of Virginia Law Library, 2.
37. Ibid., 3.
38. Orbach, *Curse on this Country*, 203.
39. Ogata, *Defiance in Manchuria*, 24.
40. Yoshihashi, *Conspiracy at Mukden*, 139.
41. Ogata, *Defiance in Manchuria*, 23-24.
42. Ishihara Kanji Dossier, Box 329, Intelligence and Investigative Dossiers Personal Files, Record Group 319, National Archives at College Park, Maryland, 65.
43. Seki, "The Manchurian Incident," 143.
44. Peattie, *Ishiwara Kanji*, 30.
45. For more on Ishiwara's understanding of warfare's spatial dimension, see Iguchi, Gerald, "Nichirenism as Modernism: Imperialism, Fascism, and Buddhism in Modern Japan," PhD Thesis, University of California San Diego, 2006, 242-243.
46. For more on Ishiwara's understanding of protracted vs. annihilation war, see Peattie, *Ishiwara Kanji,* 29.
47. To better understand the specifics of how wars of annihilation and attrition alternated in prominence according to Ishiwara's conception, see ibid., 58-59.

48. For more on Ishiwara's understanding of the quickening pace of technological progress, see Iguchi, "Nichirenism as Modernism," 244-245.
49. For more on Ishiwara's views on the centrality of aircraft, see Peattie, *Ishiwara Kanji*, 61-62.
50. Iguchi, "Nichirenism as Modernism," 246-247.
51. For more on Ishiwara's views on the east and west, see Peattie, *Ishiwara Kanji*, 55-57.
52. Ibid.
53. Godart, Clinton G., "Nichirenism, Utopianism, and Modernity: Rethinking Ishiwara Kanji's East Asia League Movement," *Japanese Journal of Religious Studies* 42, no. 2, (November 2015): 238, https://doi.org/10.18874/jjrs.42.2.2015.235-274.
54. Peattie, *Ishiwara Kanji*, 46.
55. Ibid., 47.
56. Ibid., 67-68.
57. Yamamuro, *Manchuria Under Japanese Dominion*, 19.
58. Ibid.
59. Peattie, *Ishiwara Kanji*, 68-70.
60. Ogata, *Defiance in Manchuria*, 43.
61. Seki, "The Manchurian Incident," 138.
62. Peattie, *Ishiwara Kanji*, 95.
63. Ogata, *Defiance in Manchuria*, 48.
64. Peattie, *Ishiwara Kanji*, 103.
65. Ibid., 102.
66. Seki, "The Manchurian Incident," 148.
67. Ibid., 149.
68. Peattie, *Ishiwara Kanji*, 52.
69. Seki, "The Manchurian Incident," 149.
70. Yamamuro, Manchuria Under Japanese Dominion, 15.
71. Yoshihashi, *Conspiracy at Mukden*, 138.
72. Peattie, *Ishiwara Kanji*, 100.
73. Yoshihashi, *Conspiracy at Mukden*, 140-141.
74. Seki, "The Manchurian Incident," 150.
75. Yamamuro, *Manchuria Under Japanese Dominion*, 15.
76. "Summary of Manchuria and Mongolia governance plan (1)," (JACAR), Ref.C13010350400, Educational materials (1) Sumiya Tei Historic Ma-

terials (National Institute for Defense Studies), 1, Translated by Ms. Fatima Janaghir.

77. Seki, "The Manchurian Incident," 44.
78. "Summary of Manchuria and Mongolia governance plan," 2-5.
79. Ogata, *Defiance in Manchuria*, 45-47.
80. "Sending the opinions about the policies on Manchuria and Mongolia," (JACAR), Ref.C01003764000, Mitsu Dainikki, 4 of 6, 1927 (National Institute for Defense Studies), Translated by Ms. Fatima Janaghir.
81. Yamamuro, *Manchuria Under Japanese Dominion*, 16.
82. Seki, "The Manchurian Incident," 154.
83. Yoshihashi, *Conspiracy at Mukden*, 141.
84. Yamamuro, *Manchuria Under Japanese Dominion*, 17.
85. For more on all of Ishiwara's reconnaissance trips in 1929-1930, see Peattie, *Ishiwara Kanji*, 104-105.
86. Yoshihashi, *Conspiracy at Mukden*, 137-138.
87. Peattie, *Ishiwara Kanji*, 106.
88. Seki, "The Manchurian Incident," 154.
89. Peattie, *Ishiwara Kanji*, 106.
90. For more on Komoto's trip to Manchuria, see Seki, "The Manchurian Incident," 160.
91. For more on Komoto's efforts in Japan, see ibid., 162.
92. "International Military Tribunal for the Far East (IMFTE) Judgement, Chapter V, Japanese Aggression Against China," 530.
93. For more on this situation estimate, see Seki, "The Manchurian Incident," 161.
94. Orbach, *Curse on this Country*, 196.
95. For more on Nagata's trip, see Seki, "The Manchurian Incident," 161-162.
96. Ibid.
97. Orbach, *Curse on this Country,* 198.
98. Crowley, *Japan's Quest for Autonomy*, 97.
99. Orbach, *Curse on this Country*, 206.
100. Seki, "The Manchurian Incident," 163.
101. Crowley, *Japan's Quest for Autonomy*, 98.
102. Yoshihashi, *Conspiracy at Mukden*, 87.
103. Crowley, *Japan's Quest for Autonomy*, 98-99.
104. Yoshihashi, *Conspiracy at Mukden*, 88.

105. Ibid.
106. Orbach, *Curse on this Country*, 204.
107. Ogata, *Defiance in Manchuria*, 164.
108. Yoshihashi, *Conspiracy at Mukden*, 92.
109. For more on Okawa's Efforts to secure the support of left- and right-wing political organizations, see ibid., 91-92.
110. Crowley, *Japan's Quest for Autonomy*, 100-101.
111. Orbach, *Curse on this Country*, 196.
112. Crowley, *Japan's Quest for Autonomy*, 99.
113. Orbach, *Curse on this Country*, 206.
114. Ogata, *Defiance in Manchuria*, 32.
115. Yoshihashi, *Conspiracy at Mukden*, 92.
116. Ogata, *Defiance in Manchuria*, 99-100.
117. Yoshihashi, *Conspiracy at Mukden*, 93.
118. Orbach, *Curse on this Country*, 207-208.

Chapter 11: Countdown to Zero Hour

1. Peattie, *Ishiwara Kanji*, 110.
2. For Ishiwara's views on war with all four powers, see Seki, "The Manchurian Incident," 166.
3. Ibid., 165.
4. For Itagaki's views on war with China and the Soviet Union see ibid., 167-168.
5. Yamamuro, *Manchuria Under Japanese Dominion*, 25.
6. Seki, "The Manchurian Incident," 168.
7. Ibid., 164-165.
8. Peattie, *Ishiwara Kanji*, 112.
9. For further details about how the various junior officers that took part in the Mukden Incident learned of their roles, see Seki, "The Manchurian Incident," 212.
10. Ibid.
11. Ibid.
12. Crowley, *Japan's Quest for Autonomy*, 115.
13. Seki, "The Manchurian Incident," 172.
14. Crowley, *Japan's Quest for Autonomy*, 102.

15. Seki, "The Manchurian Incident," 173.
16. For more on the Geneva Arms Limitation Conference see Crowley, *Japan's Quest for Autonomy*, 105-106.
17. "Hashimoto Urged War in Manchuria Testifies News Article," Box 2, Second Phelps Scrapbook, C.W.J. Phelps Collection, University of Virginia Law Library, 1.
18. Seki, "The Manchurian Incident," 173.
19. Ibid.
20. For more on the "Situation Estimate" Document, see ibid., 173-174.
21. Ogata, *Defiance in Manchuria*, 54-55.
22. "Doc 3029-Views Concerning Our Imperial National Defense Against the USSR," Box 11, IPS Documents 3000-3999, Tavenner Papers & IMTE Official Records, University of Virginia Law Library, 1.
23. Yamamuro, *Japanese Dominion in Manchuria*, 26.
24. Seki "The Manchurian Incident," 174.
25. For more on Hanaya's meetings in Tokyo, see Ogata, *Defiance in Manchuria*, 55-56.
26. For more on the reactions of officers to Hanaya's proposals, see ibid., 55-56.
27. Ibid., 53-54.
28. Seki, "The Manchurian Incident," 176
29. Lee, Chong-Sik, *Revolutionary Struggle in Manchuria: Chinese Communism and Soviet Interest,* (University of California Press, 2023), 24.
30. In 1909 there were roughly 82,999 Koreans in Jiandao, a part of Jilin province that borders Korea; the increase in the number of Koreans was likely due to the establishment of a Japanese protectorate over Korea. For more on this topic, see Pratt, Edward, Earl Alexander, "Wanpaoshan, 1931: Japanese Imperialism, Chinese Nationalism and the Korean Problem in Northeast China on the Eve of the Manchurian Incident," (Master's Thesis, University of Virginia, May 1983), 20-22.
31. Lattimore, *Manchuria: Cradle of Conflict*, 239.
32. Koo, Wellington, "Memorandum on Koreans in the Three Eastern Provinces" in *Memoranda nos. 1-29 Submitted by the Chinese Assessor to the Commission of Enquiry, into the Sino-Japanese Dispute*, Library of Congress, Main Reading Room, 1.
33. Ibid., 6.

34. Pratt, "Wanpaoshan," 31.
35. Lytton, "Appeal by the Chinese Government," 56-57.
36. Japanese Government, "Twenty-One Demands," April 26, 1915 USC US-China Institute.
37. Koo, Wellington, "Memorandum on the Wanpaoshan Case" in *Memoranda nos. 1-29 Submitted by the Chinese Assessor to the Commission of Enquiry, into the Sino-Japanese Dispute*, Library of Congress, Main Reading Room, 9.
38. For more on the anti-Korean discriminatory measures of local Chinese officials, see Pratt, "Wanpaoshan," 39-83.
39. Ibid., 43-44.
40. Ibid., 44-45.
41. Ibid., 56, 63.
42. Ibid., 7.
43. Koo, Wellington, "Memorandum on the Wanpaoshan Case," 7.
44. Pratt, "Wanpaoshan," 8.
45. Ibid.
46. Ibid., 8.
47. For more on the negotiations surrounding the Wanpaoshan Incident, see ibid., 9-15.
48. Koo, Wellington, "Memorandum on the Wanpaoshan Case," 8.
49. Ibid.
50. For more on the attacks on Chinese in Korea, see Pratt, "Wanpaoshan," 15-16.
51. Yoshihashi, *Conspiracy at Mukden*, 144.
52. Ibid., 145.
53. Japanese Government, "Twenty-One Demands," April 26, 1915 USC US-China Institute.
54. Koo, Wellington, "Relation of Japan's Claims Based on the So-called Fifty-Three Outstanding Cases" in *Memoranda nos. 1-29 Submitted by the Chinese Assessor to the Commission of Enquiry, into the Sino-Japanese Dispute,* Library of Congress, Main Reading Room, 49.
55. For more on Nakamura's mission and his preparations for it, see Ramsdell, Daniel S., "The Nakamura Incident and the Japanese Foreign Office," *The Journal of Asian Studies,* 25, no. 1, (November 1965): 53-54, https://doi.org/10.2307/2051040.

56. Ibid.
57. Ibid.
58. "IMFTE Judgement, Chapter V," 139.
59. Ramsdell, "The Nakamura Incident," 55.
60. "IMFTE Judgement Chapter V," 139.
61. Lytton, "Appeal by the Chinese Government," 64.
62. "The Consul at Mukden (Vincent) to the Minister in China (Johnson)," Document 1 August 20th, 1931, Papers Relating to the Foreign Relations of the United States, 1931, The Far East, Volume III.
63. Ramsdell, "The Nakamura Incident," 55-56.
64. Ibid.
65. "The Consul at Mukden (Vincent)," Document 1.
66. For more on the Chinese investigations of Nakamura's death, see Lytton, "Appeal by the Chinese Government," 64-65.
67. Yamamoto, Tsutomo, David, "The Japanese Press and Japanese Foreign Policy, 1927-1933," (PhD Thesis, School of Oriental and African Studies, May 1981), 149.
68. Wilson, Sandra, *The Manchuria Crisis and Japanese Society*, (Taylor & Francis, 2002), 19.
69. "The Consul at Mukden (Vincent)," Document 1.
70. Wilson, *The Manchuria Crisis*, 19.
71. "The Consul at Mukden (Vincent)," Document 1.
72. Ramsdell, "The Nakamura Incident," 59.
73. Yoshihashi, *Conspiracy at Mukden*, 146-147.
74. Crowley, *Japan's Quest for Autonomy*, 107.
75. Yoshihashi, *Conspiracy at Mukden*, 147.
76. "3. Measures for Manchurian-Mongolian problem," (JACAR), Ref. B02030032800, Documents relating to negotiations of Manchurian-Mongolian problem, Matsumoto Records, Vol. 1 (A.1.1.0.1_002) (Diplomatic Archives of the Ministry of Foreign Affairs), 1-6, Translated by Ms. Fatima Janaghir.
77. "1. Manchuria in general / 1. From September 19, 1931 to November 9, 1931," (JACAR), Ref.B02030206500, Manchurian Incident (Documents relating to confrontation between Japanese and Chinese armies caused by Chinese soldiers' explosion of Liutiaogou of Manchurian Railway) / Protection, Repatriation, Evacuation and Damage of Japanese Residents

/ Protection, Repatriation and Evacuation, Vol. 9 (A.1.1.0.21-1-1_009) (Diplomatic Archives of the Ministry of Foreign Affairs), 4, Translated by Ms. Fatima Janaghir.

78. Ibid.
79. Ogata, *Defiance in Manchuria*, 57.
80. "Subject No. 76-Minami, Jiro (General)," Box 8, Files on Defendants; Minami, Muto 1946, Tavenner Papers and IMFTE Official Records, University of Virginia Law Library, 16.
81. "Jiro, Minami," Box 1, Files on Defendants; Minami, Muto 1946, Tavenner Papers and IMFTE Official Records, University of Virginia Law Library, 1.
82. "Subject No. 76-Minami, Jiro (General)," 18.
83. "Reference Made to Minami in the Narrative Summary of the Record," Box 8, Files on Defendants; Minami, Muto 1946, Tavenner Papers and IMFTE Official Records, University of Virginia Law Library, 5.
84. Yamamoto, "Japanese Press and Foreign Policy," 150.
85. "Subject No. 76," 20.
86. Seki, "The Manchurian Incident," 186.
87. Ibid.
88. Ogata, *Defiance in Manchuria*, 57.
89. Ibid., 56
90. Seki, "The Manchurian Incident," 216.
91. Ogata, *Defiance in Manchuria*, 56.
92. "IMFTE Judgement, Chapter V," 540-541.
93. "Saionji-Harada Memoirs Re Manchurian Aggression (I)," Box 7, Saionji-Harada Memoirs from March-July 1947, Tavenner Papers and IMFTE Official Records, University of Virginia Law Library, 1.
94. Seki, "The Manchurian Incident," 189.
95. Ibid., 189-190.
96. For more on Wakatsuki's meeting with Minami see ibid., 190.
97. "Saionji-Harada Memoirs Re Manchurian Aggression (I)," 2.
98. For more on this dynamic, see ibid.
99. Yoshihashi, *Conspiracy at Mukden*, 149.
100. Seki, "The Manchurian Incident," 200.
101. For more on Doihara's meeting with these figures, see ibid.
102. Ibid.

103. Ogata, *Defiance in Manchuria*, 58.
104. Seki, "The Manchurian Incident," 202-203.
105. Bix, *Hirohito and the Making of Modern Japan*, 211.
106. "Saionji-Harada Memoirs Re Manchurian Aggression (I)," 3.
107. Ibid.
108. Seki, "The Manchurian Incident," 202-203.
109. Ibid., 216.
110. Ibid., 216-217.
111. Ibid., 221.
112. Ibid.
113. Ibid.
114. "Affidavit of Tanaka Ryukichi," Box 1, First Phelps Scrap Book, C.W.J. Phelps Collection, University of Virginia Law Library, 4.
115. Yoshihashi, *Conspiracy at Mukden*, 154.
116. Ibid.
117. Ibid., 155.
118. "Affidavit of Tanaka Ryukichi," 4.
119. Ogata, *Defiance in Manchuria*, 58-59.
120. Yoshihashi, *Conspiracy at Mukden*, 156.
121. For more details on the debate between the various Kwantung Army plotters, read Seki, "The Manchurian Incident," 222-223.
122. Ibid., 223.

Chapter 12: The Mukden Incident

1. For more on the KMT's war with the Guangxi Clique, see Jowett, Philip S., *The Bitter Peace: Conflict in China, 1928-37* (Amberley Publishing, 2017), 25-40.
2. To read more about the Central Plains War, see ibid., 41-59.
3. Coble, *Facing Japan*, 28.
4. For more on the first three encirclement and suppression campaigns, see Hsu, *Art of War Papers*, 76-89.
5. For more on Mao Zedong's understanding and application of asymmetric warfare in battle, see Mao, Zedong, "Problems of Strategy in China's Revolutionary War," in *Selected Military Writings of Mao Tse-Tung* (Foreign Languages Press, 1967), 77-152.

6. Coble, *Facing Japan*, 17.
7. Ibid.
8. Lytton, "Appeal By the Chinese Government," 69.
9. Mitter, *The Manchurian Myth*, 74.
10. Fenby, *The Generalissimo*, 204.
11. Coox, *Nomonhan*, 27.
12. Ibid.
13. Coble, *Facing Japan*, 12.
14. Coox, *Nomonhan* , 27.
15. Jowett, *The Bitter Peace*, 83.
16. Ibid.
17. Lytton, "Appeal by the Chinese Government," 71.
18. For more on the Kwantung Army's weaknesses in September 1931, see Coox, *Nomonhan*, 26.
19. Ibid.
20. Jowett, *The Bitter Peace*, 83.
21. Seki, "The Manchurian Incident," 212.
22. Ibid., 226.
23. Yoshihashi, *Conspiracy at Mukden*, 157.
24. Seki, "The Manchurian Incident," 226.
25. To read more about these plans, see ibid., 225.
26. Ibid., 225.
27. Ibid.
28. Yoshihashi, *Conspiracy at Mukden*, 158.
29. Seki, "The Manchurian Incident," 226.
30. For more on Itagaki and Tatekawa's conversation at Penchi, see Yoshihashi, *Conspiracy at Mukden*, 158.
31. Seki, "The Manchurian Incident," 226.
32. For details on Itagaki and Tatekawa's conversation at the inn, see ibid., 227.
33. "IMFTE Judgement, Chapter V," 550.
34. Seki, "The Manchurian Incident," 227.
35. Yoshihashi, *Conspiracy at Mukden*, 159.
36. "IMFTE Judgement, Chapter V," 550.
37. For more on Tanaka's allegations about Tanaka's views and actions on September 18th, see "Affidavit of Tanaka Ryukichi," 4-5.

38. Similar analysis can be found in Ogata, *Defiance in Manchuria*, 59.
39. Ferrell, Robert, "The Mukden Incident: September 18-19, 1931," *The Journal of Modern History* 27, no. 1, (March 1955): 70, https://doi.org/10.1086/237763.
40. "Saionji-Harada Memoirs Re Manchurian Aggression (I)," 9.
41. For more on Morishima and Kimura's views on Tatekawa's presence, see Seki, "The Manchurian Incident," 226.
42. Ibid.
43. For more details on how Imada told his subordinates of the planned attack, see ibid., 227-228.
44. Ibid., 228.
45. Ibid.
46. Ibid.
47. Toland, *The Rising Sun: The Decline and Fall of the Japanese Empire*, (Random Publishing House, 1970), 8.
48. Seki, "The Manchurian Incident," 228.
49. Lytton, "Appeal by the Chinese Government," 68.
50. "10. Record of Conversations between the Members of the Commission of Inquiry of the League of Nations and Lieutenant-General Honjo / Commander-in-Chief of the Kwantung Army 1," (JACAR), Ref. B02030457400, Manchurian Incident (Documents relating to confrontation between Japanese and Chinese armies caused by Chinese soldiers' explosion of Liutiaogou of Manchurian Railway) / Remedial measures/ Investigators of League of Nations to China / Memorandum of conversation (record of investigation) (A.1.1.0.21-12-2-5) (Diplomatic Archives of the Ministry of Foreign Affairs), 18-19.
51. Lytton, "Appeal by the Chinese Government," 71.
52. Coox, *Nomonhan*, 30.
53. For more on the 3rd Company's assault see Lytton, "Appeal by the Chinese Government," 68.
54. *Imperial Japanese Army in Manchuria*, 70.
55. For more on the 1st and 4th Company assaults, see Lytton, "Appeal by the Chinese Government," 68-70
56. Ibid.
57. Ibid.
58. Seki, "The Manchurian Incident," 229.

59. Ibid., 229
60. For more on the Japanese attack on Mukden, see "10. Record of Conversations," 24.
61. For more on these forces' coordinated move towards Mukden, see "Doc 1641-The Manchurian Incident," Box 10, IPS Documents 1500-1999, Tavenner Papers & IMFTE Official Records, University of Virginia Law Library, 2-3.
62. "10. Record of Conversations,"24.
63. "12. Record of Conversations between the Members of the Commission of Inquiry of the League of Nations and Lieutenant-General Honjo, Commander-in-Chief of the Kwantung Army 3," (JACAR), Ref. B02030457600, Manchurian Incident (Documents relating to confrontation between Japanese and Chinese armies caused by Chinese soldiers' explosion of Liutiaogou of Manchurian Railway) / Remedial measures/ Investigators of League of Nations to China / Memorandum of conversation (record of investigation) (A.1.1.0.21-12-2-5) (Diplomatic Archives of the Ministry of Foreign Affairs), 95-96.
64. Jowett, *Bitter Peace*, 82.
65. Mitter, *The Manchurian Myth*, 78.
66. One such incident occurred at Fenghuangcheng, near a district office in the city; for more on this incident, see "1. Manchuria in general," 4, Translated by Ms. Fatima Janaghir.
67. "Liutiaogou (Liu-t'iao-kou) Vol. 1 Addendum 1. From September 19, 1931 to September 21, 1931," (JACAR) Ref.B02030186800, Manchurian Incident (Documents relating to confrontation between Japanese and Chinese armies caused by Chinese soldiers' explosion of Liutiaogou of Manchurian Railway), Vol. 1 (A.1.1.0.21_001) (Diplomatic Archives of the Ministry of Foreign Affairs), 1-4, Translated by Ms. Fatima Janaghir.
68. "1. Manchuria in general," 4, Translated by Ms. Fatima Janaghir.
69. Ibid.
70. "Liutiaogou (Liu-t'iao-kou) Vol. 1 Addendum 1. From September 19, 1931 to September 21, 1931," 1-4, Translated by Ms. Fatima Janaghir.
71. Mitter, *The Manchurian Myth,* 78.
72. "12. Record of Conversations," 104.
73. For more on Hasebe's decision, see ibid., 105-106.
74. For more on the attack on the Nanling Barracks, see ibid.

75. For more on the renewed attack on the Nanling Barracks, see ibid.
76. Ibid.
77. For more on the attack on Kuanchengtze, see ibid., 106
78. Ibid., 106
79. Ibid.
80. Yoshihashi, *Conspiracy at Mukden,* 164.
81. "IMFTE Judgement, Chapter V," 552.
82. Yoshihashi, *Conspiracy at Mukden*, 164-165.
83. Ibid., 165.
84. Shimada, "The Extension of Hostilities," 229.
85. Yoshihashi, *Conspiracy at Mukden*, 165.
86. Both of these telegrams are within "File of telegrams sent by Mukden Consul-General Hayashi to Foreign Minister Shidehara, 1-5" JACAR Ref. B02030188700, Manchurian Incident (Documents relating to confrontation between Japanese and Chinese armies caused by Chinese soldiers' explosion of Liutiaogou of Manchurian Railway), Vol. 2 (A.1.1.0.21_002) (Diplomatic Archives of the Ministry of Foreign Affairs).
87. Ibid.
88. Ferrell, "The Mukden Incident", 70.
89. "International Military Tribunal for the Far East, Proceedings," 15,735.
90. The exact wording of Hayashi's telegram to Shidehara is in "IMFTE Judgement, Chapter V," 553.
91. "File of telegrams sent by Mukden Consul-General Hayashi to Foreign Minister Shidehara, 5," 1-5, Translated by Ms. Fatima Janaghir.
92. IMFTE Proceedings, 15,736.
93. Ogata, *Defiance in Manchuria*, 63.
94. Ferrell, "The Mukden Incident," 70-71.
95. "10. Record of Conversations," 11.
96. Ibid., 11-12
97. Seki, "The Manchurian Incident," in *Japan's Road to the Pacific War, Volume I* 229
98. Ogata, *Defiance in Manchuria*, 60
99. "10. Record of Conversations," 11-12.
100. Ogata, *Defiance in Manchuria*, 60.
101. "IMFTE Judgement, Chapter V," 558.
102. Seki, "The Manchurian Incident," 230.

103. Ferrell, "The Mukden Incident," 71.
104. Yoshihashi, *Conspiracy at Mukden*, 167-168.
105. Ogata, *Defiance in Manchuria*, 60.
106. Yoshihashi, *Conspiracy at Mukden*, 168.
107. Ogata, *Defiance in Manchuria*, 60.
108. Shimada, "The Extension of Hostilities," 241.
109. Crowley, *Japan's Quest for Autonomy*, 122.
110. Ibid.
111. Shimada, "The Extension of Hostilities," 242.
112. Yoshihashi, *Conspiracy at Mukden*, 172-173.
113. Shimada, "The Extension of Hostilities," 242-243.
114. Yoshihashi, *Conspiracy at Mukden*, 172.
115. "IMFTE Judgement, Chapter V," 559.
116. "Saionji-Harada Memoirs Re Manchurian Aggression (I)," 4.
117. "Reference Made to Minami in the Narrative Summary of the Record," 2.
118. Hattori, *Japan at War and Peace*, 204.
119. Shimada, "The Extension of Hostilities," 246.
120. Yoshihashi, *Conspiracy at Mukden*, 8-9.
121. Ibid., 9.
122. Shimada, "The Extension of Hostilities," 246.
123. For more on the meeting between the "big three" see ibid., 253.
124. Ibid.
125. Ibid.
126. Ibid., 254.
127. For more on this meeting, see Bix, *Hirohito and the Making of Modern Japan*, 216.
128. For more on Wakatsuki's conversation with Harada, see "Saionji-Harada Memoirs Re Manchurian Aggression (I)," 5.
129. Ogata, *Defiance in Manchuria*, 62.
130. Ibid.
131. For more on the meeting at Harada's house on the evening of September 19th, see Bix, *Hirohito and the Making of Modern Japan*, 215.
132. Ibid.
133. Shimada, "The Extension of Hostilities," 244.
134. Yoshihashi, *Conspiracy at Mukden*, 172.
135. Shimada, "The Extension of Hostilities," 244.

136. Ibid.
137. Ibid., 245.
138. Yoshihashi, *Conspiracy at Mukden*, 172.
139. Crowley, *Japan's Quest for Autonomy*, 123-124.
140. Shimada, "The Extension of Hostilities," 255.
141. For more on the conference at the Shinyokan hotel on the night of September 19th, see Coox, *Nomonhan*, 34-35.
142. Shimada, "The Extension of Hostilities," 255.
143. For more on the conference at the Shinyokan hotel on the night of September 19th, see Coox, *Nomonhan*, 34-35.
144. For more on Tatekawa's conversation with Honjo, see ibid., 35.
145. Ibid., 35.
146. Ogata, *Defiance in Manchuria*, 65.
147. Peattie, *Ishiwara Kanji*, 123.
148. "Doc 1641-The Manchurian Incident," Box 10, IPS Documents 1500-1999, Tavenner Papers & IMFTE Official Records, University of Virginia Law Library, 5.
149. Coox, *Nomonhan*, 37.
150. Peattie, *Ishiwara Kanji*, 124.
151. Ibid.
152. Ogata, *Defiance in Manchuria*, 64.
153. Ibid.
154. For more on Itagaki and Miyake's appeals to Honjo, see Coox, *Nomonhan*, 36.
155. Peattie, *Ishiwara Kanji*, 124.
156. Ogata, *Defiance in Manchuria*, 64-65.
157. "Doc 1641-The Manchurian Incident," 5.
158. Coox, *Nomonhan*, 36.
159. For reference to Xi Qia possibly being held at gunpoint by Taro Jimon while declaring independence, See Yoshihashi, *Conspiracy at Mukden*, 180.
160. Coox, *Nomonhan*, 36.
161. 4. Independence and area of influence of Zhang Haipeng," (JACAR), Ref.B02032034900, Documents relating to independence of Manchuria and Mongolia after Manchurian Incident, Vol. 1 (A.6.2.0.1_001) (Diplomatic Archives of the Ministry of Foreign Affairs), 2, Translated by Ms. Fatima Janaghir.

162. "Saionji-Harada Memoirs Re Manchurian Aggression (I)," 7
163. "From Doc 827-Biography of General Minami by Sugiyama, Ken," Box 10, IPS Documents 0-999, Tavenner Papers and IMFTE Official Records, University of Virginia Law Library, 6.
164. "Saionji-Harada Memoirs Re Manchurian Aggression (I)," 7.
165. "Summary of Military Operations of 20th Division Headquarters, Mixed 39th Brigade," (JACAR), Ref.A03023779400, Various date and information, Press releases on the Manchuria and China Incident (National Archives of Japan), 1-2, Translated by Ms. Fatima Janaghir.
166. "Reference Made to Minami in the Narrative Summary of the Record," 2.
167. "Subject No. 76-Minami, Jiro," 5.
168. Shimada, "The Extension of Hostilities," 250.
169. Ibid.
170. For a slightly different account of this meeting, see "Saionji-Harada Memoirs Re Manchurian Aggression (I)," 7
171. Shimada, "The Extension of Hostilities," 250-251.
172. "Summary of Interrogation of Baron Wakatsuki, Reijiro," Box 2, General Reports and Memoranda from 1946, Tavenner Papers and IMFTE Official Records, University of Virginia Law Library, 2.
173. Ibid.
174. Bix, *Hirohito and the Making of Modern Japan*, 216-217.
175. Shimada, "The Extension of Hostilities," 261.
176. Coox, *Nomonhan*, 37.
177. Ogata, *Defiance in Manchuria*, 67-68.
178. Shimada, "The Extension of Hostilities," 262.
179. Ibid., 262.
180. Ibid., 263.
181. Ogata, *Defiance in Manchuria*, 68.
182. Ibid.

Chapter 13: Intermission in Northeastern China: Waves Beneath the Surface

1. Lytton, "Appeal by the Chinese Government," 72.
2. "IMFTE Judgement, Chapter V," 566.

3. "The Consul at Geneva (Gilbert) to the Secretary of State," Document 80, September 26th, 1931, Papers Relating to the Foreign Relations of the United States, 1931, Far East Volume III, https://history.state.gov/historicaldocuments/frus1931v03/d80.
4. "The Charge in Japan (Neville) to the Secretary of State," Document 123, October 6th, 1931, Papers Relating to the Foreign Relations of the United States, 1931, Far East Volume III, https://history.state.gov/historicaldocuments/frus1931v03/d123.
5. Ogata, *Defiance in Manchuria*, 76.
6. Ibid.
7. Ibid., 77.
8. *From Emperor to Citizen, the Autobiography of Aisin-Gioro Pu Yi*, 219.
9. Ibid.
10. For more on the meeting with Puyi, see ibid. 219-220.
11. "Saionji-Harada Memoirs Re Manchurian Aggression (I)," 8.
12. "Saionji-Harada Memoirs Re Manchurian Aggression (II)," Box 7, Saionji-Harada Memoirs from March-July 1947, Tavenner Papers and IMFTE Official Records, University of Virginia Law Library, 2.
13. Ogata, *Defiance in Manchuria*, 79.
14. To see the declarations of independence described, see ibid., 90-91.
15. Mitter, *The Manchurian Myth*, 80-81.
16. Ibid.
17. Shimada, "The Extension of Hostilities," 267.
18. Mitter, *The Manchurian Myth*, 79.
19. Lytton, "Appeal by the Chinese Government," 72.
20. Shimada, "The Extension of Hostilities," 267.
21. "4. Independence and area of influence of Zhang Haipeng," (JACAR), Ref.B02032034900, Documents relating to independence of Manchuria and Mongolia after Manchurian Incident, Vol. 1 (A.6.2.0.1_001) (Diplomatic Archives of the Ministry of Foreign Affairs), 2, Translated by Ms. Fatima Janaghir.
22. "IMFTE Judgement Chapter V," 573.
23. Shimada, "The Extension of Hostilities," 288.
24. Ibid.
25. Ibid.
26. Ogata, *Defiance in Manchuria*, 81.

27. More details about the targets and intentions of Ishiwara's decision to bomb Jinzhou can be found in Shimada, "The Extension of Hostilities," 288-289.
28. Peattie, *Ishiwara Kanji*, 126.
29. Shimada, "The Extension of Hostilities," 289.
30. Yoshihashi, *Conspiracy at Mukden*, 191.
31. Ibid.
32. Ogata, *Defiance in Manchuria*, 82.
33. Shimada, "The Extension of Hostilities," 290.
34. Ibid.
35. Peattie, *Ishiwara Kanji*, 127.
36. Wright, Quincy, "The Manchurian Crisis," 58-59.
37. Ibid., 289.
38. Crowley, *Japan's Quest for Autonomy*, 129.
39. Ibid.
40. Ogata, *Defiance in Manchuria*, 93-94.
41. Ibid., 94.
42. Peattie, *Ishiwara Kanji*, 128.
43. Shimada, "The Extension of Hostilities," 329-330.
44. Ogata, *Defiance in Manchuria*, 93.
45. Ibid.
46. For more on the meeting between Ishiwara and Imamura, see Peattie, *Ishiwara Kanji*, 129.
47. Ibid.
48. Ibid.
49. Shimada, "The Extension of Hostilities," 328.
50. Yoshihashi, *Conspiracy at Mukden*, 194.
51. As Hashimoto himself put it, the purpose of the October Incident was twofold: to "destroy the government then in power and to bring about a new government to bring about a settlement of the Manchurian Incident," according to "Affidavit of Hashimoto Kingoro," Box 8, Files on Defendants: Hashimoto, Kingoro, Tavenner Papers and IMFTE Official Records, University of Virginia Law Library, 1.
52. Orbach, *Curse on this Country*, 214.
53. Ibid., 215.
54. Yoshihashi, *Conspiracy at Mukden*, 194-195.

55. Orbach, *Curse on this Country*, 215.
56. Crowley, *Japan's Quest for Autonomy*, 132.
57. Orbach, *Curse on this Country*, 217.
58. For more on the details of these conversations, see Yoshihashi, *Conspiracy at Mukden*, 198-199.
59. "Saionji-Harada Memoirs Re Manchurian Aggression (II)," 4.
60. Orbach, *Curse on this Country*, 217.
61. "Saionji-Harada Memoirs Re Manchurian Aggression (II)," 4.

Chapter 14: To the North

1. *Imperial Japanese Army in Manchuria*, 71-72.
2. Lytton, "Appeal by the Chinese Government," 72.
3. "Doc 1641-The Manchurian Incident," 7.
4. Yoshihashi, *Conspiracy at Mukden*, 207.
5. Shimada, "The Extension of Hostilities," 269.
6. Ibid.
7. Ibid.
8. Lytton, "Appeal by the Chinese Government," 73.
9. Shimada, "The Extension of Hostilities," 270.
10. Lytton, "Appeal by the Chinese Government," 73.
11. Ibid.
12. "12. Record of Conversations," 110.
13. Ogata, *Defiance in Manchuria*, 107.
14. Ibid.
15. Shimada, "The Extension of Hostilities," 270.
16. For more on this fighting, see Lytton, "Appeal by the Chinese Government," 73-74.
17. Ibid., 74.
18. Ibid., 74.
19. "12. Record of Conversations," 111.
20. Ogata, *Defiance in Manchuria*, 108.
21. Ibid., 109.
22. Shimada, "The Extension of Hostilities," 273.
23. Ibid., 279.
24. Ibid., 274.

25. Ibid., 275.
26. Lytton, "Appeal by the Chinese Government," 74.
27. Ibid.
28. Shimada, "The Extension of Hostilities," 276.
29. Ibid.
30. Ibid.
31. Lytton, "Appeal by the Chinese Government," 74.
32. Ogata, *Defiance in Manchuria*, 111.
33. Shimada, "The Extension of Hostilities," 280.
34. Ibid., 278.
35. For more on the embrace and decreased hostility to Kwantung Army aggression in northern Manchuria, see Shimada, "The Extension of Hostilities," 280.
36. Ibid., 280.
37. For more on the Kwantung Army's alteration of the AGS's approved terms to Ma, see ibid., 280.
38. Lytton, "Appeal by the Chinese Government," 74.
39. Shimada, "The Extension of Hostilities," 281.
40. Jowett, *The Bitter Peace,* 86.
41. Ibid.
42. Other sources say these forces only numbered roughly 8,000 men; whatever the case, Ma's forces had withdrawn.
43. For more on Chinese fortifications outside Qiqihar, see ibid.
44. Ibid.
45. "Rout of Ma," *Time Magazine*, November 30th, 1931.
46. Ibid.
47. Jowett, *The Bitter Peace,* 87.
48. "Rout of Ma," *Time Magazine*, November 30th, 1931.
49. "12. Record of Conversations," 112.
50. Jowett, *The Bitter Peace,* 87.
51. "Saionji-Harada Memoirs Re Manchurian Aggression (II)," 7.
52. Ogata, *Defiance in Manchuria*, 112.
53. To read more about Ninomiya's change of heart, see Shimada, "The Extension of Hostilities," 285.
54. Ibid.
55. Ogata, *Defiance in Manchuria*, 113.

56. Shimada, "The Extension of Hostilities," 286.
57. For more on Doihara's meeting with Puyi in Tianjin, see Pu Yi, Henry, *The Last Manchu: The Autobiography of Henry Pu Yi, the Last Emperor of China*, (Skyhorse Publishing, 2010), 142-144.
58. Ibid., 143.
59. For more on Puyi's conversation with Doihara, see ibid., 143-144.
60. Behr, Edward S., *The Last Emperor*, (Bantam Books, 1987), 190-192.
61. *From Emperor to Citizen: The Autobiography of Aisin-Gioro Puyi*, (University of the Pacific Press, 2001), 228.
62. "IMFTE Judgement, Chapter V," 569.
63. Behr, *The Last Emperor,* 190-192.
64. Lytton, "Appeal by the Chinese Government," 76-77.
65. *From Emperor to Citizen,* 228-229.
66. Ibid., 230.
67. Shimada, "The Extension of Hostilities," 290.
68. Ibid., 291.
69. Lytton, "Appeal by the Chinese Government," 75-76.
70. Peattie, *Ishiwara Kanji*, 131-132.
71. Shimada, "The Extension of Hostilities," 293.
72. Ibid.
73. Lytton, "Appeal by the Chinese Government," 76—77.
74. For more on the AGS conversations on what to do about the second Tianjin Incident, see Shimada, "The Extension of Hostilities," 293.
75. Ibid.
76. Ibid.
77. Yoshihashi, *Conspiracy at Mukden*, 217.
78. Lytton, "Appeal by the Chinese Government," 77.
79. Shimada, "The Extension of Hostilities," 294.
80. Ibid., 295.
81. Ibid., 295-297.
82. Ibid., 297.
83. Crowley, *Japan's Quest for Autonomy*, 137.
84. Ibid.
85. Yoshihashi, *Conspiracy at Mukden*, 220.
86. Ibid., 220-221.
87. Ogata, *Defiance in Manchuria*, 101.

88. For more details on Adachi's various attempts to form a coalition government with the Seiyukai see Yoshihashi, *Conspiracy at Mukden*, 223-226.
89. Ibid., 226-227.
90. Ibid., 227.
91. Shimada, "The Extension of Hostilities," 298.
92. Ogata, *Defiance in Manchuria*, 116.
93. Lytton, "Appeal by the Chinese Government," 77.
94. Ibid.
95. "Manchurian Incident 9-2 Jinzhou Incident (part 2) (from December 26, 1931 to January 1, 1932)," (JACAR), Ref.B02030494900, Manchurian Incident (Documents relating to confrontation between Japanese and Chinese armies caused by Chinese soldiers' explosion of Liutiaogou of Manchurian Railway) / Jinzhou Incident (A.1.1.0.21-32) (Diplomatic Archives of the Ministry of Foreign Affairs), 1, Translated by Ms. Fatima Janaghir.
96. For more information on the redeployment of Kwantung Army units in Manchuria, see ibid., 77
97. Lytton, "Appeal by the Chinese Government," 77.
98. Shimada, "The Extension of Hostilities," 302.
99. Ibid.
100. Ibid.
101. Lytton, "Appeal by the Chinese Government," 78.
102. Ibid.
103. Jowett, *The Bitter Peace,* 93.
104. "IMFTE Judgement Chapter V," 586.
105. Shimada, "The Extension of Hostilities," 303-304.
106. Ibid.
107. Initially, only two Japanese battalions were able to deploy towards Harbin according to Lytton, "Appeal by the Chinese Government," 79.
108. Ibid.
109. Shimada, "The Extension of Hostilities," 304.
110. Lytton, "Appeal by the Chinese Government," 80.
111. "11. Record of Conversations between the Members of the Commission of Inquiry of the League of Nations and Lieutenant-General Honjo, Commander-in-Chief of the Kwantung Army 2," (JACAR), Ref. B02030457500, Manchurian Incident (Documents relating to confron-

tation between Japanese and Chinese armies caused by Chinese soldiers' explosion of Liutiaogou of Manchurian Railway) / Remedial measures/ Investigators of League of Nations to China / Memorandum of conversation (record of investigation) (A.1.1.0.21-12-2-5) (Diplomatic Archives of the Ministry of Foreign Affairs).

112. Lytton, "Appeal by the Chinese Government," 80.
113. Jowett, *The Bitter Peace,* 94.
114. Lytton, "Appeal By the Chinese Government," 80.
115. Ibid., 80.

Chapter 15: Japan and the World Reacts to Kwantung Army Aggression

1. Purport Lists and Cards, Records of the Department of State Relating to Political Relations Between China and Japan, 1930-1944 (National Archives Microfilm Publication M976, roll 3, General Records of the Department of State, Record Group 59.
2. Lytton, "Appeal by the Chinese Government," 4.
3. Purport Lists and Cards, Records of the Department of State Relating to Political Relations Between China and Japan, 1930-1944 (National Archives Microfilm Publication M976, roll 3, General Records of the Department of State, Record Group 59.
4. "The Minister in China (Johnson) to the Secretary of State," Document 47, September 23rd, 1931, Department of State, Papers Relating to the Foreign Relations of the United States, 1931, The Far East, Volume III, https://history.state.gov/historicaldocuments/frus1931v03/d47.
5. Purport Lists and Cards, Records of the Department of State Relating to Political Relations Between China and Japan, 1930-1944 (National Archives Microfilm Publication M976, roll 3, General Records of the Department of State, Record Group 59.
6. Coble, *Facing Japan*, 34.
7. Ibid.
8. For more on the publication of anti-Chiang articles, see Coble, Parks, M., "Chiang Kai-shek and the anti-Japanese Movement in China: Zou Tao-fen and the National Salvation Association, 1931-1937," 296-299.
9. Wilson, *The Manchurian Crisis*, 22.

10. Ibid.
11. "13. Record of Conversations between the Members of the Commission of Inquiry of the League of Nations and Lieutenant-General Honjo, Commander-in-Chief of the Kwantung Army 4," (JACAR) Ref. B02030457700, Manchurian Incident (Documents relating to confrontation between Japanese and Chinese armies caused by Chinese soldiers' explosion of Liutiaogou of Manchurian Railway) / Remedial measures/ Investigators of League of Nations to China / Memorandum of conversation (record of investigation) (A.1.1.0.21-12-2-5) (Diplomatic Archives of the Ministry of Foreign Affairs), 118.
12. "The Ambassador in Japan (Forbes), to the Secretary of State," Document 397, November 12th, 1931, Department of State, Papers Relating to the Foreign Relations of the United States, 1931, The Far East, Volume III, https://history.state.gov/historicaldocuments/frus1931v03/d397.
13. Wilson, *The Manchurian Crisis*, 20.
14. For more on this phenomenon, see Wilson, *The Manchurian Crisis*.
15. For more on the skepticism of certain members of the press regarding the Mukden Incident, see Matsuyama, "Japanese Press and Foreign Policy," 155-157.
16. For more on army censorship of the press, see ibid., 159-179.
17. "The Secretary of State to the Minister in China (Johnson)," Document 63, September 24th, 1931, Department of State, Papers Relating to the Foreign Relations of the United States, 1931, The Far East, Volume III, https://history.state.gov/historicaldocuments/frus1931v03/d63.
18. "The Secretary of State to President Hoover, Aboard the 'U.S.S. Arkansas,' Hampton Roads, Document 241, October 18th, 1931, Department of State, Papers Relating to the Foreign Relations of the United States, 1931, The Far East, Volume III, https://history.state.gov/historicaldocuments/frus1931v03/d241.
19. "The Vice Consul at Mukden (Lynch) to the Minister in China (Johnson)," Document 21, September 20th, 1931, Department of State, Papers Relating to the Foreign Relations of the United States, 1931, The Far East, Volume III, https://history.state.gov/historicaldocuments/frus1931v03/d21.
20. Purport Lists and Cards, Records of the Department of State Relating to Political Relations Between China and Japan, 1930-1944 (National

Archives Microfilm Publication M976, roll 3, General Records of the Department of State, Record Group 59.

21. Stimson, Henry L., *The Far Eastern Crisis: Recollections and Observations* (Harper and Brothers Publishers, 1974), 32.
22. Ibid., 33.
23. "Memorandum by the Chief of Division of Current Information (McDermott), of the Press Conference on September 19th, 1931," Document 17, September 19th, 1931, Department of State, Papers Relating to the Foreign Relations of the United States, 1931, The Far East, Volume III, https://history.state.gov/historicaldocuments/frus1931v03/d17.
24. For more on Stimson's views on Shidehara, see Stimson, *The Far Eastern Crisis,* 34-37.
25. Purport Lists and Cards, Records of the Department of State Relating to Political Relations Between China and Japan, 1930-1944 (National Archives Microfilm Publication M976, roll 3, General Records of the Department of State, Record Group 59.
26. Wright, "The Manchurian Crisis," 72-73.
27. Smith, Sara, R., *The Manchurian Crisis, 1931-1932, A Tragedy in International Relations*, (Greenwood Press, 1948), 191.
28. Ibid.
29. "Saionji-Harada Memoirs Re Manchurian Aggression (II)," 9.
30. Bassett, Reginald, *Democracy and Foreign Policy: A Case History of the Sino-Japanse Dispute,* (Taylor & Francis, 1952), 75.
31. "The Minister in China (Johnson) to the Secretary of State," Document 402, November 13th, 1931, Department of State, Papers Relating to the Foreign Relations of the United States, 1931, The Far East, Volume III, https://history.state.gov/historicaldocuments/frus1931v03/d402.
32. "The Charge in France (Shaw), to the Secretary of State," Document 485, November 21st, 1931, Department of State, Papers Relating to the Foreign Relations of the United States, 1931, The Far East, Volume III, https://history.state.gov/historicaldocuments/frus1931v03/d485.
33. Lytton, "Appeal by the Chinese Government," 17.
34. Ibid., 127-128.
35. Ibid., 130-131.

Bibliography

Primary Sources

Japan Center for Asian Historical Records

"1. Manchuria in general / 1. From September 19, 1931 to November 9, 1931," Japan Center for Asian Historical Records (JACAR) Ref.B02030206500, Manchurian Incident (Documents relating to confrontation between Japanese and Chinese armies caused by Chinese soldiers' explosion of Liutiaogou of Manchurian Railway) / Protection, Repatriation, Evacuation and Damage of Japanese Residents / Protection, Repatriation and Evacuation, Vol. 9 (A.1.1.0.21-1-1_009) (Diplomatic Archives of the Ministry of Foreign Affairs), Translated by Ms. Fatima Janaghir.

"10. Record of Conversations between the Members of the Commission of Inquiry of the League of Nations and Lieutenant-General Honjo / Commander-in-Chief of the Kwantung Army, 1," Japan Center for Asian Historical Records (JACAR) Ref.B02030457400, Manchurian Incident (Documents relating to confrontation between Japanese and Chinese armies caused by Chinese soldiers' explosion of Liutiaogou of Manchurian Railway) / Remedial measures/ Investigators of League of Nations to China / Memorandum of conversation (record of investigation) (A.1.1.0.21-12-2-5) (Diplomatic Archives of the Ministry of Foreign Affairs).

"11. Record of Conversations between the Members of the Commission of Inquiry of the League of Nations and Lieutenant-General Honjo, Commander-in-Chief of the Kwantung Army 2," 73, Japan Center for Asian Historical Records (JACAR) Ref.B02030457500, Manchurian Incident (Documents relating to confrontation between Japanese and Chinese armies caused by Chinese soldiers' explosion of Liutiaogou of Manchurian Railway) / Remedial measures/ Investigators of League of Nations to China / Memorandum of conversation (record of investigation) (A.1.1.0.21-12-2-5) (Diplomatic Archives of the Ministry of Foreign Affairs).

"12. Record of Conversations between the Members of the Commission of Inquiry of the League of Nations and Lieutenant-General Honjo, Commander-in-Chief of the Kwantung Army 3," Japan Center for Asian Historical Records (JACAR) Ref.B02030457600, Manchurian Incident (Documents relating to confrontation between Japanese and Chinese armies caused by Chinese soldiers' explosion of Liutiaogou of Manchurian Railway) / Remedial measures/ Investigators of League of Nations to China / Memorandum of conversation (record of investigation) (A.1.1.0.21-12-2-5) (Diplomatic Archives of the Ministry of Foreign Affairs).

"13. Record of Conversations between the Members of the Commission of Inquiry of the League of Nations and Lieutenant-General Honjo, Commander-in-Chief of the Kwantung Army 4," Japan Center for Asian Historical Records (JACAR), Ref.B02030457700, Manchurian Incident (Documents relating to confrontation between Japanese and Chinese armies caused by Chinese soldiers' explosion of Liutiaogou of Manchurian Railway) / Remedial measures/ Investigators of League of Nations to China / Memorandum of conversation (record of investigation) (A.1.1.0.21-12-2-5) (Diplomatic Archives of the Ministry of Foreign Affairs).

"Independence and area of influence of Zhang Haipeng," Japan Center for Asian Historical Records (JACAR) Ref.B02032034900, Documents relating to independence of Manchuria and Mongolia after Manchurian Incident, Vol. 1 (A.6.2.0.1_001) (Diplomatic Archives of the Ministry of Foreign Affairs), Translated by Ms. Fatima Janaghir.

"5 December 22, 1911 to December 31, 1911," Japan Center for Asian Historical Records (JACAR) Ref.B03050659300, Overseas miscellaneous news about the revolution conflicts in Qing China/Resident Generals Office of Kwantung Report Volume 4 (1.6.1.51-1_004) (Diplomatic Archives of the Ministry of Foreign Affairs), Translated by Ms. Fatima Janaghir.

"File of telegrams sent by Mukden Consul-General Hayashi to Foreign Minister Shidehara," JACAR Ref.B02030188700, Manchurian Incident (Documents relating to confrontation between Japanese and Chinese armies caused by Chinese soldiers' explosion of Liutiaogou of Manchurian Railway), Vol. 2 (A.1.1.0.21_002) (Diplomatic Archives of the Ministry of Foreign Affairs), Translated by Ms. Fatima Janaghir.

"The Fengtien-Chihli war, No. 3/ 4 From October 9, 1914 to October 24, 1914," Japan Center for Asian Historical Records (JACAR), Ref.B03050758900, Confusional relationships of Kiangsu-Chekang and Fengtian-Chihli/ A strategic situation Vol.7 (1.6.1.85-2_007) (Diplomatic Archives of the Ministry of Foreign Affairs), Translated by Ms. Fatima Janaghir.

"General Condition of War and Politics/6 November 28, 1925 to December 8, 1925," Japan Center for Asian Historical Records (JACAR) Ref. B03050778500, Occurrence of confusion in the anti-Mukden (Shenyang) groups, Vol. 1 (1.6.1.86_001) (Diplomatic Archives of the Ministry of Foreign Affairs), Translated by Ms. Fatima Janaghir.

"Invitation to Chang Tso-lin as a Japanese Officer," Japan Center for Asian Historical Records (JACAR) Ref.C03022654100, (National Institute for Defense Studies), Translated by Ms. Fatima Janaghir.

"Liutiaogou (Liu-t'iao-kou) Vol. 1 Addendum 1. From September 19, 1931 to September 21, 1931," Japan Center for Asian Historical Records (JACAR) Ref.B02030186800, Manchurian Incident (Documents relating to confrontation between Japanese and Chinese armies caused by Chinese soldiers' explosion of Liutiaogou of Manchurian Railway), Vol. 1 (A.1.1.0.21_001) (Diplomatic Archives of the Ministry of Foreign Affairs), Translated by Ms. Fatima Janaghir.

"Manchurian Incident 9-2 Jinzhou Incident (part 2) (from December 26, 1931 to January 1, 1932)," Japan Center for Asian Historical Records (JACAR) Ref.B02030494900, Manchurian Incident (Documents relating to confrontation between Japanese and Chinese armies caused by Chinese soldiers' explosion of Liutiaogou of Manchurian Railway) / Jinzhou Incident (A.1.1.0.21-32) (Diplomatic Archives of the Ministry of Foreign Affairs), Translated by Ms. Fatima Janaghir.

"Message to be conveyed to Chang Tso-lin," Japan Center for Asian Historical Records (JACAR) Ref.C03022725300, Mitsu Dainikki, 4 of 6, 1925 (National Institute for Defense Studies), Translated by Ms. Fatima Janaghir.

"Preface, Japanese National Defense Resources August, 1917," Japan Center for Asian Historical Records (JACAR) Ref.C12121558200, Japanese national defense resources August, 1917 (National Institute for Defense Studies), Translated by Ms. Fatima Janaghir.

"Sending the Opinions About the Policies on Manchuria and Mongolia," Japan Center for Asian Historical Records (JACAR) Ref.C01003764000, Mitsu Dainikki, 4 of 6, 1927 (National Institute for Defense Studies), Translated by Ms. Fatima Janaghir.

"Summary on Manchuria and Mongolia Governance Plan (1)," Japan Center for Asian Historical Records (JACAR) Ref.C13010350400, Educational materials (1) Sumiya Tei Historic Materials (National Institute for Defense Studies), Translated by Ms. Fatima Janaghir.

"Summary of Military Operations of 20th Division Headquarters, Mixed 39th Brigade," Japan Center for Asian Historical Records (JACAR) Ref. A03023779400, Various date and information, press releases on the Manchuria and China Incident (National Archives of Japan), Translated by Ms. Fatima Janaghir.

"War and political situations in general (from October 21, 1925 to November 24, 1925)," Japan Center for Asian Historical Records (JACAR) Ref. B03050778100, Occurrence of confusion in the anti-Mukden (Shenyang) groups, Vol. 1 (1.6.1.86_001) (Diplomatic Archives of the Ministry of Foreign Affairs), Translated by Ms. Fatima Janaghir.

Library of Congress

Koo, Wellington. "Memorandum on Japan's Plots and Schemes Against the Unification of China." Essay. In *Memoranda Nos. 1-29 Submitted by the Chinese Assessor to the Commission of Enquiry, into the Sino-Japanese Dispute*. Washington D.C.: Library of Congress, Main Reading Room.

Koo, Wellington. "Memorandum on Koreans in the Three Eastern Provinces." Essay. In *Memoranda Nos. 1-29 Submitted by the Chinese Assessor to the Commission of Enquiry, into the Sino-Japanese Dispute*. Library of Congress, Main Reading Room.

Koo, Wellington. "Memorandum on the Wanpaoshan Case." Essay. In *Memoranda Nos. 1-29 Submitted by the Chinese Assessor to the Commission of Enquiry, into the Sino-Japanese Dispute.* Washington D.C.: Library of Congress, Main Reading Room.

Koo, Wellington. "Relation of Japan's Claims Based on the So-Called Fifty-Three Outstanding Cases." Essay. In *Memoranda Nos. 1-29 Submitted by the Chinese Assessor to the Commission of Enquiry, into the Sino-Japanese Dispute, Library of Congress.* Washington D.C.: Library of Congress, Main Reading Room.

Lytton, Robert. Appeal by the Chinese Government: Report of the Commission of Enquiry, Library of Congress, Washington D.C., Library of Congress, Main Reading Room https://www.loc.gov/item/2021666890/.

National Archives at College Park

"Ishihara Kanji Dossier," Box 329, Intelligence and Investigative Dossiers Personal Files, Record Group 319, National Archives at College Park.

"Records of Proceedings of International Military Tribunal for the Far East (IMFTE)," Records of Trial Files, Record Group 331, National Archives at College Park.

"2nd Report on Progress in Manchuria to 1930," Box 261, Organizational History Files, Record Group 550, National Archives at College Park.

"Japan in Manchuria," Box 2, Organizational History Files, Record Group 59, National Archives at College Park.

Purport Lists and Cards, Records of the Department of State Relating to Political Relations Between China and Japan 1930-1944 (National Archives Microfilm Publication M976, roll 3, General Record of the Department of State), Record Group 59, National Archives at College Park.

War Department Map Collection, Record Group 77, National Archives at College Park.

Papers Relating to Foreign Relations of the United States

"The Ambassador in Japan (Forbes), to the Secretary of State." Document 397, Papers Relating to the Foreign Relations of the United States, 1931,

Far East Volume III, November 12, 1931. https://history.state.gov/historicaldocuments/frus1931v03/d397.

"The Charge in France (Shaw), to the Secretary of State." Document 485, Papers Relating to the Foreign Relations of the United States, 1931, Far East Volume III, November 21, 1931. https://history.state.gov/historicaldocuments/frus1931v03/d485.

"The Charge in Japan (Neville) to the Secretary of State." Document 123, Papers Relating to the Foreign Relations of the United States, 1931, Far East Volume III, October 6, 1931. https://history.state.gov/historicaldocuments/frus1931v03/d123.

"The Consul at Geneva (Gilbert) to the Secretary of State." Document 80, Papers Relating to the Foreign Relations of the United States, 1931, Far East Volume III, September 26, 1931. https://history.state.gov/historicaldocuments/frus1931v03/d80.

"The Consul at Mukden (Vincent) to the Minister in China (Johnson)." Document 1, Papers Relating to the Foreign Relations of the United States, 1931, The Far East, Volume III, August 20, 1931. https://history.state.gov/historicaldocuments/frus1931v03/d1.

"Memorandum by the Chief of Division of Current Information (McDermott), of the Press Conference on September 19th, 1931." Document 17, Papers Relating to the Foreign Relations of the United States, 1931, Far East Volume III, September 19, 1931. https://history.state.gov/historicaldocuments/frus1931v03/d17.

"The Minister in China (Johnson) to the Secretary of State." Document 402, Papers Relating to the Foreign Relations of the United States, 1931, Far East Volume III, November 13, 1931. https://history.state.gov/historicaldocuments/frus1931v03/d402.

"The Minister in China (Johnson) to the Secretary of State." Document 47, Department of State, Papers Relating to the Foreign Relations of the United States, 1931, The Far East, Volume III, September 23, 1931. https://history.state.gov/historicaldocuments/frus1931v03/d47.

"The Secretary of State to President Hoover, Aboard the 'U.S.S. Arkansas,' Hampton Roads." Document 241, Papers Relating to the Foreign Relations of the United States, 1931, Far East Volume III, October 18, 1931. https://history.state.gov/historicaldocuments/frus1931v03/d241.

"The Secretary of State to the Minister in China (Johnson)." Document 63, Papers Relating to the Foreign Relations of the United States, 1931, The Far East, Volume III, September 24, 1931. https://history.state.gov/historicaldocuments/frus1931v03/d63.

"The Vice Consul at Mukden (Lynch) to the Minister in China (Johnson)." Document 21, Papers Relating to the Foreign Relations of the United States, 1931, Far East Volume III, September 20, 1931. https://history.state.gov/historicaldocuments/frus1931v03/d21.

University of Virginia Law Library

"Addition to Doc 684, Asia, Europe, Japan, by Okawa, Shumei," Box 10, IPS Documents 0-999, Tavenner Papers and IMFTE Records, University of Virginia Law Library.

"Affidavit of Hashimoto Kingoro," Box 8, Files on Defendants: Hashimoto, Kingoro, Tavenner Papers and IMFTE Official Records, University of Virginia Law Library

"Affidavit of Tanaka Ryukichi," Box 1, First Phelps Scrap Book, C.W.J. Phelps Collection, University of Virginia Law Library.

"Amplified Report on Dr. Okawa, Shumei," Box 8, Files on Defendants; Okawa, Oshima, 1946, Tavenner Papers and IMFTE Official Records, University of Virginia Law Library.

"Brief-Hashimoto Kingoro." Box 8, Files on Defendants: Hashimoto-Hata, 1946, Tavenner Papers & IMFTE Official Records, University of Virginia Law Library.

"Doc 1641-The Manchurian Incident," Box 10, IPS Documents 1500-1999, Tavenner Papers & IMFTE Official Records, University of Virginia Law Library.

"Doc 3029-Views Concerning Our Imperial National Defense Against the USSR," Box 11, IPS Documents 3000-3999, Tavenner Papers & IMTE Official Records, University of Virginia Law Library.

"From Doc 827-Biography of General Minami by Sugiyama, Ken," Box 10, IPS Documents 0-999, Tavenner Papers and IMFTE Official Records, University of Virginia Law Library.

"Hashimoto Kingoro-Miscallaneous Notes," Box 8, Files on Defendants: Hashimoto-Hata, 1946, Tavenner Papers & IMFTE Official Records, University of Virginia Law Library.

"Hashimoto Urged War in Manchuria Testifies News Article," Box 2, Second Phelps Scrapbook, C.W.J. Phelps Collection, University of Virginia Law Library.

"Jiro Minami." Box 1, Files on Defendants; Minami, Muto 1946, Tavenner Papers and IMFTE Official Records, University of Virginia Law Library, University of Virginia Law Library.

"Reference Made to Minami in the Narrative Summary of the Record," Box 8, Files on Defendants; Minami, Muto 1946, Tavenner Papers and IMFTE Official Records, University of Virginia Law Library.

"Saionji-Harada Memoirs Re Manchurian Aggression (I)," Box 7, Saionji-Harada Memoirs from March-July 1947, Tavenner Papers and IMFTE Official Records, University of Virginia Law Library.

"Saionji-Harada Memoirs Re Manchurian Aggression (II)," Box 7, Saionji-Harada Memoirs from March-July 1947, Tavenner Papers and IMFTE Official Records, University of Virginia Law Library.

"Subject No. 76-Minami, Jiro (General)." Box 8, Files on Defendants; Minami, Muto 1946, Tavenner Papers and IMFTE Official Records, University of Virginia Law Library.

"Summary of Interrogation of Baron Wakatsuki, Reijiro," Box 2, General Reports and Memoranda from 1946, Tavenner Papers and IMFTE Official Records, University of Virginia Law Library.

Other Primary Sources

Abend, Hallett, "Japan Wins Tsinan; Drives out Chinese in Week's Fighting," *The New York Times*, May 11, 1928. https://www.nytimes.com/1928/05/11/archives/japan-wins-tsinan-drives-out-chinese-in-weeks-fighting-nippons.html.

Aisin-Gioro, Puyi. *From Emperor to Citizen: The Autobiography of Aisin-Gioro Puyi*. Honolulu, HI: University of the Pacific Press, 2001.

Aisin-Gioro, Puyi. *The Last Manchu: The Autobiography of Henry Pu Yi, Last Emperor of China*. New York, NY: Skyhorse Publishing, 2010.

"Avalon Project—Foreign Relations of the United States: Treaty Between the United of States of America, Belgium, the British Empire, China, France, Italy, Japan, the Netherlands, and Portugal. Signed at Washington February 6, 1922." The Avalon Project. Accessed June 26, 2025. https://avalon.law.yale.edu/20th_century/tr22-01.asp.

Boxer Protocol, Peking 7. September 1901, June 26, 2025. https://reparations.qub.ac.uk/assets/uploads/Boxer-Protocol.pdf.

"CHANG TO GIVE CHINA NEW CONSTITUTION; Unity Is Possible Under the Right Kind of Government, General Declares." *The New York Times*. December 30, 1921. https://www.nytimes.com/1921/12/30/archives/chang-to-give-china-new-constitution-unity-is-possible-under-the.html.

"Convention for the Lease of the Liaotung Peninsula," in *Treaties and Agreements with and Concerning China, 1894-1919*. Oxford University Press, 1921.

Ikki, Kita. *Outline Plan for the Reorganization of Japan*. Monee, IL: Adept Press, 2024.

"International Military Tribunal for the Far East (IMFTE) Judgement Chapter V: Japanese Aggression Against China Sections I and II." Hyperwar: International Military Tribunal for the Far East [Chapter 5]. Accessed August 4, 2025. https://www.ibiblio.org/hyperwar/PTO/IMTFE/IMTFE-5.html.

Ito, Miyoji. "The Constitution of the Empire of Japan." 国立国会図書館—National Diet Library. Accessed March 7, 2025. https://www.ndl.go.jp/constitution/e/etc/c02.html.

Itō, Takeo, and Joshua A. Fogel. *Life Along the South Manchurian Railway: The Memoirs of Itō Takeo*. Abingdon, Oxon: Routledge, 2015.

James, H E.M. *The Long White Mountain: A Journey in Manchuria with Illustrations and a Map*. London: Longmans, Green, & Co, 1888.

The Japan-Manchukuo Yearbook, 1934: Cyclopedia of General Information and Statistics on the Empires of Japan and Manchukuo. Tokyo: The Japan-Manchukuo Yearbook Co., 1934.

"Japanese Government, 'Twenty-One Demands,' April 26, 1915." | USC China Institute. Accessed August 4, 2025. https://china.usc.edu/node/20643.

Japanese Intelligence Planning Against the USSR. Japanese Research Division, Japanese Studies on Manchuria, (1955).

Logistics in Manchuria. Japanese Research Division, Japanese Studies on Manchuria, (1959).

The Imperial Japanese Army in Manchuria, 1894-1945. Japanese Research Division, Japanese Studies on Manchuria, (1959).

"Rout of Ma." *Time Magazine*, November 30, 1931.

"Russo-Japanese Peace Treaty." The World and Japan Database. Accessed August 10, 2025. https://worldjpn.net/documents/texts/pw/19050905.T1E.html.

Strategic Study of Manchuria Military Geography and Topography, Part I, Japanese Research Division, Japanese Studies on Manchuria, (1955).

Tikowara, Hesibo, and Robert Francis Sidney Grant. *Before Port Arthur in a Destroyer: The Personal Diary of a Japanese Naval Officer.* London, UK: John Murray, 1907.

Treaty of Shimonoseki, August 10, 2025. https://reparations.qub.ac.uk/assets/uploads/Treaty-of-Shimonoseki.pdf.

"Treaty of Tientsin Between the Queen of Great Britain and the Emperor of China." The World and Japan' Database. Accessed June 26, 2025. https://worldjpn.net/documents/texts/pw/18580626.T1E.html.

"Ultimatum Orders Peking Cabinet Out, Wu Pei-Fu, with Troops Already Moving, Gives Liang Shih-Yi Three Days to Resign, Civil War Believed Near, Rival Threatens Attack Unless Chang Tso-Lin Dictatorship Ends Within a Week." *The New York Times.* January 16, 1922, https://www.nytimes.com/1922/01/16/archives/ultimatum-orders-peking-cabinet-out-wu-peifu-with-troops-already.html.

"Union of All China Is Sought in Peking; Chang Tso-Lin Manchurian Chief Who Caused Cabinet's Resignation, Takes Control. Delegation Here Hopeful Statement on Rehabilitation Move at Home Says Stay Still." *The New York Times.* December 20, 1921, https://www.nytimes.com/1921/12/20/archives/union-of-all-china-is-sought-in-peking-chang-tsolin-manchurian.html.

Secondary Sources

Allen, George Cyril. *A Short Economic History of Modern Japan: 1867-1937, by G.C. Allen.* London: G. Allen and Unwin, 1950.

Asada, Sadao. "Between the Old Diplomacy and the New, 1918-1922: The Washington System and the Origins of Japanese-American Rapprochement." *Diplomatic History* 30, no. 2 (April 2006): 211-30.

Asada, Sadao. "Japan's 'Special Interests' and the Washington Conference." *The American Historical Review* 67, no. 1 (October 1961): 62-70. https://doi.org/10.2307/1846262.

Auslin, Michael R. *Negotiating with Imperialism: The Unequal Treaties and the Culture of Japanese Diplomacy.* Cambridge, MA: Harvard University Press, 2022.

Barnhart, Michael A. *Japan Prepares for Total War: The Search for Economic Security, 1919-1941.* Ithaca, NY: Cornell University Press, 1987.

Barnhart, Michael A. "The Origins of the Second World War in Asia and the Pacific: Synthesis Impossible?" *Diplomatic History* 20, no. 2 (April 1996): 241-60. https://doi.org/10.1111/j.1467-7709.1996.tb00624.x.

Bassett, Reginald. *Democracy and Foreign Policy: A Case History of the Sino-Japanese Dispute, 1931-1933.* Abingdon, Oxon: Taylor & Francis, 1952.

Beasley, W. G. *The Meiji Restoration.* Stanford, CA: Stanford University Press, 1972.

Beasley, W. G. *The Modern History of Japan.* New York: Frederick A Praeger, 1969.

Behr, Edward, and Peary, Gerald. *The Last Emperor.* Toronto: Bantam Books, 1987.

Bickers, Robert. *The Scramble for China: Foreign Devils in the Qing Empire, 1832-1914.* New York: Penguin Books, 2016.

Bix, Herbert P. *Hirohito and the Making of Modern Japan.* New York: HarperCollins Publishers, 2000.

Bix, Herbert P. "Japanese Imperialism and the Manchurian Economy, 1900-31." *The China Quarterly* 51 (July 1972): 425-43. https://doi.org/10.1017/s0305741000052231.

Boot, Max. *War Made New: Technology, Warfare, and the Course of History, 1500 to Today.* New York: Gotham Books, 2006.

Büttner, Clemens. "The National Protection War and the Intellectual Foundations of Chinese Warlordism." *Modern China* 49, no. 6 (April 28, 2023): 679-708. https://doi.org/10.1177/00977004231153331.

Cameron, Craig M. "Race and Identity: The Culture of Combat in the Pacific War." *The International History Review* 27, no. 3 (September 2005): 550-66. https://doi.org/10.1080/07075332.2005.9641072.

Carter, James. "The Rise of Nationalism and Revolutionary Parties." Essay. In *Oxford Illustrated History of Modern China*, edited by Jeffrey Wasserstrom. Oxford: Oxford University Press, 2016.

Cha, Victor D. *Japan's Grand Strategy on the Korean Peninsula: Optimistic Realism*, 2001, 227-66.

Chan, Anthony. *Arming the Chinese: The Western Armaments Trade in Warlord China*. Vancouver: UBC Press, 2010.

Chi, Hsi-sheng. *The Chinese Warlord System: 1916 to 1928*. Washington D.C.: American University, Center for Research in Social Systems, 1969.

Chi, Madeleine. "Bureaucratic Capitalists in Operation: Ts'ao Ju-Lin and His New Communications Clique, 1916-1919." *The Journal of Asian Studies* 34, no. 3 (May 1975): 675-88. https://doi.org/10.2307/2052549.

Chu, Samuel C., and Kwang-Ching Liu. *Li Hung-chang and China's Early Modernization*. Abingdon, Oxon: Routledge, 2015.

Clodfelter, Micheal. *Warfare and Armed Conflicts: A Statistical Encyclopedia of Casualty and Other Figures, 1492-2015*. Vol. II. Jefferson, NC: McFarland & Company Publishers, 2017.

Coble, Parks M. "Chiang Kai-shek and the Anti-Japanese Movement in China: Zou Tao-Fen and the National Salvation Association, 1931-1937." *The Journal of Asian Studies* 44, no. 2 (February 1985): 293-310. https://doi.org/10.2307/2055924.

Coble, Parks M. *Facing Japan: Chinese Politics and Japanese Imperialism, 1931-1937*. Cambridge, MA: Council on East Asian Studies, Harvard University, 1991.

Connaughton, Richard. *Rising Sun and Tumbling Bear: Russia's War with Japan*. Orion, 2007.

Conroy, Hilary. "Government Versus 'Patriot': The Background of Japan's Asiatic Expansion." *Pacific Historical Review* 20, no. 1 (February 1, 1951): 31-42. https://doi.org/10.2307/3634474.

Coogan, Anthony. "Northeast China and the Origins of the Anti-Japanese United Front." *Modern China* 20, no. 3 (July 1994): 282-314. https://doi.org/10.1177/009770049402000302.

Coox, Alvin D. *Nomonhan: Japan Against Russia, 1939*. Stanford, CA: Stanford University Press, 1985.

Coox, Alvin. "The Kwantung Army Dimension." Essay. In *Japanese Informal Empire in China*, edited by Peter Duus, Mark Peattie, and Ramon Myers. Princeton University Press, 2014.

Costello, John. *The Pacific War*. New York: Perennial, 2002.

Crean, Jeffery. *The Fear of Chinese Power: An International History*. London: Bloomsbury, 2024.

Crowley, James B. *Japan's Quest for Autonomy: National Security and Foreign Policy 1930-1938*. Princeton, NJ: Princeton University Press, 1966.

End, Nicholas. "Chapter 7: Japan During the Interwar Period: From Monetary Restraint to Fiscal Abandon." Essay. In *Debt and Entanglements Between the Wars*, edited by Era Dabla-Norris. Washington, D.C: International Monetary Fund, 2019.

Darity, William. "A Model of 'Original Sin': Rise of the West and Lag of the Rest." *American Economic Review* 82, no. 2 (May 1992): 162-67.

Davis, Malcolm W. "Railway Strategy in Manchuria." *Foreign Affairs* 4, no. 3 (1926): 499-502. https://doi.org/10.2307/20028472.

Drea, Edward J. *Japan's Imperial Army: Its Rise and Fall*. Lawrence, KS: University of Kansas Press, 2016.

Drea, Edward. "The Imperial Japanese Army (1868-1945): Its Origins Evolution and Legacy." Essay. In *War in the Modern World Since 1815*, edited by Jeremy Black. Oxford: Taylor & Francis, 2003.

Dull, Paul S. "The Assassination of Chang Tso-Lin." *The Far Eastern Quarterly* 11, no. 4 (August 1952): 453-63. https://doi.org/10.2307/2942022.

Dunscomb, Paul E. "A Great Disobedience Against the People: Popular Press Criticism of Japan's Siberian Intervention, 1918-1922." *Journal of Japanese Studies* 32, no. 1 (Winter 2006): 53-81. https://doi.org/10.5771/9780739146026.

Duus, Peter. *The Abacus and the Sword: The Japanese Penetration of Korea, 1859-1910*. Berkeley, CA: University of California Press, 1995.

Eastlake, F. Warrington, and Yamada Yoshi-Aki. *Heroic Japan: A History of the War Between China and Japan*. Washington, D.C: University Publications of America, 1979.

Eckstein, Alexander, Kang Chao, and John Chang. "The Economic Development of Manchuria: The Rise of a Frontier Economy." *The Journal of Economic History* 34, no. 1 (March 1974): 239-64. https://doi.org/10.1017/s0022050700079729.

Esherick, Joseph, and C. X. George Wei. *China: How the Empire Fell*. London: Routledge, 2014.

Esherick, Joseph. *The Origins of the Boxer Uprising*. Berkeley, CA: University of California Press, 1987.

Esposito, Gabriele, and Giuseppe Rava. *Armies of the First Sino-Japanese War 1894-95*. London: Bloomsbury Publishing, 2022.

Evans, David C., and Mark Peattie. *Kaigun: Strategy, Tactics, and Technology in the Imperial Japanese Navy, 1887-1941*. New York, NY: Naval Institute Press, 2015.

Fairbank, John King, Kwang-Ching Liu, and Denis Crispin Twitchett. *The Cambridge History of China, Volume 11, Part 2, Late Ch'ing, 1800-1911*. Cambridge: Cambridge University Press, 1980.

Fairbank, John King, and Wilbur, Martin, *The Cambridge History of China, Volume 12 Part 1, Republican China*, Cambridge: Cambridge University Press, 1980.

Fenby, Jonathan. *Generalissimo: Chiang Kai-shek and the China He Lost*. London: Simon & Schuster UK, 2003.

Ferrell, Robert H. "The Mukden Incident: September 18-19, 1931." *The Journal of Modern History* 27, no. 1 (March 1955): 66-72. https://doi.org/10.1086/237763.

"The Financial Cutting Edge in the Partition of China." *Pacific Affairs* 9, no. 2 (June 1936): 165-76. https://doi.org/10.2307/2751404.

Fung, Edmund S. "Military Subversion in the Chinese Revolution of 1911." *Modern Asian Studies* 9, no. 1 (January 1975): 103-23. https://doi.org/10.1017/s0026749x00004893.

Godart, G. Clinton. "Nichirenism, Utopianism, and Modernity: Rethinking Ishiwara Kanji's East Asia League Movement." *Japanese Journal of Reli-*

gious Studies 42, no. 2 (November 2, 2015). https://doi.org/10.18874/jjrs.42.2.2015.235-274.

Goldman, Stuart D. *Nomonhan, 1939: The Red Army's Victory that Shaped World War II*. Annapolis, MD: Naval Institute Press, 2013.

Gray, Jack. *Rebellions and Revolutions: China from the 1880s to 2000*. Oxford: Oxford University Press, 2003.

Gu, Edward X. "Who Was Mr. Democracy? The May Fourth Discourse of Populist Democracy and the Radicalization of Chinese Intellectuals (1915-1922)." *Modern Asian Studies* 35, no. 03 (July 2001): 589-621. https://doi.org/10.1017/s0026749x01003043.

Hacker, Barton C. "The Weapons of the West: Military Technology and Modernization in 19th-Century China and Japan." *Technology and Culture* 18, no. 1 (January 1977): 43. https://doi.org/10.2307/3103204.

Hall, Simon. Blinded by the Rising Sun: Japanese Military Intelligence from the First Sino-Japanese War to the End of World War II. PhD Thesis, University of Adelaide, December (2016). https://digital.library.adelaide.edu.au/bitstream/2440/109796/2/02whole.pdf.

Harada, Yutaka. "Why Was Wilsonian-Taisho Moment Lost in Japan in Spite of Its Economic Success?" *Japanese Journal of Political Science* 19, no. 4 (December 2018): 571-86. https://doi.org/10.1017/s1468109918000348.

Harries, Meirion, and Susie Harries. *Soldiers of the Sun: The Rise and Fall of the Imperial Japanese Army*. New York: Random House, 1991.

Hattori, Ryūji. *Japan at War and Peace: Shidehara Kijūrō and the Making of Modern Diplomacy*. Canberra, ACT, Australia: Australian National University Press, 2021.

Havens, Thomas R.H. *Farm and Nation in Modern Japan: Agrarian Nationalism, 1870-1940*. Princeton, NJ: Princeton University Press, 1974.

Hendrix, Henry. "The Roots of Japanese Militarism." Master's Thesis, Naval Postgraduate School, 1994. https://apps.dtic.mil/sti/pdfs/ADA293660.pdf.

Hibbert, Christopher. *The Dragon Wakes: China and the West, 1793-1911*. Harmondsworth, UK: Penguin, 1988.

Hosoya, Chihiro. "Origin of the Siberian Intervention, 1917-1918." *The Annals of the Hitotsubashi Academy* 9, no. 1 (October 1958): 91-108.

Hsu, Immanuel Chung-Yueh. *The Rise of Modern China*. Hong Kong: Oxford University Press, 1983.

Hsu, Wilbur W. *Survival Through Adaptation: The Chinese Red Army and the Extermination Campaigns, 1927-1936*. Fort Leavenworth, KS: Combat Studies Institute Press, US Army Combined Arms Center, 2012.

Hu, Shizhang. *Stanley K. Hornbeck and The Open Door Policy, 1919-1937*. Westport, CT: Greenwood Press, 1995.

Humphreys, Leonard A. *The Way of the Heavenly Sword: The Japanese Army in the 1920s*. Stanford, CA: Stanford University Press, 1995.

Iguchi, Gerald. Imperialism, Fascism, and Buddhism in Modern Japan, PhD Thesis, University of California San Diego, (2006). https://escholarship.org/content/qt53m639t5/qt53m639t5_noSplash_ef03389b9358fa889f151640254ec12b.pdf.

Ikle, Frank W. "The Triple Intervention. Japan's Lesson in the Diplomacy of Imperialism." *Monumenta Nipponica* 22, no. 1/2 (1967): 122-30. https://doi.org/10.2307/2383226.

Iriye, Akira. *After Imperialism: The Search for a New Order in the Far East, 1921-1931*. Cambridge, MA: Harvard University Press, 1965.

Iriye, Akira. "Chang Hsueh-Liang and the Japanese." *The Journal of Asian Studies* 20, no. 1 (November 1960): 33-43. https://doi.org/10.2307/2050070.

"Ishihara Kanji: Portraits of Modern Japanese Historical Figures: National Diet Library, Japan." Portraits of Modern Japanese Historical Figures. Accessed August 20, 2025. https://www.ndl.go.jp/portrait/e/datas/355/.

Iyenaga, Toyokichi. "Japan in South Manchuria." *The Journal of Race Development* 2, no. 4 (1912): 373-98. https://doi.org/10.2307/29737926.

Jacob, Frank. *East Asia and the First World War*. Oldenbourg: De Gruyter, 2022.

Jacob, Frank. "Secret Societies in Japan and Preparation for the Russo-Japanese War (1904-1905)." *Diacronie*, 28, no. 4, (December 29, 2016). https://doi.org/10.4000/diacronie.4738.

Jansen, Marius. "Japanese Imperialism: Late Meiji Perspectives." Essay. In *The Japanese Colonial Empire, 1895-1945*, edited by Ramon Myers and Peter Duus. Princeton, NJ: Princeton University Press, 1984.

Jaundrill, D. Colin. *Samurai to Soldier: Remaking Military Service in Nineteenth-Century Japan*. Ithaca, NY: Cornell University Press, 2017.

Jordan, Donald A. *The Northern Expedition China's National Revolution of 1926-1928*. Honolulu, HI: University of Hawai'i Press, 2019.

Jowett, Philip S. *China's Wars: Rousing the Dragon, 1894-1949*. Oxford, UK: Osprey Publishing, 2013.

Jowett, Philip S. *The Bitter Peace: Conflict in China, 1928-37*. Gloucester, UK: Amberley Publishing, 2017.

Jowett, Philip. *Chinese Warlord Armies, 1911-30*. Oxford, UK: Osprey, 2010.

Jukes, Geoffrey. *The Russo-Japanese War*. Oxford, UK: Osprey, 2022.

Kang, David C. "International Order in Historical East Asia: Tribute and Hierarchy Beyond Sinocentrism and Eurocentrism." *International Organization* 74, no. 1 (September 30, 2019): 65-93. https://doi.org/10.1017/s0020818319000274.

Karl, Rebecca E. *Mao Zedong and China in the Twentieth-Century World: A Concise History*. Durham, NC: Duke University Press, 2010.

Kawamoto, Bondon. "How Old Regimes Help New Economic Growth." Honors Thesis, Portland State University, (2013). https://doi.org/10.15760/honors.25.

Kawamura, Noriko. "Wilsonian Idealism and Japanese Claims at the Paris Peace Conference." *Pacific Historical Review* 66, no. 4 (November 1, 1997): 503-26. https://doi.org/10.2307/3642235.

Kitaoka, Shin'ichi. "The Army as a Bureaucracy: Japanese Militarism Revisited." *The Journal of Military History* 57, no. 5 (October 1993): 67-86. https://doi.org/10.2307/2951803.

Kitaoka, Shin'ichi. "The Army's China Experts." Essay. In *The Japanese Informal Empire in China, 1895-1937,* edited by Peter Duus, Ramon Myers, and Mark Peattie. Princeton, NJ: Princeton University Press, 1989.

Koda, Yoji. "The Russo-Japanese War-Primary Causes of Japanese Success." *Naval War College Review* 58, no. 2, (2005): 12-44. https://digital-commons.usnwc.edu/nwc-review/vol58/iss2/3/.

Kowner, Rotem. *Historical Dictionary of the Russo-Japanese War*. Lanham, MD: Rowman & Littlefield, 2017.

Kublin, Hyman. "The 'Modern' Army of Early Meiji Japan." *The Far Eastern Quarterly* 9, no. 1 (November 1949): 20. https://doi.org/10.2307/2049123.

Kwok, Tat Wai Dwight. "A Translation of Datsu-A Ron: Decoding a Prewar Japanese Nationalistic Theory." Master's Thesis, University of Toronto, 2009. https://tspace.library.utoronto.ca/handle/1807/18797.

Kwong, Chi Man. *War and Geopolitics in Interwar Manchuria: Zhang Zuolin and the Fengtian Clique During the Northern Expedition*. Leiden: Brill, 2017.

Lattimore, Owen. *Manchuria: Cradle of Conflict*. New York: Macmillan, 1932.

Lauren, Paul Gordon. *Power and Prejudice: The Politics and Diplomacy of Racial Discrimination*. London: Routledge, 2019.

Lee, Chong-Sik. *Revolutionary Struggle in Manchuria: Chinese Communism and Soviet Interest, 1922-1945*. Berkeley, CA: University of California Press, 2023.

Lenox, Bertam L., and Putnam Weale. *The Fight for the Republic in China*, New York: Dodd, Mead and Co, 1917.

Lewis, Michael. *Rioters and Citizens: Mass Protest in Imperial Japan*. Berkeley, CA: University of California Press, 1990.

Lockwood, William W. "Japanese Silk and the American Market." *Far Eastern Survey* 5, no. 4 (February 12, 1936): 31-36. https://doi.org/10.2307/3021472.

Lone, Stewart. *Japan's First Modern War Army and Society in the Conflict with China, 1894-5*. London: Palgrave Macmillan, 1994.

Luo, Zhitian. "National Humiliation and National Assertion: The Chinese Response to the Twenty-One Demands." *Modern Asian Studies* 27, no. 2 (May 1993): 297-319. https://doi.org/10.1017/s0026749x00011501.

MacMillan, Margaret. *Paris 1919: Six Months that Changed the World*. London: John Murray Publishers, 2019.

Maddison, Angus. *Development Centre Studies the World Economy: Historical Statistics*. OECD Publications, 2003.

Maeda, Kiyotaka. "Dynamics of the Rice Market During the Interwar Period," (Undergraduate Thesis, Keio University, 2022) June 2022. https://www.sanken.keio.ac.jp/publication/KEO-dp/170/KEO-DP170.pdf.

Mainardi, Benjamin E. "The Russo-Japanese War: Origins and Implications." *James Madison Undergraduate Research Journal* 7, no. 1, 2020: 6-14.

Makeham, John. *China: The World's Oldest Living Civilization Revealed*. Thames & Hudson, 2009.

Mallory, Walter H. "The Open Door in China: A Reappraisal." *Foreign Affairs* 26, no. 1 (October 1947): 155-68. https://doi.org/10.2307/20030095.

Manela, Erez. "Imagining Woodrow Wilson in Asia: Dreams of East-West Harmony and the Revolt against Empire in 1919." *The American Historical Review* 111, no. 5 (December 1, 2006): 1327-51. https://doi.org/10.1086/ahr.111.5.1327.

Mao, Zedong. "Problems of Strategy in China's Revolutionary War." Essay. In *Selected Military Writings of Mao Tse-Tung*, 77-152. Beijing, China: Foreign Languages Press, 1967.

Matsusaka, Yoshihisa Tak. *The Making of Japanese Manchuria, 1904-1932*. Cambridge, MA: Harvard University Asia Center, 2001.

McClain, Kendra. "Explaining Meiji Japan's Top-Down Revolution." Master's Thesis, Naval Postgraduate School, 2017. https://apps.dtic.mil/sti/citations/AD1053365.

McCormack, Gavan. *Chang Tso-lin in Northeast China, 1911-1928*. Stanford, CA: Stanford University Press, 1977.

Metzler, Mark. *Lever of Empire: The International Gold Standard and the Crisis of Liberalism in Prewar Japan*. Berkeley: University of California Press, 2006.

Millard, Thomas F. *The ABC's of the Twenty-One Demands*. New York: Chinese Students' Committee on Washington Conference, 1921.

Mitter, Rana. *The Manchurian Myth: Nationalism, Resistance and Collaboration During the Manchurian Crisis, 1931-1933*. Berkeley, CA: University of California Press, 2000.

Morley, James William, ed. "The Extension of Hostilities, 1931-1932." Essay. In *Japan's Road to the Pacific War Volume I*, 241-335. New York: Columbia University Press, 1984.

Morse, Hosea Ballou. *The International Relations of the Chinese Empire*. London: Longmans, Green & Co, 1910.

Myers, Ramon. "Japanese Imperialism in Manchuria: The South Manchuria Railway Company." Essay. In *The Japanese Informal Empire in China, 1895-1937*, edited by Peter Duus and Mark R. Peattie. Princeton, NJ: Princeton University Press, 2016.

Nish, Ian. *The Iwakura Mission to America and Europe: A New Assessment*. London: Routledge, 2008.

Nish, Ian. *The Origins of the Russo-Japanese War*. New York: Longman Group, 1985.

Noguchi, Masayoshi. "The South Manchuria Railway Company: An Accounting and Financial History 1907-1943." University of Kobe, Research Institution of Economics and Business Administration, March 15, 2013. https://www.rieb.kobe-u.ac.jp/academic/ra/dp/English/DP2013-08.pdf.

Norman, E. Herbert, and Lawrence Timothy Woods. *Japan's Emergence as a Modern State: Political and Economic Problems of the Meiji Period*. UBC Press, 2007.

Norman, E. Herbert. "The Genyosha: A Study in the Origins of Japanese Imperialism." *Pacific Affairs* 17, no. 3 (September 1944): 261. https://doi.org/10.2307/2751627.

Ogata, Sadako N. *Defiance in Manchuria the Making of Japanese Foreign Policy, 1931-1932*. Berkeley, CA: University of California Press, 1964.

Olender, Piotr. *Sino-Japanese Naval War: 1894-1895*. Hampshire, UK: Stratus, 2014.

Orbach, Danny. *Curse on this Country: The Rebellious Army of Imperial Japan*. Ithaca, NY: Cornell University Press, 2017.

Orbach, Danny. "The Military-Adventurous Complex: Officers, Adventurers, and Japanese Expansion in East Asia, 1884-1937." *Modern Asian Studies* 53, no. 2 (August 13, 2018): 339-76. https://doi.org/10.1017/s0026749x17000543.

Otterstedt, Charles. "The Kwantung Army and the Nomonhan Incident: Its Impact on National Security," April 4, 2000. https://doi.org/10.21236/ada378202.

Paine, S.C.M. *The Sino-Japanese War of 1894-95: Power, Perceptions, and Primacy*. Cambridge, UK: Cambridge University Press, 1951.

Peattie, Mark R. *Ishiwara Kanji and Japan's Confrontation with the West*. Princeton, N.J: Princeton University Press, 1975.

Perry, J. C. "The Battle off the Tayang, 17 September 1894." *The Mariner's Mirror* 50, no. 4 (January 1964): 243-59. https://doi.org/10.1080/00253359.1964.10657787.

Pesek, Pauline. "The Second Chinese Banking Consortium: The Open Door and Finance Diplomacy." Master's Thesis, University of Nebraska, May 1, 1976. https://digitalcommons.unomaha.edu/cgi/viewcontent.cgi?article=1451&context=studentwork.

Platt, Stephen R. “New Domestic and Global Challenges, 1792-1860.” Essay. In *The Oxford Illustrated History of Modern China*, edited by Jeffrey Wasserstrom, 37-62. Oxford, UK: Oxford University Press, 2016.

Pratt, Edward Earl. “Wanpaoshan, 1931: Japanese Imperialism, Chinese Nationalism, and the Korean Problem in Northeast China On the Eve of the Manchurian Incident.” Master’s Thesis, University of Virginia, May 1983.

Railway and Locomotive Engineering. New York: Sinclair, 1913.

Ramsdell, Daniel B. “The Nakamura Incident and the Japanese Foreign Office.” *The Journal of Asian Studies* 25, no. 1 (November 1965): 51-67. https://doi.org/10.2307/2051040.

Roberts, Stephen F. “The Imperial Chinese Steam Navy.” *Warship International* 11, no. 1 (1974): 19-57.

Sabey, John Wayne. “The Gen’yōsha, the Kokuryūkai, and Japanese Expansionism.” PhD Thesis, University of Michigan, (1975).

Schlitz, Michael. “Nishihara Kamezo and the Abortive China Loans, 1917-18.” Thesis, Catholic University of Leuven, Munich Personal REpEC Archive, 2007. https://mpra.ub.uni-muenchen.de/7100/1/MPRA_paper_7100.pdf.

“Secretary of State John Hay and the Open Door in China, 1899-1900.” United States State Department, Office of the Historian. Accessed June 26, 2025. https://history.state.gov/milestones/1899-1913/hay-and-china.

Seki, Hiroharu. “The Manchurian Incident, 1931.” Essay. In *Japan’s Road to the Pacific War: Japan Erupts The London Naval Conference and the Manchurian Incident 1928-1932*, edited by James William Morley, 121-230. New York: Columbia University Press, 1984.

Seth, Michael J. *A History of Korea: From Antiquity to the Present*. Lanham, MD: Rowman & Littlefield, 2011.

Shimada, Toshihiko. “The Extension of Hostilities, 1931-1932.” Essay. In *Japan’s Road to the Pacific War: Japan Erupts The London Naval Conference and the Manchurian Incident 1928-1932*, edited by James William Morley, 231-335. New York: Columbia University Press, 1984.

Silbey, David. *The Boxer Rebellion and the Great Game in China*. New York: Hill and Wang, 2013.

Smith, S. A. *A Road is Made: Communism in Shanghai*. Richmond: Curzon, 2000.

Smith, Sara R., *The Manchurian Crisis: A Tragedy in International Relations, 1931-1932* Westport, CT: Greenwood Press, 1948

Spector, Stanley. *Li Hung-Chang and the Huai Army: A Study in Nineteenth-Century Chinese Regionalism*. Seattle: University of Washington Press, 1964.

Spence, Jonathan D. *The Search for Modern China*. New York: W. W. Norton, 1999.

Stewart, John R. "Manchuria: The Land and Its Economy." *Economic Geography* 8, no. 2 (April 1932): 134-60. https://doi.org/10.2307/140247.

Stimson, Henry L. *The Far Eastern Crisis: Recollections and Observations*. New York: Harper & Brothers Publishers, 1974.

Stine, Jeffrey K., and Daniel R. Headrick. "The Tools of Empire: Technology and European Imperialism in the Nineteenth Century." *The Journal of Modern History* 51, no. 2 (June 1979): 231-63. https://doi.org/10.2307/3104499.

Su, Christine. "Justifiers of the British Opium Trade: Arguments by Parliament, Traders, and the Times in the Leading Up to the Opium War." *Stanford Undergraduate Research Journal*, (2008).

Temperley, Harold V. *A History of the Paris Peace Conference*. Oxford: Oxford University Press, 1920.

Toland, John. *The Rising Sun: Decline and Fall of the Japanese Empire*. New York: Random, 1970.

Tyler, Sydney. *The Russo-Japanese War: An Illustrated History of the War in the Far East*. Sharp Ink, 2022.

Valiant, Robert. "Japanese Involvement in Mongol Independence Movements, 1912-1919." *The Mongolian Society Bulletin* 11, no. 2 (1972): 1-32.

Volpicelli, Zenone. *The China-Japan War: Compiled from Japanese, Chinese and Foreign Sources*. New York: Charles Scribner's Sons, 1896.

Wagner, Donald B. "The Traditional Chinese Iron Industry and Its Modern Fate." *East Asian Science, Technology, and Medicine* 12, no. 1 (July 5, 1995): 138-62. https://doi.org/10.1163/26669323-01201010.

Warner, Denis, and Peggy Warner. *The Tide at Sunrise: A History of the Russo-Japanese War, 1904-1905*. New York: Charterhouse, 1974.

Wasserstrom, Jeffrey N. *Twentieth-Century China: New Approaches*. Hoboken: Taylor and Francis, 2013.

Watts, Anthony John. *The Imperial Russian Navy*. London: Arms and Armour, 1990.

Weale, Putnam, and B.L. Lenox. *The Vanished Empire*. London: Macmillan, 1926.

Weland, James. "Misguided Intelligence: Japanese Military Intelligence Officers in the Manchurian Incident, September 1931." *The Journal of Military History* 58, no. 3 (July 1994): 445-60. https://doi.org/10.2307/2944134.

Whitehurst, G. William. *The China Incident: Igniting the Second Sino-Japanese War*. Jefferson, NC: McFarland & Company Publishers, 2021.

Wilbur, Clarence, and Julie-Lienying How. *Missionaries of Revolution: Soviet Advisors and Nationalist China 1920-1927*. Cambridge, MA: Harvard University Press, 1989.

"Wilson's Fourteen Points." United States State Department Office of the Historian, Accessed August 19th, 2025. https://history.state.gov/milestones/1914-1920/fourteen-points.

Wilson, Sandra. *The Manchuria Crisis and Japanese Society, 1931-1933*. New York, NY: Taylor & Francis, 2002.

Wilson, Sandra. "The Manchurian Crisis and Moderate Japanese Intellectuals: The Japan Council of the Institute of Pacific Relations." *Modern Asian Studies* 26, no. 3 (July 1992): 507-44. https://doi.org/10.1017/s0026749x00009896.

Worthing, Peter M. *General He Yingqin the Rise and Fall of Nationalist China*. Cambridge: Cambridge University Press, 2018.

Wright, Quincy. "The Manchurian Crisis." *American Political Science Review* 26, no. 1 (February 1932): 45-76. https://doi.org/10.2307/1946441.

Wright, Quincy. "The Washington Conference." *The American Political Science Review* 16, no. 2 (May 1922): 285-97.

Yamamoto, Tsumoto David. Japanese Press and Japanese Foreign Policy, PhD Thesis, School of Oriental and African Studies, May 1981.

Yamamuro, Shin'ichi, and Joshua A. Fogel. *Manchuria Under Japanese Dominion*. Philadelphia, PA: University of Pennsylvania Press, 2006.

Yoshihashi, Takehiko. *Conspiracy at Mukden: The Rise of the Japanese Military*. New Haven, CT: Yale University Press, 1963.

Young, John W. "The Hara Cabinet and Chang Tso-Lin, 1920-1." *Monumenta Nipponica* 27, no. 2 (Summer 1972): 125-42. https://doi.org/10.2307/2383715.

Young, Louise. *Japan's Total Empire: Manchuria and the Culture of Wartime Imperialism*. Berkeley, CA: University of California Press, 1999.

Youngsoo Kim. "Two Perspectives on the 1895 Assassination of Queen Min." *Korea Journal* 48, no. 2 (June 2008): 160-85. https://doi.org/10.25024/kj.2008.48.2.160.

Zachmann, Urs Matthias. "Imperialism in a Nutshell: Conflict and the 'Concert of Powers' in the Tripartite Intervention, 1895." *Japanstudien* 17, no. 1 (January 2006): 57-82. https://doi.org/10.1080/09386491.2006.11826924.

Zarrow, Peter. "Felling a Dynasty and Founding a Republic." Essay. In *Oxford Illustrated History of Modern China*, edited by Jeffrey Wasserstrom. Oxford: Oxford University Press, 2016.

Zheng, Xiaowei. *The Politics of Rights and the 1911 Revolution in China*. Stanford, CA: Stanford University Press, 2020.

Index

J

O

P

U

Made in the USA
Monee, IL
14 September 2026

62974520R00272